In the Hearts of the Beasts

In the Hearts of the Beasts

How American Behavioral Scientists Rediscovered the Emotions of Animals

ANNE C. ROSE

OXFORD
UNIVERSITY PRESS

OXFORD
UNIVERSITY PRESS

Oxford University Press is a department of the University of Oxford. It furthers the University's objective of excellence in research, scholarship, and education by publishing worldwide. Oxford is a registered trade mark of Oxford University Press in the UK and certain other countries.

Published in the United States of America by Oxford University Press
198 Madison Avenue, New York, NY 10016, United States of America.

Library of Congress Cataloging-in-Publication Data
Names: Rose, Anne C., 1950– author.
Title: In the hearts of the beasts : how American behavioral scientists rediscovered the emotions of animals / Anne C. Rose.
Description: New York, NY : Oxford University Press, [2020] |
Includes bibliographical references and index.
Identifiers: LCCN 2019038611 (print) | LCCN 2019038612 (ebook) |
ISBN 9780190935610 (hardback) | ISBN 9780190935627 (updf) |
ISBN 9780190935634 (epub)
Subjects: LCSH: Emotions in animals—Research—United States—History. |
Animal behavior—Research—United States—History. | Behavioral scientists—United States.
Classification: LCC QL785.27 .R67 2020 (print) | LCC QL785.27 (ebook) |
DDC 591.5—dc23
LC record available at https://lccn.loc.gov/2019038611
LC ebook record available at https://lccn.loc.gov/2019038612

1 3 5 7 9 8 6 4 2

Printed by Integrated Books International, United States of America

For our beagles

Contents

List of Illustrations

Preface

The history of a science is easy to overlook because it seems a record of past errors or at best incompleteness. No research paper about the emotions of animals fails to cite the latest scientific evidence, and yet similar breadth connected with time—logging the slow steps to new insights—is less common. Once in a while, a scholar references a pioneering scientist and then skips ahead to the present without reviewing protracted intervening debates about such concepts as instinct, behavior, or consciousness.

History, though, is a crucial dimension of science, at least so I argue. Formal protocols provide immediate answers, but experimentation has, to put it crudely, feet in the mud: the many and changing preoccupations of scientists, their institutions, and societies. Researchers who study animals also face the unpredictability of living creatures. These historical variables affect the way later practitioners frame scientific questions. Taking a sustained look at scientific roots facilitates understanding of the present.

Although it is shortsighted to think that our own time is exceptional, the modern world does seem unusually steeped in emotion. One-time human virtues of reasonableness and self-restraint appear less persuasive today and hence less practiced. Acting willfully does not make us more animal-like. Nonhuman and human animals are too complex to credit the old adage that passions are bestial. Still, unashamed displays of feeling across modern cultures push contemporary scholars to search for the essential human being: the latter-day product of evolution akin to numerous other organisms. The rising American science of animal emotions, with origins around 1900, echoes these twenty-first-century concerns.

My ability to reconstruct this story has depended on the excellent support of archives that house the papers and photographs of my principal historical figures. I am especially grateful to the archivists whose names appear in parentheses. The institutions include the American Philosophical Society (Charles Greifenstein); Archives, California Institute of Technology; Archives and Special Collections, Mount Holyoke College (Deborah Richards); Archives and Special Collections, Vassar College (Dean M. Rogers); Arthur and Elizabeth Schlesinger Library, Harvard University Archives, and Center for the History of Medicine of the Francis A. Countway Library of Medicine (Jessica Murphy), all Harvard University; David M. Rubenstein Rare Book and Manuscript Library, Duke University (Amy McDonald); Division of Rare and Manuscript Collections, Cornell University; Eberly Family Special Collections Library, Penn State University (Alexandra Bainbridge); Manuscripts and Archives, Yale University (Michael Frost); Rockefeller Archive Center (Nancy Adgent and Monica Blank); Special Collections Research Center, University of Chicago; Special Collections and University Archives, Stanford University; and Whitney Library, New Haven Museum. I am grateful to Penn State's Distinguished Professor stipend that has supported my travel to archives. I thank these institutions for permission to quote from manuscripts and reproduce photographs in their possession.

Fellow scholars have been extraordinarily generous in their comments on my manuscript and suggestions for further questions and sources. I heartily thank Justin T. Clark, Janet M. Davis, Marian Stamp Dawkins, Donald A. Dewsbury, Paula Droege, Amy S. Greenberg, Theodora Kalikow, and Bruce Kuklick. The late Victoria Braithwaite exemplified scientific curiosity and collegiality. My indebtedness to Nadine Weidman for sharing her thoughts and friendship is profound. Johanna N. Shields never lost faith in this project and gave me characteristically excellent advice and sustaining kindness over many years. Others who have offered invaluable encouragement and opportunities include Vincent Colapiertro, Gary Cross, Alan C. Elms, Adele Lindenmeyr, Wilson Moses, Ian Nicholson, Robert Sapolsky, Phyllis Wentworth, and Edward O. Wilson. The Mechanisms of Mind seminar at Penn State, founded and sustained by Paula Droege, was an ideal setting to interact with scientists, social scientists, and humanists who have a common interest in mental experience. My deepest intellectual debts are to my teachers: the late Sydney E. Ahlstrom, David Brion Davis, and Robert H. Elias. I hope my work honors their commitment to critical inquiry informed by ethics.

I have been fortunate to have resourceful and energetic research assistants who read more dusty pages about animals than they imagined existed. Thank you to Kate Gregory, Yokasta Lara, Meagan McDermott, Katelyn Powlus, Jenna Stern, and Daphnie Vega. Joshua Wretzel provided expert editorial support. My Penn State students have helped me sharpen my thinking. I have appreciated their kindness when they asked more than once: How's your book coming?

At Oxford University Press, I am grateful to Abby Gross for her generosity at a crucial moment. I sincerely thank Joan Bossert for her sustained commitment to my book and her excellent guidance. Her assistant editors, Lynnee Argabright followed by Phil Velinov, have been a pleasure to work with. Devasena Vedamurthi and her colleagues at Newgen greatly facilitated the book's production.

My family is my foundation. I love you Jon, Ellie, Colin, Kellan, Benny, and Buddy. Adam remains a caring supporter. My beagles and African frog have been my best friends and research assistants. Cathy Kassab and the Nittany Beagle Rescue, Dr. Renee Calvert and the staff of Centre Animal Hospital, Hansen Home and Pet Sitting (especially Jenny Dean), and Madyson Sechler have all added to our happiness. Thank you Katie, Winnie, Rosie, Kayden, and George!

Introduction

Surprising Glimpses into Animal Hearts

> It is reasonable to suppose that a rat learns more from wandering about the maze and finally arriving at the end than it does from hastily running through as best it can.
>
> —N. R. F. Maier and T. C. Schneirla, 1935

In the summer of 1917, Walter Cannon, the Harvard physiologist known for his *Bodily Changes in Pain, Hunger, Fear and Rage*, had a brief moment of insight into his dependence on animals. Arriving in France to investigate wound shock—the sudden collapse of life functions following injury—he saw "that it is hopeless to attempt to do any experimental work on animals here." "Animals are hard to get," and his laboratory in the port city of Boulogne was poor anyway. Cannon's inability to rely on familiar test subjects was part of his general disorientation. The war's "welter of agony" made "it difficult to realize that I am really living a continuation of my previous life." Suffering now exposed the limits of reason and the power of emotion in history. It was a "beastly torment," he wrote home, a "mutilation of our own kind." Soon, though, Cannon was experimenting on cats and dogs in Dijon and London, comparing results with "surgeons of extensive experience at casualty clearing stations."[1] But it took another half century for American scientists to think squarely about the emotions of animals. This book explains why attention to animal feelings gradually reappeared.

Cannon's habit of using animals as experimental proxies was based on evolutionary theory, and Charles Darwin had presented emotions as links among species in *The Expression of the Emotions in Man and Animals* (1872).[2] It seems puzzling, then, that Cannon treated animals and their sensibilities as afterthoughts. The question deepens, because American laboratories were overflowing with animals.

From the 1890s on, the strenuous empiricism of Americans who investigated behavior favored animal experiments. Precisely because animals could not be asked to introspect and explain their mental experiences, they appealed to experimenters eager to avoid speculation and instead gather facts. Organisms thought to be simple suited the streamlined method. John B. Watson sharply endorsed observation in 1913 when he denounced the notion of an intangible mind as a "yoke of consciousness" on his aspiring science. More positively, Margaret Washburn identified "inference from behavior" as a reliable substitute for verbal reports of mental states in *The Animal Mind* (1908). These affirmations of objective data echoed an enduring American preference. The new nation's scientists around 1800, one historian writes, believed that finding truth required a "simple accretion of knowledge, the heaping up of fact upon unrelated fact." In 1946 the émigré biologist Ernst Mayr similarly advised a European friend coming to lecture in the United States, "American scientists do not care for too much speculation but like thorough analyses of observed phenomena."[3]

Empiricism thus bound Americans as an ethos and a method, and animals bore the consequences of the insistent hunt for evidence. They were less likely than humans to possess something like a mind and had no language anyway to miscommunicate private information. They were also available for constant scrutiny. The habits of ants, frogs, pigeons, crows, rats, raccoons, pigs, and primates soon filled an expanding American scientific literature on behavior.[4]

If this commonsensical American science made animals a magic key to understanding organisms, intelligence was the initial focus of inquiry, and not emotions. Mental facility for the Americans nearly meant know-how: acquired skill especially admired by a people historically devoted to useful knowledge. The combination of "*thoughtful* observing" and weighing "multiple hypotheses" is "the necessary condition of progress," declared John Dewey in *How We Think* (1910). Besides, mastering a task could be measured. In *Animal Intelligence: Experimental Studies* (1911), Edward Thorndike explained how kittens or chicks gradually learned to release a door in a puzzle box or maneuver a labyrinth. Searching for calculable evidence of improvement, experimenters sidestepped emotions as elusive. Psychology would never become "an undisputed natural science," Watson warned, so long as critics might charge that "your introspection is again faulty": "Feelings are never clear." Robert Yerkes was more apologetic about the lopsidedness of research. Although study of "our feelings, emotions, and sentiments. . . is less advanced along strictly scientific lines than is that of sensation, perception, ideation, and the various aspects of cognitive consciousness," there are few areas "of more obvious importance scientifically than affection and its complexes." But skepticism that animals have feelings compounded other reasons why laboratories did not add emotional tests. Animals are less capable than humans of "emancipation" from the "tyranny of sense stimuli and habit," in Dewey's judgment.[5] Despite their many uses, animals were unlikely to advance a science of emotional experience.

This history makes it unsurprising that intelligence as a trait still inflected the words chosen by the American zoologist Donald Griffin for his 1976 title, *The Question of Animal Awareness: Evolutionary Continuity of Mental Experience*. And yet Griffin's version of knowing exceeded cleverness and made room for "animal feelings." He had a solid reputation as an experimenter, credited with discovering how bats navigate when they probe their surroundings by emitting resonant sounds. He named the process echolocation in 1944. Three decades later, generalizing more boldly, Griffin ventured onto new—or, strangely enough, old—intellectual ground. Animals have "mental images" that may be called "*consciousness*." Moreover, "the obvious fact [is] that our mental experiences include not only images and intentions, but also feelings, desires, hopes, fears." Griffin was comfortable with scientific inference and impressed by animals' natural intelligence, including emotions that may assist survival. Some American colleagues were aghast at his boldness. Had Griffin penned "the satanic verses of animal cognition," as one critic charged, or, as others whispered, was he suffering "premature senility?"[6] Griffin's book was not an aberration, however, but instead an outgrowth of the empiricism it now challenged.

I argue that American empirical practice, as applied to the field of animal behavior, was a dynamic system that not only permitted scientists to change their minds but also encouraged them to do so. During the twentieth century, successive cohorts of psychologists and biologists reintegrated emotion into their explanations of the activities of diverse species. Inherent in the empirical mindset was a productive instability: the imperative never to rest content until all the evidence was in. The down-to-earth labor of accumulating practical trials put disciplinary loyalties in the background and promoted collaboration. Emotions, physiologically grounded and yet individually felt, hovered between the biological and mental sciences. Even a single organism's life cycle was sufficiently complex to spark conversations across rising specialties ranging from genetics to ecology. Most basically, animals pushed forward revisionism. As living beings, they upset hypotheses.

Yet for all its nurture of inquisitiveness and reflection, American empiricism was a doctrine. Its practitioners mistakenly saw experimentalism as nature's method, giving them direct access to the phenomena around them. Too often this made them blind to their premises, unaware of their theorizing, and intolerant of opponents they charged with not having run enough tests. Characteristically, American psychologists cast their nature-nurture tussle of the 1920s with the British-born émigré William McDougall as a disagreement about acceptable technique. McDougall's defense of inborn "native dispositions" could not be credited, one critic wrote, because of his unfortunate tendency to "postulate" using "hypothetical" evidence. In contrast, Dewey praised empiricism as akin to the child's perfect instincts: the "unspoiled attitude of childhood, marked by ardent curiosity, fertile imagination, and love of experimental inquiry, is near, very near, to the attitude of the scientific mind."[7] This conviction of special national insight filtered—but did not block—ever-accelerating internationalism in the field of animal behavior. Professional study, collaboration, and fieldwork abroad increased at a steady pace. The Americans' faith in their monopoly on hard-headedness was nonetheless tenacious. They took pride in fact-finding as their singular skill and were eager to show it off. Predictably, there were episodes of unappealing censorship of foreign science. Empiricism itself proved mutable, however, in the long run.[8]

Because I take the Americans' mandate to investigate as an honest aspiration, I approach the rediscovery of animal emotions from the bottom up. My case for the transformative power of practice rests on two proposals: animals were influential actors, and scientists learned from them. From the outset, they spent many days together in close quarters. Imagining their subjects to be a mix of evolutionary link and human substitute, the scholars came face to face with the animals' personalities and sensed that they were more than machines. "Even those who claim that animals are automata do not treat them as such," wrote Robert Yerkes at the Harvard Psychological Laboratory in 1906. Giving a lecture across town around 1903, Walter Cannon distinguished "*hunger* as *need*" and "*appetite*. . . as desire."[9] Simple biology triggered less tangible yearnings that set off a hunt for satisfaction. The humans were open-minded enough to be surprised.

The animals were immensely varied in kind and plentiful in numbers. Although laboratories gravitated over time toward use of species that were prolific and easily cared for—memorably, the white rats of psychologists and fruit flies of geneticists—subjects ranged from unicellular organisms to the great apes. A scientist's academic discipline affected his or her preference. Psychologists who published in the *Journal of Animal Behavior* in the 1910s studied mainly mammals and birds (94% of their papers), in contrast to the zoologists' focus on invertebrates (69% of their articles). Data on medical testing, where the goal was to understand pathology and not behavior, suggest the quantity of animals. In 1908, when popular protests against vivisection led Walter Cannon to make inquiries about humane treatment, he learned that Cornell University Medical College performed tests that year on 1,124 pigs, rabbits, cats, and dogs. In 1938, when C. R. Carpenter sailed from India with 500 rhesus monkeys to launch a breeding colony in Puerto Rico, the United States was importing 20,000 monkeys annually for medical trials. Whether the focus was physiology or psychology, laboratories were the favored American site, so much that scholars abroad depended on their findings. Nearly all of the illustrations in *Investigation of Mind in Animals* (1915), issued in Britain, were borrowed from American publications.[10] But debate never ended about ideal conditions of observation, and study also flourished in zoos, field stations, natural history museums, and wild habitats as near as the backyard or remote as the jungle.

The time American researchers spent with animals was enough to force glimpses of the creatures' autonomy and the affective quality in the scientist-subject bond. Close scrutiny did not produce empathy for the animals so much as a kind of respect. Early on, when facilities were small, lead scientists had routine contact with animals. The widow of psychologist Calvin P. Stone reminisced in the 1950s that her husband had scoured "garbage heaps" at 5:00 a.m. to capture wild rats. One of his former students added, "Stone himself fed, watered, cleaned, and housed the colonies of rats, guinea pigs, canaries, pigeons, rabbits, chicks, cats, dogs, and monkeys that, at one or more times, he used experimentally." The experimenters scrupulously recorded details of animal activity and incorporated stray notes into future projects. Wallace Craig separated a young dove from its parents in 1907 to see if the bird learned normally to mate and nest. In the process, he found that Jack reacted in a "social manner" to him, "charging up and down the cage, kahing and bowing-and-cooing to me, and pecking my hand whenever it came within his cage" to deliver food. Craig was impressed that "innate nervous organization" could be so "supplemented by experience," the crux of his theory of appetites a decade later. Robert Yerkes was similarly intrigued by the human factor in 1913 when he conducted "tests of the savageness, wildness, and timidity of wild rats, tame rats, and of the first and second generation hybrids." His experiment was most likely inspired by anecdotes that laboratory-bred rats were easier to handle, making it a management issue that induced him to examine behavior resembling fear and aggression.[11]

The Americans were nonetheless doubly unprepared to name emotions, much less systematically analyze them: intelligence was their stated interest, and inherited terms were fictions to be shunned. Yerkes's 1911 phrase, "affective aspect of consciousness," soon became impossible following disuse of the concept of mind. Instincts, long recognized as innate prods of feelings, faced skeptics between the wars. The supposed "instincts and their accompanying emotions" cannot be verified "in quite the same objective way one proves the existence of a pineal gland," the psychologist Calvin Stone summarized in 1931. Edward Tolman ventured "driving adjustment" in 1922 to cover reactions such as "flight, pugnacity, [and] repulsion" in "behavioral terms." No new shared nomenclature for emotions appeared, however, because the Americans favored a plurality of concrete descriptors. Around 1960, Frank Beach detected a "system of preference" in his beagles' mate selections, and T. C. Schneirla identified an ant colony's "excitement" generated by "social stimulation" as the catalyst of its queen's fertility.[12] The words were less strictly observational than they seemed. Read closely, they downplayed untutored feelings. "System" intimated that acquired habits channeled the dogs' sex drive, and "stimulation" located the ants' motivation in their surroundings. In neither case were emotions ineffable, but instead embedded in measurable situations. Said more positively, the ad hoc American vocabulary fit emotions into a like-minded science that aimed to remain open-ended.

In experimental practice, however, the Americans adhered with remarkable constancy to the psychic dualism of nineteenth-century British utilitarianism: pleasure and pain. This scientific application of a happiness scale most likely made sense in a country keen on the pursuit of personal satisfaction. Researchers behaved as if sensations on a continuum of agreeableness animated their animal subjects to learn and grow. Early learning trials used food, presumably pleasurable, to reward animals for mastering assigned tasks. Less often, punishment, either physical or mental, painfully discouraged errors. Sexual performance tests, the absorbing work of animal laboratories in the 1920s, equated consummation with fulfillment. Animal suffering experiments of the 1930s, involving induced disturbances to bodies or habits, reflected the premise that disrupted routines bring psychic disintegration. Since the 1960s, legal initiatives to regulate human treatment of laboratory and food-producing animals have proceeded in tandem with experimental studies of pain. The enduring scientific success of the pleasure-pain model in the United States seems rooted in its usefulness: appetite and aversion are easily elicited and tracked, and the results readily applied to future science or policy. Needless to say, the categories were also a screen that hid other responses in animals while they revealed these two. No less, they conveniently skirted what emotions are: whether physical or mental, inborn or learned, or some combination. The value of the dualism was to sustain a steady output of animal work.[13]

The researchers were mirror images of the animals: complex organisms with personalities partly located within the boundaries of science. No matter how we judge Robert Yerkes's idealized view in 1943 of a chimpanzee at his field station as "servant of science," the phrase equally applied to him.[14] Four men typify stages in the rethinking of animal emotions in American science: Robert Yerkes (1876–1956), who considered temperament in relation to learning; Karl Lashley (1890–1958), who added physiological variables to psychology experiments; Frank Beach (1911–1988), who set animal psychology in an evolutionary framework; and Donald Griffin (1915–2003), who viewed subjective experience as a condition of a species' success. All began with at least one professional foot in laboratory psychology, and they remained active experimenters. All were eclectic in disciplinary terms, and although none was free of prejudices or enemies, they tended to be reconcilers. Over time, there was increasing emphasis on the intimacy of physical and mental experience, and the integration of inborn dispositions and acquired habits. All along they presumed the truth of evolution, but gradually widened their focus from the organism's single life to the biosphere's dynamics. With their peers, these leaders pursued a series of scientific problems that, in retrospect, unfolded in a clear chronological order: learning, motivation, sexuality, suffering, maturation, and consciousness. Evidence of animal emotions duly accumulated, but it was the scholars who took steps to comprehend it.

The academic training of animal researchers varied, in part because the charge to be observant made "behavior" a common watchword. This was not the nurture-only philosophy associated with "behaviorism." Rather, psychologists and biologists holding a spectrum of opinions about the nature-nurture balance all aimed to interpret actions—that is, behaviors—as signs of mental processes or bodily dispositions. In 1898 the biologist C. O. Whitman lectured on "Animal Behavior." In 1911 the psychologists John B. Watson and Robert Yerkes, respectively, began their "Behavior Monographs" series and *Journal of Animal Behavior*. In 1975 E. O. Wilson offered his comprehensive "sociobiology" as the key to "all social behavior."[15] The welcoming inclusiveness of the behavior concept produced a crazy quilt of teacher-student lineages. Whitman trained the ornithologist Wallace Craig and entomologist William Morton Wheeler, first-class analysts of animal motivation. Craig in turn taught future-psychologist Leonard Carmichael at Harvard, where Wheeler mentored Alfred Kinsey, a wasp expert before he developed human sexology. The armchair psychologist G. Stanley Hall instructed Margaret Morse Nice, who conducted fieldwork on birds in the 1930s and collaborated in Austria with Konrad Lorenz. The British émigré William McDougall, denounced in America for speculating about instincts, was the first psychology teacher of C. Ray Carpenter, a primatologist. The doctoral adviser of bat expert Donald Griffin was the neuropsychologist Karl Lashley.[16]

Individual fields also changed. Between the wars, animal psychologists supplemented learning trials with ablative procedures performed on organs. Mind may once have been thought to be immaterial, but it increasingly meant a physical brain. Similarly, if the word "biology" ever simply connoted scrutiny of living matter, by mid-century it covered diverse methodologies in genetics, biochemistry, embryology, systematics, and ecology. The "crisis in biology," an ominous phrase coined by paleontologist George Gaylord Simpson in 1967 to warn of indistinct boundaries with the physical sciences, in fact attested to biology's vast reach.[17]

In this scholarly landscape, I understand the scientists I study to be students of animal behavior. This description narrows the field more than it might seem. They were psychologists who concentrated on animals (and not humans) and biologists who focused on animal activities (and not bodies). Not all were interested in animal species for their own sake; they were not necessarily zoologists. But someone whose choice of test subject was incidental to a scientific problem oftentimes became an accidental expert on pigeons or rats.

Keen as they were about experimentation, animal scientists had lives outside laboratories and field sites. They raised families, traveled the world, served in wars, spoke about politics, and dealt with illness. They shared a philosophy that integrated their work and personal choices: the orderly perfection of nature and the capacity of science to discover its secrets. Comparing "the body physiologic and the body politic" in a speech on the eve of World War II, Walter Cannon presented the healthy organism as a social model. If the body's "evolution of a self-regulated internal economy" has "liberated [it] from deranging external influences"—making homeostasis "one of the marvels of biology"—"Can this sort of liberty be achieved in human society?"[18] Cannon answered surely yes.

For these scientists, imagining social policies based on nature's way had no single political outcome. Against the liberal mores of the 1920s, Cannon's wife Cornelia, a successful novelist, pictured responsible sex as directed by eugenics and indulged in only for reproduction. A decade later, however, her husband tirelessly organized medical aid for Spanish Republicans in their civil war against Fascists.[19] In another case, the pacifist

Edward Tolman argued in *Drives toward War* (1942) that satisfaction of psychic needs would forever end violence. This was no bland professional advice. He wrangled repeatedly with authorities. Supporting draft resistance in 1918, Tolman was "hauled before" the president of Northwestern University, where he was teaching, and feared he might be "imprisoned." Riled once again by government in 1955, he successfully sued the University of California after he was fired for refusing the loyalty oath required of faculty. Such colorful politics are important for this book to the extent that they influenced conclusions about animal emotions. Most decisively, World War I helped redirect research from learning to motivation. Clearly, Tolman—for one—was no mere lab rat, but I focus on his career as a "rat-psychologist," as one of his students put it, who dedicated his major book, *Purposive Behavior in Animals and Men* (1932), to "M. N. A. (*Mus novegicus albinus*)": the laboratory rat.[20] I treat the portal between biography and science as a half-open door and put animal behavior science in the foreground.

More narrowly, academic politics—and particularly the right attitude toward foreigners—had a clear effect on Americans investigating animal behavior. On the one hand, they wore their empiricism as a badge of superiority, encouraging isolation, and yet on the other, they searched the world for evidence for unvarying natural laws. Making facts the rule of all value could be intellectually limiting. Robert Yerkes's display of Benjamin Franklin-like pragmatism in 1904, when he advised his peers "to experiment much and speculate little," reappeared in 1966 as a gnarly attack by T. C. Schneirla on the "non-experimental anecdotal" method of the Austrian Konrad Lorenz. The phrase damningly cast Lorenz's theory that animals act out inherited behavior patterns as unscientific. At other times, the Americans neglected European sources. "In Search of the Engram," remembered as a milestone paper delivered by Karl Lashley in Britain in 1949, made no mention of the origin of the idea of engrams, or memory traces, in the German psychologist Richard Semon's *Die Mneme* (1904). Instead, more than one-third of Lashley's fifty-seven citations referenced his own experiments, along with a cluster of titles by his collaborators and just four foreign-language studies, three in German and one in French. This pattern of debts did not seem so much nativist as focused on facts. Sharing a holistic view of the brain with the German-Jewish refugee psychiatrist Kurt Goldstein, Lashley did list Goldstein's clinical summary, recently published in English. Nonetheless, Lashley presented himself to his audience at the University of Cambridge as a tester of theory, explaining the logical progression of his own laboratory trials over the past thirty years.[21]

The visible irony was that Lashley had traveled abroad for an international meeting. The Americans never doubted that truth was unitary for all nature, mandating worldwide science. Visiting Leningrad in 1935, as Ivan Pavlov's personal guest at the International Physiological Congress, Walter Cannon lauded the universalism of "the omnipotent scientific method" that he expected to "deliver man from the existing darkness." Researchers also left home to hunt specimens. Lured by exotic ant species, William Morton Wheeler sailed westward in 1914, away from war, for a collecting expedition in Australia, filling idle hours on shipboard with "Melville's interesting story" of the South Seas, *Typee*.[22]

The journeys were more than the dispassionate treks of fact-gatherers. The romance of chasing down life itself exposed the human being in the scientist. Behind the hard-headed analyst, in the words of E. O. Wilson, hovered the Alabama boy he once was, "daydreaming and aimlessly wandering" along the shore. It was a wonderment the American scientists shared with their countrymen. Brisk sales of Teddy Roosevelt's *Through the Brazilian Wilderness* (1914) coincided with the beginning of European war. Writing that summer, as he recovered from a near-fatal leg wound, Roosevelt chronicled his four-month journey from the border of Paraguay to the mouth of the Amazon, in part down the uncharted thousand-mile River of Doubt. The travelers marveled at toucans and capybaras, confronted jaguars and piranhas, and witnessed a drowning and murder. They collected 3,000 rare birds, mammals, reptiles, amphibians, and fish, dead and alive, for museums and zoos. They left a record of grit in extremis, as Roosevelt recalled his sense of "duty to go forward, if necessary on all fours," although in truth his son had to command the delirious ex-president, for days prone in a canoe or carried by porters, not to die.[23] The book seemed to transport readers to the very heart of nature. This was equally the aspiration of the scientists.

The sustained adventure was their scholarship, however, and beginning in the 1960s, there was a surge of interest in the emotions and in animals, as separate or converging subjects. American researchers of the previous half century were not the sole forerunners of this sweeping intellectual turn. Nonetheless, their empirical mandate had produced prolonged struggle with the problem of animal motivation. In the United States, the halting science of animal emotions set the stage for escalating investigation of the awareness of organisms.[24]

The new views were expansive and often rebellious. Philosophers including Susanne K. Langer, Robert Solomon, and Thomas Nagel now critiqued the explanatory power of forms of rationalism that insisted on empirical verification. They reinserted ideas of feeling and subjectivity into conversations about the mind. The "modern synthesis" forged by evolutionary theory, first promoted by Julian Huxley during World War II and accommodating specialties from genetics to ethology, came into its own and celebrated the relatedness of all animals and the utility of biological theory. Animal welfare activism, grounded in a view of organisms as complex beings, flourished.[25] Historians discovered antecedents of these rising subjects. Thomas Dixon traced past estimates of emotion. Janet Browne examined Darwin's own evolution, Peter Bowler presented non-Darwinian evolutionists, and Vassiliki Betty Smocovitis traced modern neo-Darwinians. James Turner and Janet Davis wrote accounts of humane movements. Donald Dewsbury charted advances in comparative psychology, and Otniel Dror looked cross-culturally at animals' emotional physiology.[26]

In the rush forward, the uneven steps of American empiricists and the animals they chose as mute witnesses to elemental behavior were nearly forgotten. The neglect was not accidental. As if converted, the American successors of Yerkes, Lashley, and Beach embraced international mentors and eclectic vocabularies. Expecting experimentalism to yield truth now appeared naïve, and its one-time champions narrow-minded. In Donald Griffin's words, an evolutionary animal science encompassing awareness and feeling rightfully displaced "the behavioristic tradition in psychology or the comparable reductionism in biology." The revolt was indeed liberating. "What Is it Like to Be a Bat?" This title communicated Thomas Nagel's curiosity in 1974 about what "it is like to *be* that organism—something it is like *for* the organism." His hypothetical creature was the lowly bat, in fairy tales the unpleasant companion of witches on broomsticks and yet now esteemed for a sophisticated directional system that helped sustain colonies. Bats had become fit subjects for a philosopher. Science had made them more than thinking machines.[27]

Today scientific prudence has curbed earlier enthusiasm for a theory of animal consciousness, and yet there is a redoubled conviction that emotions are powerful life forces, even in animals. Marian Stamp Dawkins, the Oxford scholar long devoted to understanding animal suffering, recently worried that excessive claims for nonhuman minds risk "confusion, misunderstanding, and even antagonism to animal welfare." Careful experiments to discern "what animals want" better serve science. At the same time, the past half-century has witnessed the "ascent of affect" as a cultural preoccupation, Ruth Leys writes, and Americans obsessively wonder how they can become, in the words of Daniel Horowitz's book title, *Happier?* Differing opinions about animal subjectivity do not erase a basic consensus that animals feel. In a public exchange over the decision to shoot two birds for a field experiment, Marc Bekoff and Stephen T. Emlen—the first focused on animal suffering and the second on adaptive behavior–concurred that the "animal cost" for "scientific gain" should be small. "If asked, everyone would agree that unnecessary and unnatural pain and suffering in animals should be minimized," Emlen wrote, despite disagreements about where to draw the line.[28] It was Griffin and his contemporaries who made it reasonable to imagine emotional animals.

Before them were the generations of American empiricists and their animals dating back to the turn of the twentieth century. Setting up rudimentary tasks, perhaps then handicapping the animals in successive trials, comparing them with wild strains, and all the while watching and generalizing: intellectually, this was a powerful process. From World War I on, American psychologists paid attention to physiological components of mental activity in the nervous, endocrine, and digestive systems. At the same time, American biologists turned from physical bodies toward unlearned activities such as homing and mating. Emotions crept into American animal study, because feelings seemed located at the interstices of mind and matter, where they helped explain motivations. They had no settled names, and their importance was nearly lost in the recurrent litany of method: test, test, test. Yet slowly but surely animal emotions found their way into American behavior science before Griffin's explosive declaration: to be aware is also to feel, and animals do both. How that happened is the subject of this book.

1
Conversations with the Animals

> Few persons have the time to spare and also the necessary interest and enthusiasm to rise and go out at three o'clock in the morning to hear a bird sing. If you, kind reader, are one of these few persons, please plan to make some records of the twilight song during the coming summer.
>
> —Wallace Craig, "The Twilight Song of the Wood Pewee," 1926

Language distinguishes humans from other animals, but the natural world is still a noisy place. Circumstances such as danger or migration elicit sound patterns in a species, as do reproductive activities such as mating and nurture. In a marathon fifteen-hour session in Massachusetts in 1931, the naturalist Margaret Morse Nice and her husband recorded 1,313 songs of a Black-throated Green Warbler, the father of newly hatched nestlings.[1] In earlier years, by contrast, the first American students of animal behavior were less interested in repertories of wild expressions than answers to experimental questions. Their technical vocabularies in effect gave their animal subjects a human voice. The most frequent question was, how do you learn? Fortunately for the study of emotion, however, experimentation was not fully regulated. A steady, cacophonous undertone, produced in part by the affective life of animals and the humans who studied them, found its way into the record.

John B. Watson's fieldwork from 1907 to 1915 on homing patterns in terns was one case of nonroutinized science. Its origin was serendipitous. The Carnegie Institution was looking for projects for a new Florida field station, and Watson needed a summer research salary. Ranging geographically from the birds' nesting ground in the Dry Tortugas to a release point in Texas, the experiments were in every way distant from his winter rat tests at Johns Hopkins. Watson began by treating homing as a kind of intelligence, regulated by instinct, vision, and memory. His tests quickly led, though, to mating behavior. In one trial "I obliterated the old nest and made another only a few inches to the right or left" to determine if the bird could still find it. Another experiment examined something more than information processing: he removed one mate from the nest to see if the remaining parent stayed with the egg. He also recognized individual temperaments. Trapping birds to transport by ship to Galveston, where they were banded and tracked, Watson noted that "bold birds," less readily frightened away by the roaming experimenters, "are the ones always captured."[2]

Throughout, he presented himself as "the human observer," and he came across as curious, ingenious, and a bit heartless, but also empathetic. It was "a life-and-death matter" for the tern, Watson realized, to leave the nest to find food and then return on a "greatly congested" strip amid "quarrelsome" neighbors.[3] For bird and man alike, emotion figured in the investigation.

Watson recalled in the 1930s that he "always was uncomfortable" with "the stuffy, artificial instructions" typically given by psychologists in their laboratories to human subjects, but "with animals I was at home." Raised in a rural South Carolina family with few social advantages, he seemed happy to avoid the feeling that he had "acted unnaturally" with the people he encountered at top-tier universities.[4] Animal research, however, could not in truth be called natural. Especially in the laboratory, but also in the field, conventions grew up hand-in-hand with behavioral study after 1890 that shaped the work itself.

What made a test animal, or the scientist devoted to understanding it? Published advice and practical behavior alike aimed at an ideal of precision in experimental performance and scientific reporting. Method thereby became the great unifier across a range of animal species and human training. It hardly seemed worth discussing that the substantive question ought to be how animals know: how do raccoons learn to unhook a latch, fish acquire skill at distinguishing colors, or crustaceans use their senses to locate mates? Freedom to invent diverse tasks evoking animal cleverness grew alongside standard procedural values of planning and routine. Not all behavioral scientists conceded that animals were surer subjects than humans; but animal work attracted many psychologists, who found they had much to say to the ant, bird, or primate expert down the hall. In these disciplined communities of animals and humans, there was room for unpredictability. Animal behavior was too variable to conform entirely to protocols, and the reason for any pattern of action was obscure, inducing scholars repeatedly to revise their thinking. Perhaps the tern depends for navigation, Watson honestly speculated, "upon the muscular sense somewhat as does the blind man."[5]

The emotions entered American scientific conversations about animal behavior by this circuitous route. I begin with how experimental animals and their observers came into being. What were their origins, places of labor, and work schedules? Parallel lives did not mean equality: humans made the initial choices, although the animals were willful actors. Nor were the scholars' estimates of emotion ever simple. Quietly edged aside by the investigators' interest in intelligence, the emotions were damned again by their prominence in nineteenth-century theories of mind, now sidelined as unverifiable by American empiricists. The near-universal use of food rewards to motivate animal learning, however, assumed the test subjects' deep-rooted preference for comfort. In all, emotions were laced through animal research that remained focused elsewhere.

Life under the Microscope

At a time when many American researchers imagined a valid animal experiment to be measuring reactions to sensory cues or timing trials in puzzle boxes, the informal observations of the past were roundly censored. Rigorous method, versus intimate knowledge or shrewd insight, was the avenue to truth. In 1908 Margaret Washburn dismissed any amateur who notices "the singular behavior of an animal, and at his leisure writes an account of it," especially if "the animal is a pet." Neither "scientifically trained" nor "acquainted with the habits of the species," the person would be easily misled by "personal affection" and "the desire, common to all humanity, to tell a good story."[6] Descartes, Montaigne, and Darwin were among her offenders. Although her judgments gave voice to the new zeal for controlled conditions and objectivity, her professional colleagues still interacted informally with both amateur naturalists and their own animals. Washburn's line between sloppiness and precision was blurred in practice. As noteworthy, despite her goal to explain nonhuman minds, her critique oddly ignored the animals' experiences. It was the human capacity for pride, carelessness, and self-promotion that damned "the method of anecdote."[7] The push for well-managed procedures in American behavioral study nonetheless gradually changed the lives of animals swept up in the science.

Although Washburn identified the prevailing source of bias as the owners' feelings for their animals, she was also reacting to haphazard data-gathering. Either way, variable contacts between animals and naturalists had typified nineteenth-century science. C. Lloyd Morgan, considered the founder of comparative psychology in the British academy, held out a future hope in 1894 for "an experimental station" at "some university in Great Britain." Meanwhile he raised chicks at home. "Presently I was roused from my writing by sounds of excitement and little pattering feet," as "Blackie" consumed the strip of fabric Morgan had substituted as an experiment for the usual worm.[8] Charles Darwin, praised by Konrad Lorenz in the 1960s for his "uncanny ability to reason on the basis of hypotheses," more seriously violated Washburn's standard.[9] For *The Expression of the Emotions in Man and Animals* (1872), he drew on notes about his pets and children, along with replies to inquiries he sent across the empire to missionaries and colonial officials. In one experiment at the London Zoo, "I placed a looking-glass on the floor before two young orangs" and watched as they "protruded their lips toward the image, as if to kiss it."[10]

From the animals' viewpoint, whether their proximity to humans polluted the science was assuredly less crucial than captivity itself. Domestication had not been uniformly ruled by empathy. Darwin coolly described "a female terrier of mine [that] lately had her puppies destroyed."[11] Washburn was correct, however, that the inherited literature about animal behavior was in no simple sense about nature.

Today, a rat in a maze constitutes our impression of early behavior experiments, but there was no clean historical leap from unregulated curiosity to replicable testing. Field observations of living animals and the collection of dead specimens coexisted with the new laboratory ideal. In 1910 William Morton Wheeler excitedly publicized "Small Artificial Ant-Nests of Novel Patterns," made from glass, aluminum, or plaster, that permitted indoor scrutiny of social insects without "untidy" dirt.[12] The sociology of communal feeding was Wheeler's lifelong interest, and he tested the comparative appetite of his captive ants for sugar, eggs, and flies in his laboratory at the University of Texas around 1900. Experimentation did not preclude other scientific activities, however. Wheeler tramped the countryside around Austin looking for ant species "under stones or logs." His notebook for a 1923 trip to the Canal Zone identified the 347 kinds of ants he observed. "Spent the morning in Zetek's lab labelling specimens" that had been asphyxiated and pinned.[13] On these journeys, Wheeler was systematic in a way that might have satisfied Washburn. Increasingly, though, controlled experiments became the norm.

Ironically, developing specialized skills made many animal experimenters all the more dependent on amateurs who had common-sense knowledge of animals. This human chain jeopardized scientific purity. Acquiring test subjects was an unexpected chore, and although some specimens purchased from go-betweens came directly from the wild, others had histories that perhaps left incalculable physical or psychological marks. One supplier ran a business on Lake Erie in 1903 that provided aquatic salamanders to Robert Yerkes at Harvard; the man also sold turtles as pets. Another middleman marketed sparrows to trap shooters in Indiana; he also caught birds for a psychology professor.[14] Academic demand seemed to encourage mass production. Harvard Medical School used 5,000 frogs a year by 1901, and Walter Cannon ordered them from a " 'frog farm' " near Chicago that bred the amphibians in " 'frog trenches.' "[15] The forty-one dogs tested in 1908 in the medical laboratory of Syracuse University had varied origins. "Some are sent by owners to be disposed of," stated a report by a faculty member, who also listed, "Purchase from owner. Purchase from men & boys who catch stray dogs. Students catch a few."[16] Clearly, grassroots entrepreneurs were exploiting the budding academic market that emerged side by side with unconventional pets, sportsmen's clubs, and the dislocations of urban life. It seems safe to call the animals caught in this unregulated business network generally unlucky.

Once animals arrived on campus, behavioral researchers made at times heroic efforts to house and feed them. The experimenters were limited by space and money, however, as well as their own ignorance, which reinforced their reliance on amateurs. At the University of Chicago, Charles O. Whitman never secured funds for a proper facility for his pigeons. He died during a cold snap in 1910 after working outside to save the birds that lived in cotes around his house. His story explains why Willard Small literally thought small for his doctoral research at Clark University around 1900. He kept four rats in a 16-by-16-by-20-inch ventilated box with sawdust for nesting, modestly adding a "squirrel revolver for exercise" for a later foursome.[17]

By contrast, the new Harvard Psychological Laboratory, opened in 1905, was magnificent. On the third floor of Emerson Hall, above a lecture room connected by a "monumental staircase" to a library upstairs, were cages in the "vivarium-room," an eighteen-foot "aquarium table," and electricity for apparatus. The layout copied the spatial organization of zoology research at German universities, praised by Whitman in 1887 as nearly "ideal." The irony was that the proud author of the Harvard report was Hugo Münsterberg, to be sure a German-trained psychologist and the laboratory's director, but also a believer that the proper subject of mental experiments was human consciousness.[18] Because Emerson Hall was a place for both introspective reporting by people and tests on animals, the building was a compromise.

More practically, elegant quarters did not guarantee expertise in animal care. Robert Yerkes, the force behind Harvard's animal work, kept handwritten notes on the habitats, diet, and reproduction of test species such as snails and newts. He learned from his salamander trapper that amphibians prefer "a dark place" and eat fish.[19] Good intentions went just so far. "Alack! You will believe me a fine custodian," confessed Yerkes's coworker Edwin Holt, like Yerkes a recent Harvard PhD. "Four of your large frogs have died," Holt explained to Yerkes in 1904: "I don't see how I have managed to do this."[20] Mere survival in the laboratory, even for animals nurtured for the purpose of demonstrating healthy behavior, required knowledge of the species.

At the same time that behavior investigators were setting up their animal laboratories, even they had doubts about them. In repeated calls for field stations, they expressed their worry as a concern for science, where the well-being of the animals was one, and only one, component. Whitman, the founding director in 1888 of the Marine Biological Laboratory at Woods Hole on Cape Cod, championed "a biological farm" where "the forms selected for study can be kept under natural conditions" and exhibit "natural behavior." He envisioned a pastoral middle ground between wilderness and civilization. He did not endorse amateurish Sunday treks into the woods, but, emphatically, pure research. Locate the site, Whitman declared in 1902, "at a safe distance from any large city or summer resort" for "*complete isolation from the public*" and yet near "a laboratory holding the position of a national center of biological research" and a government wildlife station.[21]

When Watson and later Yerkes took up the cause for field stations in 1906 and 1914, they, too, were disturbed by existing constraints on the scientist. Noisy laboratory animals in cities put experimentalists at risk of "arrest," Watson warned, "for keeping a nuisance." Weary of campus politics that transferred his animal quarters in Emerson Hall to the educational psychologists in 1913, Yerkes aimed to use his own 150 acres in New Hampshire "both as a summer home and as a reservation." All the proposals brimmed with faith in science, if freed of restraints. Experimentation was not the problem, because the men reaffirmed strict methods. Rather, they were disappointed by the distractions of academic laboratories. In Watson's view, field sites "will give to the animals naturalness in environment and to the investigator ample space for experimentation."[22] The discussion did not seem really to be, however, about the animals.

Whether testing took place in the city or country, the day-to-day life of the animals consisted of a work routine shaped by the investigators' own drive. Yerkes, for one, was notorious for "how stubbornly you stick to your work," in a colleague's words, and his professional influence raised the bar for animal labor.[23] In the late 1890s Edward Thorndike, credited with inventing the puzzle box where an animal learned to get in or out, required his cats, dogs, or chicks to repeat a single task about one hundred times. By 1914, Yerkes's two-month-old pigs, prompted to locate food behind multiple doors, responded to each pattern of food placement up to six hundred times. Purchased from a New Hampshire neighbor, who presumably bred the pigs for market, they instead spent their summer with a graduate student who counted their errors.[24]

L. W. Cole, another of Yerkes's students who became a professor in Oklahoma, learned that raccoons absorbed just so much of the human work ethic. He gave six raccoons fastening devices every day for nearly a year beginning in July 1905 and tabulated hundreds of results. "In each day's or half-day's work," a raccoon's initial "eagerness" gave way to "hunger without too great eagerness," after which efforts "gradually became slower" and the subject demonstrated "little more than a native desire to be occupied."[25] Attention and reaction waned more sharply in animals other than mammals, and for reasons less easy to infer. Yerkes found that over time his frogs grew "less responsive," and Margaret Washburn and Madison Bentley, collaborators at Cornell, attributed their "fish's diminishing appetite" to "confinement." In other situations Washburn felt more certain that the work itself was taxing. She felt "sympathy for the harassed animals," now rats, forced repeatedly to make trials that she believed exceeded their intelligence.[26]

Behavioral scientists did not intend, however, to work their animals to a point where reproduction stopped. Establishing colonies was a goal from the outset. Curious from the first about hereditary variation, researchers excitedly projected multigenerational studies. "In five years," Watson predicted, a field station "should have an unbroken

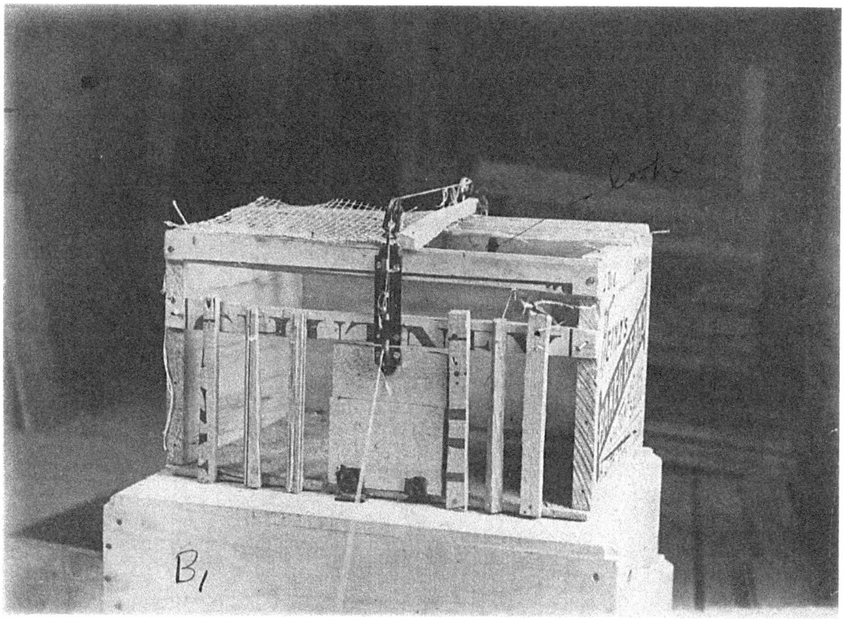

Figure 1.1 One design of Edward Thorndike's puzzle box, adaptable for cats, dogs, and chicks, ca. 1898. Courtesy of Robert Mearns Yerkes Papers (MS569), Manuscripts and Archives, Yale University Library.

succession of 20 generations of a rat family."[27] Cost-saving considerations favored plans for natural increase, too, especially for exotic species such as monkeys. Even so, the advantages of maintaining lineages did not mean that the survival of a single animal was a priority. Experimental protocols made individuals dispensable for a variety of reasons. Investigating the physiology of learning in 1901–1902 for his dissertation, John B. Watson dissected the brains and spinal cords of rats at incremental ages, beginning at birth. Nor did he let unlucky accidents jeopardize his research. One of his rats used in color discrimination experiments in 1911–1912 "worked faithfully until the day before its death," which was "due to cancerous growth upon neck; during the past month had increased enormously in size."[28]

Despite individual losses, loans and gifts of animals from scholar to scholar indicate that laboratory colonies stabilized. Wallace Craig studied pigeons belonging to "the large collection" of his mentor, C. O. Whitman, for his dissertation at the University of Chicago. A Harvard colleague who bred rats to analyze "the heredity of structural characteristics" suggested to Robert Yerkes in 1910 that Yerkes compare the psychology of wild and tame strains. Of Yerkes's "nearly three hundred individuals," the wild ones were caught on local streets, while the tame animals descended from a line developed in the "Zoological Laboratory for at least ten years."[29] Actual births in captivity of more complex animals were noteworthy events. An article by Watson and his student Karl Lashley on the unplanned birth of a monkey at Johns Hopkins in 1913 focused on maternal and infant behavior, and not the biology of reproduction.[30] The underlying message, though, was that laboratories could sustain animals through the lifecycle.

Understandably, critics outside the academy cared less about research populations than the treatment of individuals. As laboratories expanded, there was debate among reformers, journalists, lawmakers, and scientists over what constituted cruelty. On the surface, medical experiments might be expected to be more hurtful than observations of behavior. Walter Cannon's private notes on his tests of digestive processes in frogs, roosters, geese, rabbits, dogs, and cats are indeed hard reading. In one case where the animal was permitted to survive a "bi subnitrate capsule" followed by x-rays, he wrote in 1897: "Cat not anaesthetized. Tied so head free to move up & down between bars for 2–3 centimeters. Board laid flat." Behavior studies could also involve physical harm, however. To determine how male amphipods locate their mates, S. J. Holmes "blinded" some with "asphalt varnish" and in others removed "both pairs of antennae" that enable smell.[31]

As controversy swirled around creatures that evoked more empathy than these tiny crustaceans, it made sense perhaps that attention turned from the animals' suffering to human behavior. It was the experimenters, after all, who might be persuaded. In 1907 the *Nation* magazine put science on trial when it censored Watson, now depriving his rats of normal sight, smell, and touch, for "the torture of animals for merely trivial investigation."[32] William James, the nation's preeminent psychologist when he wrote for the *Boston Transcript* in 1909, blasted the "callous individual experimenter" who forgets "that there is 'a God in Israel' to whom he owes account." Although James also spoke for "the rights of the helpless," he seemed distracted by his colleagues' irresponsibility and the antivivisectionists' equal "idiocy, bad temper, untruth, and vexatiousness."[33] Whether test practices were in fact scientifically justified or wanton, the animals were unavoidably silent in these verbal battles.

Walter Cannon reasonably hoped for more humane conditions through good teaching of upcoming scholars. In a questionnaire for laboratories that he devised in 1908 for the American Medical Association, now on the defensive, he asked if students did "independent experimentation" on animals, as if poor guidance from teachers and hence ignorance were the problems.[34] In fact the quality of instruction was mixed. Undergraduate courses were broad, graduate training narrow, and some of the best practical supervision occurred at small colleges that did not educate researchers. During the same years when Yerkes came to exemplify laboratory rigor, his classes at Harvard were in every sense scattered. In 1910–1911 he taught courses in educational psychology, general psychology for extension students, and "Ontogeny of Mind," an overview of organisms that facilely relied on contested concepts such as "consciousness."[35] Because accompanying practicums in the life sciences often consisted of "taking dead and more or less smelly crayfish, earth-worms, starfish and cockroaches to pieces," in the words of the entomologist William Morton Wheeler, now at Harvard, it was no wonder that undergraduates who continued to graduate school were "obsessed with problems" and blind to nature as "an inexhaustible source of spirit and aesthetic delight."[36]

Students outside the academic mainstream seemed to acquire research skills in a more balanced way. The nearly seventy articles about animal and human experiments coauthored by Margaret Washburn and her female psychology majors at Vassar College between 1903 and 1937 were products of a deliberate teaching strategy. "The problem and method of a study having been determined by me," Washburn explained, "the experimenting was done by the students, who also formulated the results; the interpretation and writing of the reports fell to me and the paper was published under our joint names."[37] Even here, the care of animals merited no comment. It was nonetheless true that the teachers who apparently were most devoted to instruction may have been the same ones whose students had least professional contact with animals later on. In contrast to the tens and hundreds of animals acquired each year by medical schools that mainly educated men, the New York Medical College and Hospital for Women used a total of four cats and three dogs in 1908, Cannon learned, "sent from the men's homeopathic college, none are kept here."[38] There was little chance for the teaching mission of a women's college, such as Vassar, to affect the many research animals handled by men.

Apart from episodes of public criticism or population disaster, experimenters seemed half-aware of animals' traits that were unrelated to adaptive learning. Slowly, though, comments about temperamental differences between wild and captive members of a species entered the literature. The laboratory routine was somehow transformative for animals across generations, and the difference involved the emotions. Not until near the end of his published dissertation did Wallace Craig inform readers that his pigeons from Whitman's colony grew less quiet after egg-laying than did "the wild species." Treating his observation as simple fact, he did not elaborate, nor did others who took a strictly practical interest in tame behavior. In response to well-publicized reports about uncooperative monkeys, Cole, for example, "took pains . . . to handle the young raccoons as much as possible."[39] Rather than wonder what made the animals ornery, he tried to make them compliant, so as to get on with his work.

When rats were the experimental subjects, scholars agreed that the white variant that quickly became commonplace in testing was comparatively calm and friendly. Whether this trend was due to nurture or nature, however, was weakly analyzed. In 1903 Watson explained briefly that his "exceedingly tame" rats had been "raised by the writer." A decade later Yerkes attributed contrasting temperaments in wild and laboratory strains to "heritable behavior-complexes," yet still awaited "more definite

Figure 1.2 Lawrence Cole training a raccoon, 1905. Courtesy of Robert Mearns Yerkes Papers (MS569), Manuscripts and Archives, Yale University Library.

results concerning the modes of transmission."[40] This muted discussion of causes most likely reflected widespread scientific uncertainty about whether Mendelian genetics, rediscovered just after 1900, successfully disproved the heritability of acquired characteristics. Yerkes's statement to Harvard students around 1910 that "now everything is altered" because "we have Mendelism" seems more a classroom truism than a confident assertion of scientific law. As late as 1923 William Morton Wheeler remained "much interested in Lamarckism," in the words of a colleague who sent him a citation arguing for the inheritance of learning.[41] Experientially, the experimenters witnessed changes over lifespans and generations in the animals they worked with. They were not sure, however, why transformation occurred.

Irrespective of the mysteries of heredity, they also seemed to lack a common-sense understanding of just how unnatural the experimental routine was for animals. In place of days and nights spent attempting to satisfy hunger, avoid danger, and produce offspring, the animals now often performed repetitive tests, with food offered or withheld and babies permitted or removed as protocols dictated. Set in a positive light, the experimental animal was a "servant of science," words of praise coined by Robert Yerkes.[42] Researchers deliberately pinpointed specific behaviors as if putting them under a microscope, expressing only moderate worry that the technique might distort results. The emotions as a research subject were initially set aside. Scholars of animal behavior may have had few second thoughts because they themselves were busy changing. Behavioral science affected the researchers, albeit voluntarily, much as it did their animal partners.

Professional Observers

Behind the confident tone of the animal investigators' publications were careers that were far from simple. In a workmanlike way, experimental reports explained tests, pictured apparatus, logged observations, and tabulated measurements. These details were a clue that the writers were not the natural philosophers of the past, whose main occupation was authorship. Rather, they were academic workers with ongoing needs for the money, space, and living subjects required to produce empirical results at a steady pace. They felt the discomfort of having to invent procedures to make the system function.

In one awkward situation, the elder philosopher Josiah Royce loyally promoted Robert Yerkes's efforts to establish animal study in their common academic unit, the Harvard Philosophy Department. Royce made a strenuous case to one university president in 1906 that "Comparative psychology is no 'frill.'" Just months before he died in 1916 he dickered again with the next president over Yerkes's salary. In the meantime he got in a heated exchange at a faculty meeting with their colleague, George Herbert Palmer, "for your sake," he explained to Yerkes, "and that of the other psychologists." Nothing in these circumstances was unusual—neither the uncertain institutional fit of animal behavior, the financial demands, nor the divided opinions. Yerkes and his peers did not develop their science in ivory-tower laboratories. Beginning around 1900 they made themselves into animal researchers, and it was, among many things, an emotional process. When Yerkes wrote to Royce to apologize for exposing the senior scholar to

Palmer's unpleasantness, Royce returned cordial thanks "for your kind sympathy."[43] Alliances, conflicts, and manners, good and bad, were part of the business.

In terms of background, the men and sole professional woman, Margaret Washburn, who began animal behavior investigations around 1900 represented a kind of American average: comfortable without being wealthy and bright but not overly philosophical. Earlier scientists in the United States, in contrast, had often come from privileged families and wrestled painfully with the implications of science for religious faith. James McKeen Cattell, for example, who in 1891 became the first American to bear the academic title professor of psychology, considered suicide in the 1880s as he pondered "agnosticism and materialism" amid pressures from a pious mother and father, the latter a college president and Presbyterian minister.[44] The younger scholars were children of more modestly placed ministers, farmers, teachers, doctors, merchants, and manufacturers. Although a few grew up quite poor, such as John B. Watson, the middling status of most made their admission to elite universities a matter of luck and grit. Expanding its social base, Harvard recruited Walter Cannon from a public high school in Minnesota for its freshman class in 1892. He was already twenty-one years old, having worked as a teenager for a railroad. Yerkes was the same age when he arrived in Cambridge in 1897. Because Harvard would not count Yerkes's AB degree from Ursinus College as equivalent to its own training, he spent his first year there repeating undergraduate courses.[45]

Such uncomplaining steadiness also characterized the animal researchers' religious reflections. Their later memoirs routinely mentioned childhood religion, and the fathers of Edward Thorndike and Margaret Washburn were ministers. Without recorded trauma, however, devotion to science displaced faith. Like young Benjamin Franklin, who made Sundays his "Studying-Day" and "absented myself" from church, William Morton Wheeler "regularly spent my Sunday mornings" in the 1880s "drawing the palpi and epigynia of spiders." "Xmas day" 1914, on shipboard home from Australia, similarly drew no special comment from Wheeler except "a fine luncheon & dinner." As casually, Ada Yerkes informed "husband mine" that their daughter had attended Sunday school and she herself went to church.[46] The news did not seem a disputed issue in the family, but rather an offhand remark about a personal habit.

Easygoing about metaphysics, the animal researchers instead invested themselves in their work, as their test animals would learn during laborious experiments. Their education was an intellectual hodgepodge, however, both exciting and frustrating. As they completed academic degrees in various areas of psychology and biology, they received little useful guidance on experimental techniques or animal husbandry. The psychologists, mainly educated by philosophers, felt especially deficient. "The training which I have most keenly missed," Edward Thorndike reflected in 1936, was in "standard physiological and psychological apparatus" and also "mathematics." John B. Watson took the initiative to repair gaps in his instruction. More interested in neurology than most psychologists, he spent the summer of 1904 at the Johns Hopkins Hospital "working up an operative technique on animals."[47]

As people who generally grew up in towns and cities, these students were similarly weak in firsthand knowledge of animals. It was easy for them to flounder between the theorizing of armchair evolutionists, on the one hand, and the specialization of empiricists, on the other hand, researchers like C. O. Whitman, who devoted the first half of his career to leeches and the second half to pigeons. At Clark University, wide-ranging lectures delivered six days a week by psychologist and president G. Stanley Hall on "religion, child development, sex, abnormal psychology, and social problems" were obligatory for Margaret Morse Nice as she worked on her masters thesis, "The Food of the Bobwhite." She was intimate with her bird subject, Loti, who lived with her even after her marriage in 1909. But even Nice saw why a critic of her published paper "ridiculed the value of discovering that a Bobwhite would eat clothes, moths, mosquitos, and house flies."[48] Experimental technique was not Hall's forte.

Altogether, these scientists of animal behavior found their advanced studies exhilarating and uneven. Thorndike revered William James, a brilliant speculative psychologist and his first teacher, sufficiently to dedicate a book to James in 1913. It was simply outside the range of James's talents, however, to help Thorndike overcome his "distaste for using machinery and physical instruments."[49]

The unsystematic quality of the experimentalists' specialty training did not mean that they were simply one step up from amateur naturalists. Distinctive intentions, roughly stated as investigating principles versus disseminating information, separated would-be scholars from knowledgeable observers. It was nonetheless true that the aspiring scientists worked in intellectual territory already occupied by animal experts of another kind.

Frank M. Chapman, for one, was a powerful figure, and largely self-taught. Curator of birds at the American Museum of Natural History from 1888 until 1942, Chapman took up the utilitarian task of educating the public. Writing for birders in *The Auk* in 1893, he rated exhibits of "avian taxidermy" at the Chicago World's Fair, good-naturedly swallowing his distaste for one "theatric display" of Costa Rican fauna set against "a large, painted background."[50] He realized that in a nation increasingly composed of city-dwellers, reading was the surest way to engage wildlife, and few nature writers were as adept at their craft as Chapman. In books ranging from *Handbook of Birds of Eastern North America* (1895) to *Life in an Air Castle* (1938), narrating his recent Central American trip, he demonstrated, in a colleague's words, a "well-rounded mind" and "flair for journalism."[51] He comprehended more than he showed. Examining records on the locations of capture of mounted Bobolinks dating back to the 1840s, he produced a short article about migratory patterns. It was an impressive analysis.[52] Chapman was not an experimentalist, however, or a college graduate.

By contrast, Margaret Washburn was arguably among the lowliest of the animal behavior researchers, given her post at a women's college, but she little resembled Chapman. Oblivious to the public's recreational pursuit of birds or its curiosity about kinkajous and coatis, she conducted experiments on fish, rabbits, mice, and, indeed, undergraduate students. She oversaw production of four editions of her textbook, *The Animal Mind*, reissued in part to update a bibliography that finally numbered 1,683 citations. When Columbia University refused to admit her, as a woman, to its doctoral program in psychology in 1892, she doggedly turned to Cornell, where she was accepted.[53]

Both Chapman and Washburn were serious analysts of animal behavior. No mere cataloger or tour guide, Chapman crossed paths with university researchers. Against this background of a well-informed popular interest in animals and their habits, though, the new experimentalists seemed to agree that proper scholarship required advanced education.

Figure 1.3 Margaret Washburn on the Vassar College campus. Courtesy of Archives and Special Collections (Ph.f 8.93), Vassar College Library.

More immediately, the up-and-coming behavior researchers distinguished themselves by developing a common literary form, the experimental report of animal trials. The texts were intensely observational. The first step was to note down the animals' habits, thereafter systematically altering circumstances to probe research questions. Studies that began as dissertations were especially apt to use a daily-log format, chronicling steps and confessing missteps with individual animals. Whether explaining apprentice work or ongoing projects, however, the accounts included illustrations in the form of sketches and tables. There were hand-drawn pictures or photographs of mazes, puzzle boxes, and geometric shapes that the animals learned to identify. Raw numbers representing each animal's correct and incorrect responses to a problem or the time required to master it were fully listed. Although Yerkes was one of the few who supplemented these counts and measurements with statistics such as means, standard deviations, and probabilities, even he published long columns of data gathered in successive trials. In 1915 one half of a forty-page article describing his tests of pigs' intelligence consisted of series of numbers.[54]

The papers' typical lack of notes or a bibliography further focused attention on the experiment itself. Authors discussed precedents within their narratives, and readers consequently did not have to sift through appended material to grasp the implications for scholarship.[55] In all, because these stylistic conventions became commonplace quickly and without fuss, it seems safe to infer that animal experimenters saw the format as a convenience and not rigid rules. Its components comfortably matched their routines involving apparatus, animals, and patient watching. A writer could trust that a colleague who picked up the paper would know how to read it.

Having a workable template, however, did not automatically get findings into circulation. Then as now, scholars fretted about publishing arrangements. At one extreme, C. O. Whitman's fastidiousness largely kept him from issuing his records on pigeons during his lifetime. His insistence on "long-continued patient observation, experiment, and reflextion" meant that John B. Watson, Whitman's colleague at the University of Chicago, was unaware of Whitman's research even after Watson surveyed the bird literature in 1907 to prepare for his first season of fieldwork with the Florida terns.[56]

Wallace Craig, who revered Whitman as his teacher, took a different tack by placing his work in a variety of journals. Although the *Journal of Comparative Neurology and Psychology* was a fine outlet for Craig's dissertation in 1909, the *American Journal of Sociology* was an odd choice for an animal researcher when "The Voices of Pigeons Regarded as a Means of Social Control" appeared there the previous year. In 1911 the passenger pigeon's "peculiar voice," its organ for "expressions of emotion," was his subject in *The Auk*, a serious but not an academic periodical published by the American Ornithologists' Union.[57]

Whether intellectual fertility or professional desperation led Craig to disperse his papers is hard to say. He was a vivid writer who enlivened standard reporting. A boy, he asked readers to imagine, might fail to identify a bird if instructed to search by appearance. "But tell him to look for a pigeon that shrieks and chatters and clucks instead of cooing, and the boy will be less likely to make a mistake."[58] Craig, in spirit, was this eager boy, and yet also the sole psychologist at the University of Maine, "too poor" in 1913 "to buy the pigeon cages which I ought to have" and with a family so hard up for "clothes that we look hardly respectable."[59] As solid a scholar as Whitman, and with much to say, Craig was a prolific young researcher. His publishing history nonetheless underlines the lack of tried-and-true routes from the animal laboratory to readers' desks.

In this light, the publication of the *Journal of Animal Behavior* beginning in 1911 seemed a perfect solution.[60] It welcomed "field studies" and "laboratory studies" of "the naturalists and the experimentalists" pertaining to "the behavior and mental life of organisms." Its inconsistent use by contemporaries is a reminder, however, that they saw scientific success as something more than access to data. Regular contributors alternately appropriated the journal for their promotional pieces or decided to bypass it for popular magazines. Either way, they sought reputation. When Robert Yerkes, the journal's founder and managing editor, used its pages in 1914 to press his case for a rural field station, he simply redirected his habit of publicizing favorite causes. He had touted "recent progress" in animal work for the *Journal of Abnormal Psychology*, for example, in 1908.[61] His own outlet made building the field that much easier.

He was equally keen, however, on wide public visibility. He tried to persuade *Harper's Magazine* and the *Atlantic Monthly* in 1916 to accept either of two papers written by his wife Ada, "Animal Individuality and Genius" and "How Ideas Look: A Glimpse of the Monkey Mind." As his co-researcher, and not just coauthor, Ada was assuredly disappointed that her analysis of rats, crows, and primates "falls short," in the words of the *Harper's* reply to her husband, "of what we seek to give our readers in such popular science articles." The couple settled for placing one essay in the magazine of the American Museum of Natural History, with Robert now listed as lead author.[62]

At least Ada had succumbed to worthy competitors, notably John B. Watson. *Harper's* welcomed clever adaptations of his lengthy tern reports. He talked easily to "the average man" in "How Animals Find Their Way Home" in 1909 and humorously debunked scientists in 1915 who claimed that migratory birds navigate by sight. If a bird is "nine hundred miles away" from the cote and relying on vision to find home, it has to be "eighty-five miles high!"[63] Watson respectably covered the latest theories and experiments in conversational prose for middlebrow readers. Whether he earned the envy or gratitude of fellow specialists, he clearly judged time devoted to communication with the public as well spent.

Although Watson likely expected his chatty essays to attract attention to his field, outreach was perhaps also a respite from professional conflicts. Running tests on animals was just one part of scholarly work, and when a professor was not the mainstay of a tiny department of psychology or biology, intellectual battles became interpersonal. Hugo Münsterberg was not the only psychologist in the United States who so firmly equated experimentation with human introspection that he ruled animal trials "no longer psychology."[64] The issue hit home for Robert Yerkes, however, the recipient of the letter containing this judgment. It strained their relations for years after 1902 while he served under Münsterberg in the Harvard Psychological Laboratory. The "Chief" nickel-and-dimed Yerkes about "animal work." A thirty-dollar outlay seemed "disproportionate" in their budget, Münsterberg chided in 1909, and he warned in 1910 that one hundred dollars to help launch the *Journal of Animal Behavior* was a one-time gift. The next year they contended over Philosophy 20A, the laboratory practicum deemed "your course" by Yerkes, which he wished to make "our course." Münsterberg professed "astonishment" at the request, "as you have never shown the slightest willingness to ask my advice or to consult with me as to any of your animal researches."[65]

Another time, Yerkes politicked to hire John B. Watson. Edwin Holt, a colleague, doubted "the Professor" would accept someone "reputed to be an animal psychologist." Besides, Holt added, if Watson is the "only person in charge" at Johns Hopkins, "I think him more fortunate than either you or I."[66] These exchanges were made all the more painful by enduring cordiality between Münsterberg and Yerkes. "Tante Selma" Münsterberg wrote to two-year-old Roberta Yerkes about Christmas candy in 1910, and after Münsterberg died suddenly in 1916, Yerkes helped the widow move her residence. Afterward "Frau Selma" gratefully wrote, "I shall never forget your helpfulness and kindness and friendship."[67] Most noteworthy, both men remained exceptionally productive in this tense environment.

The wearying interplay of principled commitments and limited resources did not unusually affect animal experimenters. Across the disciplines, founders of research fields witnessed competitiveness erode gentlemanly scholarship. James McKeen Cattell, the Columbia psychologist and a prime mover in the establishment of the American Association of University Professors in 1915, compared his employer to "a department store" favoring "raw commercialism" and operating by "intrigue."[68] The fact that a football coach earned twice the salary of a professor, as reported at Harvard during the war, added insult to injury. The common response to academic discontent was to move if you could. Some tried to distance themselves from frustration by switching specialties. Edward Thorndike, the pioneer of animal puzzle boxes who taught for decades at Columbia, slid over to educational psychology, where healthy public demand for expertise made life a bit easier.[69]

Yerkes, in contrast, moved from Harvard to Yale by a serpentine route suggesting the career costs of having a driving interest, in his case animal behavior. In 1913 he took his first step away from the departmental bottleneck when he began to work halftime devising psychological tests for the new Harvard-sponsored Boston Psychopathic Hospital. Two years later he ran intelligence experiments on primates in California during a sabbatical, and he subsequently coordinated IQ tests for the Army and sexuality research for a national agency, before negotiating a post at Yale in 1924.[70] Through all, Yerkes held onto a basic idealism. There was a utopian quality to his idea of a research farm, and it was probably no accident that he bought his first acreage at the height of his difficulties with Münsterberg in 1911. Franklin, the New Hampshire town, was also the summer residence of colleagues Walter Cannon and the philosopher Ralph Barton Perry. It was Harvard intellectually, without the institutional problems of Cambridge. In imagination and in fact, it was a refuge that made Yerkes's pursuit of science more tolerable.[71]

Whether the prewar years were the "golden age" of animal behavior studies has been a matter of debate.[72] It may be safer to say that experiments on the minds of living things became firmly, and yet variously, embedded in individual careers. Unglamorous assessments of fine points about animal learning, exemplified by Washburn's laboratory tests at Vassar, seemed the norm outside elite universities. Tedium aside, steady plodding perhaps served the field. Here and there investigators successfully wooed the public. Walter Cannon's six Lowell Lectures in Boston during the winter of 1914–1915 deftly integrated his animal experiments, some nearly twenty years old, into a comparative overview of nervous systems, adrenal glands, and digestion. Animals, he in effect told his listeners, held answers for human stress and digestive problems. Inside research communities, the most fortunate senior scholars trained exceptional protégés. Watson, as a young professor at Johns Hopkins, connected with Karl Lashley through their common interest in neurology, although Lashley was formally a graduate student in genetics. Their lively coauthored papers on monkeys and terns just before the war bespoke their personal synergy.[73]

So much of these professional efforts were human-oriented that the animals studied seemed incidental. Work, money, and reputation were the conditions of science, however, but not its substance. When behavioral scientists reasoned seriously about the activities of organisms, they came closer to animals as whole beings that, among other things, had emotions.

Random Feelings

Although the emotions were not a prime research subject or a compelling practical worry for the first Americans who examined animal behavior, their experience made them aware of affect. Researchers do not treat test animals as simple "automata," Robert Yerkes noted, no matter how they conceive of them.[74] Working at the time with salamanders, snails, newts, turtles, and frogs, Yerkes appreciated creaturely sentience. His observation of a gap between assumptions and experience in fact understated the amount of theoretical confusion. The premise that animals are machine-like beings without sensitivity, to which Yerkes referred, had a long history in Western thinking. Yet the same tradition pictured animals as organisms mainly of sense, and for Yerkes's Anglo-American predecessors, having weak minds meant that they were ruled by emotions. Without reasoning to impose order, there is "nothing but a perpetual kaleidoscopic change of feelings," Herbert Spencer judged in 1855. Willard Small concurred in 1900 that his rats functioned by means of emotion. "That the rat *feels* 'why' or 'what' is certain," he reported from Clark University, but "that she thinks 'why' or 'what' is both doubtful and unnecessary."[75]

The persistence of belief that animals operate at an emotional level makes it arresting that early American behavior experiments focused so strongly on learning. This emphasis on animal understanding was new, and the approach succeeded because researchers now downgraded intelligence to mean gradual mastery of a task, without precise ideas or logic. Defining knowledge as a kind of efficiency was perhaps not so different from the old notion of emotion-driven instincts. But if the words used by scientists matter, mental progress stood out as the rising subject, and the intellectual heyday of affect was over.

Indeed, the coherence of nineteenth-century Anglo-American emotional theory made it especially fragile. Any inconsistency in the Western tradition did not seem to trouble recent English-speaking thinkers. For these speculative philosophers, feelings were allied with purpose, instinct, consciousness, and will. These concepts were poised to fall together like a house of cards. Giants of the British organic sciences, such as the naturalist Darwin, psychologist Alexander Bain, and anthropologist Edward Tylor, were in basic agreement: instincts facilitate adaptation, and, in higher organisms, effort and intelligence supplement native endowment.

Emotion was foundational. "The phenomena of excitability" are equivalent to "the phenomena of life" itself, wrote Darwin's protégé George John Romanes in 1884. Romanes offered nearly identical definitions of instincts, as "highly differentiated nervous mechanisms," and emotions, as "high wrought nervous mechanisms." In 1871, the year before Darwin published his book on bodily signs of emotions, Tylor had traced the ethnographic roots of human languages to "sounds merely expressing the speaker's actual feelings." The American William James explained the psychic intimacy of inclination and purposive striving in his 1884 essay "The Feeling of Effort." That feelings were building blocks of the moral will was already a commonplace, however, circulated by Bain, for one, in the 1850s.[76]

John B. Watson confessed in 1913 that his "animal work" made him feel like an outsider when he overheard such discussions of "speculative questions concerning the elements of mind." Claiming to speak for the "behavior men," he rejected abstractions as an excision needed to make psychology "an undisputed natural science."[77] Emotion survived this iconoclasm for two reasons: first, militant empiricism mandated respect for observations, including evidence of feelings, and second, Watson's dramatic strike neither began nor ended conceptual debates. In fact, emotion was a pervasive, albeit tangential, topic in animal experiments before World War I.

Despite scruples about making inferences about animals' hidden lives, the Americans did not underestimate the temptation. "A dog may apparently be *looking* at *you* intently," Watson warned half-humorously in 1906, but "the odor of your boots" is his real interest, and "when you get to the most interesting point of your discourse with him he drops his eyes, walks up to the boot, and investigates it."[78]

Caution aside, the experimenters' devotion to "careful description," using Small's phrase, frequently led them to apply affective terms to their observations.[79] After Thorndike watched a chick peck at the bars of the "pen" that separated it from both food and its fellows, he chose the words "confronted by loneliness" to describe the situation. He had noticed that his chicks were calmed by each other's presence, and his likely motive for noting their clannishness was to forewarn colleagues who might later handle them. Sharing tips was certainly James Porter's reason for discussing the "wariness" of English Sparrows. His puzzle-box trials "took much more time than had been expected" because the birds' skittishness in confinement kept them from working steadily to unlatch the door. L. W. Cole's raccoons were natural tinkerers, in contrast, with "a tendency to unfasten latches and set themselves free from the mere pleasure of performing the act." Cole told readers that he had to invent additional fastening devices to keep the animals out of mischief. Even ant species were distinctively "cowardly" or "aggressive," in language chosen by William Morton Wheeler. Wasps that consumed ants "learned to discriminate" among potential prey and attack the more passive.[80]

Experimenters were aware of variations among individuals as well as these species differences. Unequal intelligence did not always seem to be the cause. One of Small's rats "did no digging" for a food reward in a sawdust pile and instead "tried to steal her industrious comrade's hard-earned food." If descriptors of states of feeling occasionally seemed excessive or even silly, on the whole they contributed to the science. Porter called his sparrows "gregarious, quarrelsome, and pugnacious," for example, to make the point that their sociability made them excellent subjects for tests of learning by imitation.[81]

Legitimate as these comments were from a scientific viewpoint, it is surprising how much investigators uncritically relied on traditional ideas. The simple rule that organisms seek pleasure and avoid pain long infused Anglo-American philosophy alongside more detailed postulates about how the mind works. The first animal experiments in the United States adhered to this logic by relying on food as a motivator. For strictly empirical purposes, researchers recorded what their animals ate, such as the "mixed seed, grit, oyster shells, salt, and plenty of green food" consumed by C. O. Whitman's pigeons. There was also consensus that some aspects of food acquisition were unemotional. The pecking of chicks was a mere "*motor response*" unrelated to any expectation of gratification, in the estimate of British contemporary C. Lloyd Morgan.[82] The behavior was nothing more than an inborn capacity to satisfy a biological need.

Beyond these caveats, food lures to induce learning quickly dominated American experiments, and the presumption that eating equals pleasure was expressed in an offhand manner, as if everyone agreed. Margaret Washburn and Madison Bentley dangled "a young live grasshopper" from a pair of forceps to train fish to distinguish colors. Only on the final page of their article did they imply that their method depended on the fishes' association of the insect with "pleasurable consequences." Edward Thorndike believed a chick gradually mastered escape from a box, permitting it to share "food and drink" in company, because "the resulting pleasure stamps" in the memory. A Canadian critic, Wesley Mills, attacked Thorndike's premise "that hunger is an adequate stimulus" and, implicitly, comfortable fullness the animal's sole object. "Any one [*sic*] really acquainted with the nature of animals must know" they respond to multiple cues.[83] The American conviction that animals anticipate emotional satisfaction when they fill their stomachs seemed unshaken, however, in light of an expanding menu that included apples, sugar lumps, cracked corn, and milk-soaked bread, adapted by experimenters to the tastes of each species.[84]

This preference for food-based tests speaks volumes about the early researchers' approach to the emotions: that they were less attentive to pain than pleasure, assumed satiation was a core psychic goal, and treated theory as an offshoot of method. Each point contains a story. On pain, these scientists rarely aimed to motivate animals by discomfort. Only Robert Yerkes instilled learning by the disincentive of electric shock, termed "electric stimulation," in a few trials before 1910 using frogs and mice. No one, like Norman Maier in the late 1930s, trapped rats on an electrified grill to make them "neurotic" by inducing "tense or excited movements, tremors and chattering, and partial food refusal."[85]

Perhaps the turn-of-the-century scholars were blindsided by public charges of animal cruelty because they so strongly emphasized positive rewards. In practice, however, they grasped the intimacy of pleasure and pain. Time and again animals became overwrought and prone to error if experimenters let them become, as Porter said of his birds, "too hungry" before presenting a food lure. The sight or smell of a favorite food stirred happy expectations, but only to a point. This American insight did not particularly advance existing theories of sensibility. Although pain on the surface is "the contrast or antithesis of pleasure," Alexander Bain had declared in 1859, the states may overlap. Witnessing a tragic play, for example, can "touch so many springs of our emotional nature as to yield a large excess of actual pleasure." It was the American aim, however, to keep satisfaction simple. Willard Small tried and rejected a test involving removal of his rats' food reward because it was "deprivation of the fruits of their labor" and, as such, "bad method."[86]

Food-centered experiments also injected the Americans into international controversy about which emotions were most fundamental, although debate was not their intent. In a spirit of academic competitiveness the British psychologist W. H. R. Rivers, for one, cited the experiences of soldiers to contest Freud's claim that the sex drive was primal. The recent war proved the power of "another instinct, one even more fundamental than that of sex," he argued in 1920, "the instinct of self-preservation."[87] Darwinian ideas, shared across the Atlantic, implied that an equally good case might be made for the preeminence of danger responses, reproductive drives, or hunger relief as the key to survival.

The intellectual sparring in any case underlined how much the Americans before World War I focused on pleasurable eating, and how little on sexual appetite. It was mainly behavioral scientists with biological training who investigated mating systematically, and often the species they chose could not reasonably be said to have any emotion resembling desire. S. J. Holmes went only so far as to say in 1903 that a "vigorous spouse" carried a female amphipod "like a helpless burden" before copulation. W. M. Wheeler referred in passing in 1910 to an ant colony's "fecundated queen." Studying pigeons, Wallace Craig was understandably more attuned to possibly sexual feelings. After raising male birds in isolation to determine whether they knew instinctively how to mate and brood, he observed Jack "trying long and hard to accomplish the sexual act on one of the perches of his cage."[88] Craig's observation was most noteworthy because it was so unusual. The popularity of food-induced learning made it seem that American experimenters thought hunger and its attendant emotions controlled organisms.

The casual attitude toward theorizing visible here was, counterintuitively, a matter of principle. Yerkes stated the prevailing empiricism when he urged his peers "to experiment much and speculate little."[89] The liveliest questions were consequently the practical ones, and animal researchers were especially concerned about anthropomorphism. Attributing human traits to animals was a grievous error. The issue jeopardized animal experimentation. For decades armchair psychologists had repeated that a person only truly knows himself. "That the mind of each human being forms a region inaccessible to all save its possessor," Margaret Washburn began *The Animal Mind*, "is one of the commonplaces of reflection." Animal subjects, unable to self-report and also unlike people, were in this view problematic. She resolved the dilemma by endorsing "inference from behavior," at the same time reminding her coworkers not to project their own experiences.[90]

This was the standard logic, and it encouraged a measure of scientific restraint. Over time there were fewer accounts, such as Adele Fielde's study of ants in 1903, relying on human metaphors. "Feud" was the word Fielde chose for competition between ant colonies, although she otherwise shrewdly identified differing odors as the sensory trigger of "great violence."[91] Assigning numbers to animals became increasingly common, presumably to make the work more dispassionate. Inconsistent practices, however, suggest that researchers felt closer to some animals than others. Craig and Cole gave each of their subjects a number and sometimes also a name, including Craig's pigeons Jack, Billy, and Frank, and Cole's raccoons Jack and Dolly. Other numbered individuals remained nameless.[92] Bible quotations similarly indicated indistinct human-animal boundaries. One of Small's rats "tore wildly around the cage—like the wicked when no man pursueth" (Proverbs 28:1), and a lecture by Cannon about digestion in organisms included God's promise, "Behold I have given you every herb bearing seed" (Genesis 1:29).[93] In all these ways anthropomorphism, thoroughly vetted as a research flaw, nonetheless persisted and invited emotion-laden comments.

True as it was that the emotions continued be found willy-nilly in animal learning studies, the philosophical edifice that once honored feeling was weakened not only by the Americans' day-to-day practicality but also, on occasion, by their paper wars against theory. Ideas about emotion did not wholly fade from these polemics. The classic logic that instincts drive organisms to seek satisfaction broke down into a series of contested concepts. The "instinct controversy" memorably consumed psychologists during the 1920s. Purpose, deemed speculative around 1900, also regained scientific notice after World War I.[94]

Consciousness, in contrast, was the great prewar question. It was German-trained proponents of human introspection who generally contended with empiricists. But even William James, whose notion of a "stream of thought" ruled American psychology in the 1890s, issued a skeptical essay in 1904 declaring that consciousness as an "*entity is fictitious*," although single thoughts "*are fully real*."[95] Yerkes's *Introduction to Psychology* of 1911 exemplified the perplexing consequences for emotion of the consciousness debate. In a book dedicated to Hugo Münsterberg, Yerkes asserted that any "feeling which primarily interrupted the train of thought is not a product of bodily sensations." Affections, in other words, belong to consciousness. Yet in his final chapter he turned eagerly to the eugenic propositions being popularized by his former teacher Charles Davenport and attributed deviant acts such as prostitution and crime, containing emotional elements, to genetic inheritance.[96] To say the least, what and where the emotions were, or even what to call them, lacked clarity.

With the topic of emotion living a borrowed life, it is striking that two leading American experimenters, Yerkes and Cannon, responded enthusiastically to the British psychologist William McDougall's bold endorsement of affect in his *An Introduction to Social Psychology* (1908). During his years in America after 1920 as a professor at Harvard and later at Duke, McDougall became a maligned symbol of retrograde ideas as he defended instinct, purposiveness, parapsychology, and Lamarckian genetics. His *Social Psychology*, containing the seeds of these positions, systematically restated older ideas: "Each of the principal instincts conditions, then, some one kind of emotional excitement" to achieve a survival-related goal.[97]

This was a book that "deserves to be read, instead of quoted," Yerkes encouraged readers of his *Introduction*. There were points of affinity between the two scholars. Like Yerkes, McDougall rejected "the older introspective and descriptive psychology" because, he argued, organisms were inherently social and hence best studied empirically. Like Cannon, McDougall held an MD degree and appreciated physiology. When Cannon reported that an "increase of sugar in the blood" accompanied excitement induced in the laboratory, he referred for corroboration to McDougall's claim of a resemblance among fight, flight, fear, and anger.[98]

Both Americans cultivated McDougall professionally. After Yerkes sent him a copy of his newly published *Introduction*, they corresponded for two decades. On his way to France in 1917, Cannon looked up McDougall in Oxford. After dinner they had "a fine talk" and "found ourselves in agreement, at least in theories," about shell shock.[99] To say too much more about either American scientist's eager reaction to McDougall would be speculative. It seems inescapable, however, that McDougall declaimed on subjects hardly noticed or indeed suspect in America. In that sense, McDougall's certainty about the emotions was perhaps for Yerkes and Cannon a breath of fresh air.

Beyond these intellectual crosscurrents was the inescapable fact that the researchers were emotional creatures. They navigated the Anglo-American heritage on feeling in the capacity of interested parties, and their feelings in the laboratory were, if not the sole catalyst of their thinking, at least the most immediate. "I enjoyed the years in neurology and physiology and my research on the rats," Watson happily remembered in 1936, adding, "I worked night and day."[100] His recollection was accurate. Watson was not a neutral presence in the "rat quarters" as he worked on his dissertation. Testing how long it took a newborn to learn the path back to mother, he was swayed by the mother's "frantic endeavors to reach it" across a barrier and "thought best to return it." He described his animals as "little fellows," and he declared "it is a pleasure to watch them fly from place to place" as they ran the labyrinth. Using sawdust to disguise doors leading to food was, for him, "an extremely pretty test." He no less admired "the beauty of the adult cortex" when he dissected rat brains.[101] Alternately fascinated and frustrated by the daily unpredictability of animal work as a new scientific niche, Watson reacted in a far richer way than the simple pleasure-pain dichotomy used time and again in animal trials. Human emotions were part of the process of creating animal behavior studies, although the experimenters only dimly, and uneasily, recognized their influence.

Serious discussion of drives, motives, and purposes—all words somehow suggesting affect—began little by little in the United States around World War I. The origins were practical and intellectual. After years of experimenting, researchers began to see that animals may acquire skills even if a food lure is delayed or absent. The subjects were something more, in other words, than happy eaters. Needless to say, the scientists were eager to talk about the discovery, reviving old language and inventing new. At the same time, world events outside the laboratory made their equation of behavior with learning seem thin. There is an important element of truth to the proposition that the war disturbed faith in rationality, a tenet held most vigorously all along by well-educated people. Deadly combat and the politics that drove it nonetheless remained relatively abstract for most animal researchers. More immediately, they temporarily left laboratories and classrooms for public service and, not surprisingly, encountered the intense emotions stirred by wartime mobilization. Thereafter, they began to see all organisms with eyes more sensitive to the experience of feelings.

2
Animal Appetites Unleashed
World War I

> Especially during the recent war . . . for a considerable period after the reception of severe wounds, patients were to all appearances in good condition and only after several hours did the state of restlessness, anxiety and final depression begin to be evident.
>
> —Walter Cannon, 1923

> The suggestion which I offer, therefore, is that "voodoo death" may be real, and that it may be explained as due to shocking emotional stress—to obvious or repressed terror.
>
> —Walter Cannon, 1942

Like other Americans, animal behavior scientists were latecomers in 1917 to the war abroad and distant from its worst tragedies. The war nonetheless pushed forward a re-estimate of willfulness in organisms, and human beings did not benefit from the review. Adopting the voice of the king of the "Bellicose Termites," the entomologist William Morton Wheeler denounced unruly passions in a satiric paper delivered to the American Society of Naturalists in 1919. Wheeler was a flamboyant rhetorician and exploited the premise that the patriarch of what in reality are nature's largest termites—*Macrotermes bellicosus*, indigenous to Africa and up to four inches long—had answered his questionnaire. Head of a social order one hundred million years old, the monarch felt justified in instructing human beings, "gifted animals" but also "strangely immature." His advice was simple: relinquish "individualistic appetites" and "start society." Some termite habits might "shock your sensibilities," the king admitted, including relying on slave-like beetles and flies to nourish the larvae, eugenically limiting fertility to an egg-laying queen, and cannibalizing weak members. The termites' staggering record of evolutionary endurance still proved the animals' mastery of "basic problems of nutrition, reproduction and protection." In light of the recent war, the king's human audience could not claim as much.[1]

Wheeler's address soon appeared in the *Scientific Monthly* as "The Termitodoxa, or Biology and Society," newsworthy not only for its wit but also its fresh assessment of emotion. In keeping with past science, Wheeler's termites were learners: they knew how to solve problems. Mental acuity now contended, however, with a capacity for waywardness. Although Wheeler did not quite assert that termites have emotions, he imagined that their "integrated social organization" was a response long ago to "behavior toward one another," the king recalled, "most shocking to behold." This picture of a brutal termite state-of-nature was Wheeler's fantasy, but his readers may well have related to the insect's memory of "deterioration." Mixing "startling revelations in animal behavior" with a social critique, Wheeler did not pick up exactly where prewar science left off. He was sharply aware of emotional drives.[2]

Scholars on both sides of the Atlantic understood that a rethinking of the mind was in progress. "The war has been a vast crucible" consuming old ideas, declared the British psychologist W. H. R. Rivers. Before, "intellect and reason were the salt of knowledge"; now we see how much "human behavior in the mass is determined by sentiments." In the words of the Harvard philosopher Ralph Barton Perry, lecturing in California in 1918, "man is a bundle of miscellaneous impulses." An "enlightened psychology has discredited the former view that man is a mechanism that can be moved only by the expectation of pleasure or the fear of pain."[3] Each observer accurately identified elements of prewar mental science: Rivers, its focus on intelligence, and Perry, wide use of a pleasure-pain emotional model. The scientific future seemed to belong to "sentiments" and "impulses."

The path forward was not clear at the war's end, however, and American investigations of animal emotions faced both opportunities and limitations. By no means did all experts leave their laboratories in 1917 for "our great psychological experiment," as Robert Yerkes called military mobilization. Still, the war was a magnet pulling American experimenters toward innovation and internationalism. Once they left professional routines, their own unfamiliar emotions contributed to the shifting science. At the same time, the conflict reignited evolutionary thinking. Whether predator or prey, combatants behaved in ways contemporaries thought primitive. The soldiers treated by William McDougall for "regression" induced by shell shock showed "dog-like devotion to me." In this climate, king-termite's remark that "human beings are after all only animals like myself" seemed sadly obvious.[4] Scientifically speaking, evidence of human irrationality encouraged comparisons among species.

These unplanned benefits of war for animal study were accompanied by drawbacks: specifically, disrupted research and disagreeable findings. Attention to the human animal by scholars busy with military deployment was a distraction from controlled testing of other organisms. Philosophical reflection flourished, epitomized by Wheeler's bold termite-human analogies, but experimentation lagged. Rediscovery of destructive emotions was as unpleasant, moreover, as being lectured by a supercilious king-termite. Eager debate about how affect works—the fresh insights and nervous laughs one imagines in Wheeler's audience—did not change prevailing scientific opinion that emotions are dangerous.

This revived awareness of inborn goal-driven impulses occurred in a broad group of American psychologists, biologists, and philosophers. Many had been pioneers of animal study. Because the catalyst was human war, however, the conversation tended to steer away from animals. After all, human bestiality was the immediate worry, and the speculative tone of discussion invited general statements applicable to all life forms. Animals, though, were never absent from consideration, and animal experiments gradually adapted to emerging insights. Indeed, test animals and their emotions were a steadying presence in the changing science.

Out of the Cage

Although it makes perfect sense that behavior researchers involved in the war effort could not be home maintaining animal laboratories, useful information gleaned from classic learning experiments was largely exhausted anyway by World War I. To be sure, the war did reorient research. Once Robert Yerkes entered the Army as a major in 1917, he wrote exclusively about intelligence testing, mental deficiency, and personnel management until 1925, when he issued two books on primates. It is less obvious why Margaret Washburn conducted no publishable animal experiments between papers on rabbit vision in 1912 and mouse learning in 1926.[5] In fact, as established protocols yielded repetitive or problematic results, experimenters began varying procedures. The process took time, and the war was just one factor.

"A Simple Maze" (1918) by Karl Lashley was one of the rare laboratory studies of the immediate war period, and although it was theoretically brilliant, it also suggested why animal testing was at an impasse. In line with convention, Lashley expected rats to learn a maze. He did not simply offer food and then measure decreases in the time required to run the course, however. Instead, he permitted some rats to investigate the maze without food present, and he disrupted the eating of others by picking them up and returning them to the start. In short, he introduced curiosity and fear. A devoted experimentalist, wary of untestable concepts, Lashley described the first behavior as an "exploratory impulse" and the second as "emotional disturbance."[6] Although there was more to this intricate experiment than examining the role of emotions in learning, he had departed from a simple pleasure-pain paradigm when he declined to start food rewards right away or reached his hand in the maze.

These intriguing innovations were a sign of Lashley's impatience with the status quo in animal behavior tests. A young man of twenty-eight, just beginning a faculty position at the University of Minnesota, he was in a hurry to discover the truth of learning, and the word "simple" in his article title represented the crux of his reform. He peppered his report with criticisms of current methods as practically and substantively wrong-headed. Problem-solving experiments were bloated, because researchers favored teaching their animals "complex habits" and hence had to invest "in caring for them between training periods." Putting a premium on "economy of the experimenter's time," Lashley thought a "simple maze" as "dependable" as an elaborate one. As crucial, extended drilling was psychologically misguided, because comprehension was not a " 'gradual process of fixation.' " He advocated "distributed practice"—periods of effort punctuated by rest—to avoid fatigue and boredom and to encourage "variability of response" and absorption of insights.[7] In all, Lashley's paper aimed to retrieve animal experiments from the blind alley of excessive complication, but it did not challenge the laboratory as the preferred site or intelligence as the proper subject. He seemed nearly to stumble on emotional variables in an effort to demonstrate how small changes could enhance modest projects.

Whether or not due to Lashley's intervention, controlled animal experiments flourished in the long run. He was altogether correct, however, that the initial response to stagnation was to think bigger—adding organisms, apparatus, and locations to research agendas. Although he respectfully noted that "the Yerkes discrimination box" was his inspiration, Yerkes's recent animal work in fact epitomized the excesses that irked Lashley.[8] In one case, Yerkes and a graduate student spent the summer of

1914 testing two crows in an outdoor "crow house, fly, and apparatus." An operator manipulated the birds' flight through the structure to train them to recognize varying sequential patterns among nine doors opening on separate enclosures. The crows easily grasped the concept first-door-on-the-right, but second-door-on-the-left baffled them after five hundred trials. Frustrated by the animals' irregular work ethic, termed "backsliding," the experimenters adjusted pain levels by decreasing food rations and increasing the time spent confined in enclosures behind incorrect doors. Yerkes did not question that vertebrates across species uniformly rely on the ability to discern spatial order and count, or that his scientific peers considered learning an incremental process. "Doubtless many readers will object that longer training would almost certainly have enabled our subjects to solve this problem," he wrote defensively. He gave up on the crows, however, judging them comparable to a "human subject of a low level of intelligence."[9]

In Lashley's rejection of such expansive experimental designs, he overlooked how much other animal specialists also sought to revive the field, just in different ways. For Yerkes, new sites—such as the rural setting of his crow training—were one key. Arguing for the research advantages of field stations in the *Journal of Animal Behavior* in 1914, he conducted psychological tests in three controlled environments—all outside laboratories—between 1913 and 1915: his New Hampshire farm, Boston Psychopathic Hospital, and a privately owned primate collection in California. In the first case, crows, pigs, and, after the war, chimpanzees, shared the "Franklin field station" with the Yerkes family. The young daughter of Walter Cannon, a summer neighbor, remembered taking "the chimps for walks in the fields, holding onto their leathery little hands." But the country was a serious workplace. "On top of every hill, someone seemed to be writing a book," she added, not to mention the labors of the experimental animals.[10]

Yerkes's summer programs were part of a burst of procedural innovation. Without giving up his Harvard laboratory, he began half-time in 1913 as the psychologist designing intelligence tests for mental patients at the new Boston Psychopathic Hospital. The next year he planned a sabbatical at the primate research station in the Canary Islands directed by the German psychologist Wolfgang Köhler. Derailed by the war, Yerkes adapted his multiple-choice apparatus, already used on his farm for crows and pigs, to accommodate primates housed on the Santa Barbara estate of millionaire Stanley McCormick.[11]

These efforts represented a mixture of experimental replication and revision, and also self-affirmation and disappointment. In his sabbatical application, Yerkes named "ideational problems" as the common subject, and he logically sought access to diverse populations, all to be given the same test, for "comparative study." One also senses that these residential communities came closer to being "in nature," in Yerkes's estimate, than his Cambridge laboratory. By now he was shifting his emphasis toward "naturalistic observation" to supplement experimentation.[12] It was surely debatable whether the best way to acquire informative data was simply to expand the number of organisms studied. Although it was equally questionable whether testing captive animals and sequestered people qualified as the "field work," he at least was rethinking his methods.

Yerkes's influence as founding editor of the *Journal of Animal Behavior* in 1911 made his procedural changes especially important. He had a peer audience for his self-corrections. His colleagues were also generalists rather than specialists, which favored trying new strategies. Although historians rightly note that psychologists in particular lacked enduring loyalty to animal study, the open border between animal and human investigations encouraged flexible applications of established methods.[13]

Yerkes profited opportunistically, for example, from the heterogeneous pursuits of his one-time student G. V. Hamilton. Most obviously, after Yerkes's work with Köhler fell through in 1914, Hamilton hosted Yerkes at the refuge where he kept twenty monkeys and apes. As the personal psychiatrist to the estate's wealthy owner since 1908, Hamilton spent his free time analyzing the animals' sexual habits, at his employer's expense. Earlier, when Hamilton was a doctor around 1905 at McLean Psychiatric Hospital outside Boston, and simultaneously studying with Yerkes, he built a multiple-choice apparatus at the institution to test sensory discrimination in dogs. In retrospect, each man displayed professional good manners when he credited the other with inspiring the mechanical device. Whoever conceived the response box, however, Hamilton was consistently ahead of Yerkes in practical terms, working with dogs, primates, and mental patients, at the same time that Yerkes was still experimenting on crustaceans and mice in Emerson Hall. The tangible debt seemed all one direction—that is, to the student—if the goal was to get out of the laboratory.[14]

Scholars of behavior trained in biology were decisively taking up fieldwork at this time, too, and they may have influenced lab-based colleagues. Yerkes, for one, was "gratified and delighted" to welcome William Morton Wheeler to Harvard in 1908 to develop applied biology at the university's Bussey Institution. They were friendly enough in 1915 for Wheeler to fill in as editor of the *Journal of Animal Behavior* during Yerkes's sabbatical.[15] These were the same years when Wheeler began traveling often and far in search of ants. By 1918 he had undertaken research trips across the United States, as well as to Central America, Cuba, and Australia. Although zoologists, broadly defined, had long engaged in naturalistic study in exotic places, the ambition of Wheeler's treks sharply expanded the practice of his late mentor, C. O. Whitman, who made a single annual migration with his pigeons from Chicago to the Woods Hole summer field station on Cape Cod. Besides the straightforward motive of acquiring knowledge, Wheeler most likely made these journeys because he could: rail and steamship transportation were efficient, and Harvard science well funded. The war shaped his travel as well, pushing it away from combat in a westward direction.[16]

Yerkes and Wheeler visited each other's offices, and although there is no record of their conversations, Yerkes may have learned from Wheeler that fieldwork can be rigorous. Wheeler began a new notebook for each trip, and in precise handwriting he recorded when, where, and in what soil and climate conditions he found the ants he identified by the binomial system. These were strictly scientific journals, with almost no random observations about world events or personal reactions. Rare instances of Wheeler's excitement had to be inferred from the facts. In August 1917 he was so fascinated by the burrowing behavior of unusual larvae that he packed up two hundred of them in California, carried them to his summer home in Connecticut, and nurtured them in Boston until they matured the following spring. His research methods in part achieved a high standard because he had no wish to rough it. Whenever possible, he stayed in hotels, ate in restaurants, brought along family members, and took day trips by car. If Yerkes was looking for a fieldwork model at once scientifically respectable and personally comfortable, this was it.[17]

The circulation of new ideas about field study of animal behavior did not mean, however, that researchers who wished to do it actually could. Yerkes and Wheeler, after all, were established figures at an elite university where administrators willingly accommodated release from routine teaching. Even flagship programs in the mental and life sciences were experiencing institutional difficulties, however, that may have affected fresh initiatives. A generation of pioneering scholars was disappearing: William James and C. O. Whitman died in 1910, as did Hugo Münsterberg and Josiah Royce in 1916. Other faculty had been fired in policy disputes by the end of the decade, including psychologists James McKeen Cattell at Columbia and John B. Watson at Johns Hopkins.[18]

These disruptions contributed to the failure of at least some doctoral students who had an interest in animals to connect with faculty busy forging new paths. The major book-length work by Edward C. Tolman, for example, was a comparative study, *Purposive Behavior in Animals and Men* (1932). His contact with Yerkes at Harvard, where Tolman earned a PhD in psychology in 1915, was limited, however, and he never did fieldwork. Late in life, Tolman credited a course he took with Yerkes in 1911 with inspiring him to become a psychologist. But he subsequently worked with philosophy-oriented mentors—Ralph Barton Perry, Edwin Holt, and Hugo Münsterberg—who, if they did experiments at all, favored human subjects. Tolman dedicated his book to the white rat, the experimental mainstay of his laboratory at Berkeley during a long career. It seems likely that Yerkes was not around the Harvard department enough to attract the young Tolman to fieldwork. A teacher who changed direction at mid-career risked not training protégés.[19]

Nonetheless, the public expectation on the eve of the war was that travel served science, and although researchers were free to ignore popular views, as citizens they could not fully escape them. English-speaking culture was awash in narratives that mixed adventure with discovery. One survivor of Robert Falcon Scott's Antarctic exploration (1910–1913) attributed the tragic deaths of Scott and the polar party to "division of energy" on "a great scientific expedition." In retrospect it seemed foolish to this writer to race for the Pole after performing more than one "inhumanly exhausting feat" to investigate the polar environment. Teddy Roosevelt similarly described his trek through Amazonia (1913–1914) as a "zoogeographic reconnaissance." He nearly died, but "zoologically the trip had been a thorough success," yielding "over twenty-five hundred birds, about five hundred mammals, and a few reptiles, batrachians, and fishes," dead or alive, for museums and zoos.[20] Among works of fiction, *Tarzan of the Apes* (1914) was just a few steps beyond these fantastic real-life sagas. Edgar Rice Burroughs echoed evolutionary science when he placed young Tarzan as a "god-like white man" in a "simian" hoard facing conditions replicating those of "our ancestors of the dim and distant past." The novel's characters included a professor, Archimedes Porter, one of "these impractical theorists" now abandoned on the African coast by his ship's mutinous crew.[21] While it may be too much to see Porter as the symbolic scholar stumbling into wild nature, Burroughs's readers would not to be surprised to find a scientist in the jungle.

If varied types of outdoor animal behavior study were pushed forward by the limitations of learning experiments and pulled along by new expectations, there was still no simple connection between fieldwork and scientific awareness of the emotions. After all, without leaving his laboratory Lashley recognized that rats both eagerly explore and succumb to anxiety. Nor was attention to affective behavior in less regulated animals the straightforward result of observation. G. V. Hamilton's graphic report on his primates' heterosexual, homosexual, and autoerotic habits in 1914 included a methodological argument for "detailed description" of "'natural' situations." He mentioned Freud in his first sentence, however, and knew psychoanalytic theory. Hamilton's discovery of animal sexuality was the outcome of what he read and saw, probably in that order.[22]

The utter absence of sexual commentary in Yerkes's publications about the very same primates underscores the influence of the researcher's intellectual frame of reference. Yerkes was one of the field's most dogged empiricists, though, and the capacity of fieldwork to reveal emotion showed up in his work. In contrast to Hamilton's nearly risqué account, Yerkes issued a brief description of "Maternal Instinct in a Monkey" before his major California intelligence experiments appeared as *The Mental Life of Monkeys and Apes* (1916). His difficulty testing mental patients at Boston Psychopathic also drove home the psychic importance of feelings. Having no test-givers sufficiently trained to understand the patients' "extremely difficult affective problems" caused an "enormous waste of time," Yerkes complained to the hospital's medical director in 1916. He proposed "immediately, to undertake, in serious fashion, the study of instinctive and affections functions" in order to develop a testing regimen. He carried this urgent, expansive mood into the war. "These are great days for American science," he wrote excitedly in his diary in Washington in early 1918.[23]

Whatever the war meant for animal behavior science, it influenced methods that were already changing. Once it became clear that over time an animal learns, as long as the problem does not exceed the species' capacities, researchers began to vary training patterns, compare organisms, and weigh the value of locations other than laboratories. The mindset that the wider world could be appropriated for science encouraged intellectuals to think that they could productively engage in the war effort. They did, although the outcome was unexpected: they developed new ideas about motivations.

Pugnacity and Other Emotions

Despite the subtlety of Darwin's idea of natural selection, his popular readers saw survival as the reward of successful combat. Their view followed Anglo-American mental science, where fight was a basic reaction. "The Instinct of Pugnacity," one chapter in *An Introduction to Social Psychology* (1908) by the British psychologist William McDougall, epitomized this tradition.[24] Virtually no American scholar of animal behavior directly witnessed the war-making spirit during World War I, however. Only Walter Cannon went abroad, and political differences among those at home, ranging from strenuous patriotism to equally fervent pacifism, underscore that their reactions were not simple. What they had in common, though, were private emotions that were war-related. Life experience was one element in the redirection of scientific interest toward affective behavior.

It is difficult to know how attentive these intellectuals were to the protracted European war before the entry of the United States in April 1917. In retrospect, they emphasized their responsible citizenship. Yerkes's Harvard colleague Ralph Barton Perry later identified the sinking of the *Lusitania* in May 1915 as the catalyst of the military preparedness movement that he advocated. Yerkes similarly characterized the psychologists who met with him to discuss mobilization in the days after the American war declaration as being eminently civic-minded.[25] But academic life went on quite normally even then. A month after this conference Yerkes accepted a job offer for the fall from the University of Minnesota, "National demands permitting," he added in his resignation letter to Boston Psychopathic. Perry delivered a series of lectures at Berkeley in the spring of 1918, and Yerkes attended his usual professional meetings in December, although, he wrote in his diary, "I am not fresh enough to enjoy anything."[26] Both men approached the war as experts whose professional routines—mercifully for them—remained largely intact.

Yet these American scholars of mental processes also anticipated that the war experience would be momentous. They shared "the literary instinct of the troops" explored by scholar Paul Fussell in his classic study of British war writing: they kept diaries and wrote letters because it was an adventure, albeit a painful one. Yerkes bought a bound notebook that he titled "Journal 1917–1918 Especially Important Events in the Life History of Robert Mearns Yerkes." The first page recorded his birth in 1876, and each page thereafter displayed a successive year at the top. Most leaves were blank before page 83, however, where he began in earnest with the events of 1917, leading to his commission as "Major in the Sanitary Corps U.S. National Army."[27] It was a war journal, with the next one hundred pages documenting the year and a half he spent testing the new soldiers' intelligence.

The nearly one hundred letters sent by Walter Cannon from France and England, where he studied traumatic shock, were likewise a personal testament, but not as private.

Figure 2.1 Committee on Psychological Examination of Recruits, Robert Yerkes in the middle of the back row, 1917. Courtesy of Robert Mearns Yerkes Papers (MS569), Manuscripts and Archives, Yale University Library.

"I marvel at your skill as a typist," he responded to typed correspondence coming from his wife Cornelia, and yet it was her transcription into typescripts of his handwritten communications, later bound in four books, that muddied the waters as to what his letters were.[28] Knowing the tight community of Cambridge, Cannon would not have been surprised that the widow of William James hosted a reading of his letters in her parlor, or that there was talk of publishing them, which he refused. Cannon's descriptions of war conditions, and his reactions to them, were thorough from the outset, as if he anticipated an audience of family and friends. He shared what his social circle implicitly agreed was a life-altering event, to the point that he may have felt unusually exposed.[29]

The specialists in animal behavior in fact uniformly came from white Protestant families where reticence may have been customary, as it was for the Cannons. It was "a complicated household," their daughter Marian recalled—parents, five children, and two maiden aunts—who kept the peace by "the old-fashioned virtue of good manners among adults and a built-in reserve on the part of each, an emotional distance maintained instinctively, which was something distinct from the obvious warmth and humor that they shared."[30] A scientist's devotion to reasonableness most likely reinforced self-restraint. The war tested this balance.

Self-control, on the one side, and a confusing mix of distress and excitement, on the other, dominated the expansive logs of Cannon and Yerkes. Chronically subject to "depressions," in Marian's words, and consoled by the "order of his laboratory," Cannon took a risk going abroad.[31] "I have not, personally, dared to let myself get to thinking about home," he declared a month into an almost two-year sojourn. Sobriety, labor, and obligation supported him. He and fellow officers took the temperance pledge; he performed surgeries and experiments; and, as he informed Cornelia, "Duty holds me here, therefore, no matter what my wishes may be." Part of the emotional threat was his recoil from conditions that rendered him "speechless." The language he fell back on was rudimentary ("Blood was everywhere") and stereotyped ("despicable Huns"). Worse, though, was that he enjoyed himself and profited, too. "What extraordinary wounds they have!" was an exclamation of horror and eagerness, and a year later he was pleased to say "the research on shock is in a most promising stage." By Cannon's own count he published twelve scientific papers in 1918–1919 based on "work done during the war."[32] Loneliness, revulsion, and reaping rewards while others suffered were a heavy load for a man who prized equanimity.

Yerkes was disturbed by the emotional intensity of Cannon's letters at a time when he was fighting a similar psychic battle. Hearing one letter read aloud in the professors' enclave in Franklin, New Hampshire, where Yerkes was vacationing in 1918, he noted Cannon's "bitterness" in his diary. Indeed, one problem was that Cannon made the war "painfully real" for Yerkes, whereas Yerkes, stationed in Washington, kept combat at arm's length, framed in philosophic generalities: "Man seems the victim of his own craft, cunning, science and discovery." His emotional rollercoaster was no less difficult, however, though triggered by bureaucracy. Hopes and doubts hinged on the "great task" of testing the intelligence of the entire Army. "Oh joy!" and "happy indeed" were his feelings following administrative concessions. "I am depressed but not discouraged for I know that we are doing useful work," he reacted after a setback.[33] Yerkes seemed less anxious than Cannon about being emotional—whether because his diary was private, his sense of mission more naïve, or his standard of self-disclosure less rigid is hard to say. His war diary was nonetheless an eyebrow-raising record of emotionalism in a researcher hitherto devoted to analyzing thought processes.

A younger cohort of animal behavior specialists had less to gain or lose than these high-profile figures, but the war was not easy for them either. In the 1920s, Edward Tolman, Calvin Perry Stone, and Leonard Carmichael, all psychologists, and Warder Clyde Allee, a biologist, brought new perspectives to animal investigation, including neurology, embryology, and ecology. Born in the 1880s and 1890s, and earning doctorates after 1910, they anticipated a more or less straight path to becoming teacher-researchers in universities. The war shook up a simple expectation of progress from experimentation to publication and advancement.[34]

This said, Calvin Stone found much to be grateful for in his war service. It meant enough for him to jot down memories on note cards as a gift to his son on the toddler's first birthday in 1919.[35] Here Stone described the war as an opportunity, and no thought of massive hardship darkened his feeling of good fortune. His hardscrabble background, professional ambition, and stateside service helped him focus on how lucky he was. Newly married, he reluctantly volunteered for the Army, only to be pulled from training camp, honorably discharged, and reassigned as a civilian psychological examiner under Robert Yerkes. One of eight children of a widow in rural Indiana, Stone knew how to find the bright side and exploit it. In boyhood, "the nearest thing to a hobby in Calvin's life was work," his brother recalled. Stone's postwar notes identified the wartime intelligence testing as fortuitous.[36] Having earned a masters degree, he was working at the Indiana State Reformatory in 1917, so intent on "my own selfish ambition" that he scarcely read the newspapers and confessed "ignorance of the deeper significance of the world conflict." The war in effect made Stone professionally. After heading a team of psychologists that examined 75,000 recruits in less than a year, he entered a doctoral program under Karl Lashley at the University of Minnesota.[37] In 1919 his mood, looking back, was satisfaction.

By no means was the same true for pacifists. Outrage was the reaction of James McKeen Cattell, senior powerbroker in psychology, who was fired in 1917 by Columbia University for denouncing the draft. "They have hid behind the flag to stab" dissenters, he reflected acidly: "They feed upon the dead as well as the living."[38] Young scientists who opposed war, Warder Allee and Edward Tolman, both from Quaker families, could not afford to snarl similarly if they wanted careers.

It is likely that Allee, a University of Chicago PhD in biology in 1912, floated among positions during the war for reasons that Tolman's difficulties at Northwestern make clear. The "limelight" of pacifism "was not the kind one would have chosen," Allee recalled quietly. Tolman, though, left a record of turmoil. As a junior member of the psychology faculty, Tolman was "hauled before" the university president in 1918 to explain why his name appeared on the cover of *War*, "the organ of the Collegiate Anti-Militarism League." His letter to his older brother Richard, a physicist in even more political jeopardy at the University of Illinois, was filled with confusion. A woman who was accused along with Edward—"a real pacifist," he called her—planned to break her radical ties, and he would, too, if it was true, as the president claimed, that the publication was legally barred from the US mail. Twice he repeated, "I still believe in free speech." But "I do not think the issue is important enough to break the law over (no matter how foolish a law)." Even then he was ready to be "imprisoned or interned."[39] Irrespective of these subtleties of conscience, Tolman's contract was not renewed.

The tense wartime climate made it impossible for these pacifists to separate principles and ambitions. Nor did Allee, who became a lifelong leader in the Quaker community, attempt to segregate his values and science. As he began a long career at the University of Chicago in 1921, he argued in a new journal, *The Quaker*, that a cooperative approach to world food production would forever end war. "Hungry animals fight whether they be cats or cultured men."[40] For both men, touched personally by wartime conflicts, it made sense to take up affective behavior in their scholarship.

All these stories should not lead to an overestimate of the influence of wartime experiences on American professors who focused professionally on the behavior of organisms. Academic activities, I repeat, provided a steadying continuity, to the point that men who disagreed politically still collaborated. Although it is astounding to read the names of both Yerkes, waiting for his officer's commission, and Cattell, the draft critic, among the six psychologists present at a June 1917 meeting in Philadelphia to plan military intelligence testing, to exclude Cattell would have been insulting, and to attend may have seemed to him an obligation.[41]

But there was also a detachment from the war that seemed peculiarly American and indeed exaggerated in privileged families such as these. Geographic distance, the late US war declaration, and lucky birthdates contributed at times to an almost shocking disregard of international events. "Went shopping with Ralph & to Y.M.C.A. where he had a swim," William Morton Wheeler noted in his diary in Chicago on the same day in 1917 that Yerkes and Cattell got together. Just sixteen years old that summer, Ralph was his only son and the family member who most often accompanied his father on trips to collect ants, in this case touring the West until September. His father made no mention of the war in his field journal, and the following summer Ralph entered Harvard. To be sure, Wheeler waxed passionate about the war in public, using it as his reference point to praise termite-like cooperation in 1919. The following year he denounced it as the cause of corporatism in science that threatened "the investigative genius."[42] Yet as a father, Wheeler did not let the war get too close. Only Cattell among these scholars had a son abroad, an irony he drove home to his critics. His oldest "was one of the first to enlist in the army and go to France, where he is in charge of sanitation in the Harvard Hospital" under Walter Cannon.[43] Sons of the other men were all too young to force the issue on these families.

The war had an internationalizing effect on the animal behavior researchers despite this protective emotional distance. Becoming political allies connected them more firmly with British colleagues, who, the Americans sensed, learned something of scientific value during their country's prolonged military engagements since 1914. Walter Cannon traveled to Oxford on his way to France in 1917 to meet the physiologist C. S. Sherrington and psychologist William McDougall, hitherto his correspondents. In the spring of 1920, McDougall's one-time teacher at Cambridge, W. H. R. Rivers, gave a series of lectures at the Phipps Psychiatric Clinic at Johns Hopkins. That fall McDougall took a faculty position at Harvard.[44]

Professors before the war, Rivers and McDougall were among the British intellectuals transformed into therapists by the crisis. More than four years spent in military hospitals with patients suffering from shell shock led them to a sharply revisionist theory of psychoanalysis: fear of annihilation, unrelated to sex, triggered regression to an infantile mentality, and dream analysis in non-Freudian terms was a means of recovery. As the more senior of the two men, Rivers bullishly informed American readers of popular magazines including *Science* and *Scribners* that wartime therapy had dethroned the philosopher-psychologists, who obsessively studied intellect, and the Freudians, fixated on sex.[45] Both wrote about their treatment of soldiers as if they had witnessed the authentic core of human nature. Whether or not this was true, their war work was a far cry from the intelligence testing of their American counterparts.

Although there is no clear record of the steps that led to McDougall's hire by Harvard, one component for the Americans who made the offer may have been his understanding of emotion, both as a scholar and as a man who had seen a horrid war. Those who felt taxed themselves perhaps naïvely identified McDougall as a kindred spirit. This made the war a likely catalyst of the American invitation, and it made McDougall's move to the New World's Cambridge a window on who his hosts had become.

Contemporaries inconsistently described McDougall as filling vacancies left by James, Münsterberg, and Yerkes, all of which made sense: he was a polymath, both theorist and empiricist.[46] More certainly, he had become the main conduit for British thinking about emotion. Citations of his work rose steeply after Yerkes's published advice in 1911 to read McDougall. By 1918 Cannon, Wheeler, and Perry referenced McDougall on the emotions. No one, with the exception of the ever-unconventional Wallace Craig at the University of Maine, mentioned Alexander Shand, whose masterwork on feeling, *The Foundations of Character*, appeared in London in 1914. "The emotions are forces," Shand began boldly, and continued, "every primary emotion is an instinct, because innately organised to pursue a certain end."[47] Shand's essays of the late 1890s had shaped McDougall's position in *An Introduction to Social Psychology* a decade

Figure 2.2 William McDougall and family, ca. 1931, including his wife Anne and children Lesley, Duncan, Angus, and Kenneth. Courtesy of Duke University Archives.

later, and Shand now gratefully acknowledged their ongoing conversations. McDougall was university-based and hospitable to experimentation, however, whereas Shand was a gentleman scholar. This made McDougall the logical mediator for British emotional theory in America.

As a person, McDougall superficially resembled his American peers who studied the minds of organisms. Like them, he engaged the war in ways that predictably changed him. There was an all-out intensity about McDougall, however, that distinguished him. In 1914 he entered the French army as a private and drove an ambulance, until the British military recruited him as a therapist. After he demobilized in 1919 he traveled to Zurich to be psychoanalyzed by Carl Jung. More painfully, his youngest child among five died of rheumatic fever toward the end of the war, perhaps a special blow for a father who kept minute records of each child's development.[48]

Others in McDougall's British circle were more forthright than he was about wartime strain. Shand explained in 1920 that "the distractions caused by the late war made it hard to give the concentration of mind" to his book's second edition needed to complete it. Charles S. Myers, arguably the preeminent military therapist, delayed publishing material from his World War I diary until 1940, when it was still "extremely unpleasant" to recall situations where "the tolerable or controllable limits of horror, anxiety, etc. are overstepped."[49] McDougall never made a similar confession about the war's personal toll. His restlessness alone, however, including relocating in the United States, suggests that for him, too, World War I was a private watershed.

Although scholars later stressed the ill fit of McDougall in his new home, in 1920 he resembled his American peers, even if the mirror was more distorted than either party anticipated. McDougall "failed miserably," writes one historian, because the "professional climate was experimentalist and positivistic." In fact, the war spurred a virtual renaissance of American theorizing about one of McDougall's favorite subjects: affective behavior. Although he turned out to be brash and dogmatic by American standards, the war had better acquainted his new colleagues with their own emotions. They were poised to experiment with vocabulary, concepts, and methods in the interest of harnessing feeling for science.[50]

Why Do They Run?

Despite modifications in the sites, techniques, and organisms used in American behavior study by 1917, animal experimentation largely retained its classic features: it was learning-oriented, food-driven, laboratory-centered, and empirical. Military intelligence testing, proudly explained by Yerkes in 1920 in *Army Mental Tests*, represented the pinnacle of this science, based on assumptions that humans are complex animals and mobilization a vast experiment.[51] Yet alongside the usual question—how do they learn?—another appeared: why do they do it? Perhaps rats do not run the maze simply to enjoy milk-soaked bread. Although this interest in motives seems a logical extension of ongoing experiments, it entered professional discussions as a series of assertions in search of a method—much at odds, in other words, with the routinized way Americans normally investigated behavior. The catalyst, I argue, was mainly extra-scientific: reflections set off by the war about the springs of action.

The year 1918 by itself was a moment of bold revision. Appetite, drive, instinct, and purpose—all concepts mentioned in passing in recent years—were systematically analyzed in interdisciplinary forums. Animal specialists contributed to these wider conversations. In March, Karl Lashley's dismissal of mindlessly elaborate learning trials, "A Simple Maze," appeared in a new journal, *Psychobiology*. Although stimulating emotions in rats was part of his reform, it remained for Wallace Craig in February and William Morton Wheeler in April to promote the general idea of appetite.

The very title of Craig's "Appetites and Aversions as Constituents of Instincts" was a sign that he was addressing issues either neglected or contested in standard reports of task mastery. His declarations sounded straightforward: "The end action of the series, the consummatory action, is always innate." This statement was a minefield, however, if his readers were empiricists. Craig posited hidden impulses toward biologically sustaining goals such as eating and reproduction. His organisms were self-stimulating, in motion, and emotional. A female dove "settles herself on the eggs with many movements indicative of satisfying emotion (complete consummatory reaction)." He expected objections. Although he was consciously writing a major paper, citing literature dating back to the 1840s, he incorporated evidence from his steady stream of bird experiments over the past decade. His experience convinced him that appetites were "externally *observable*."[52] Whether or not motivations were scientifically accessible, however, Craig's animals were not simply responsive, adapting themselves to experimental protocols. They had purposes of their own.

Wheeler had read Craig's essay by the time he delivered "A Study of Some Ant Larvae" at the American Philosophical Society on April 19. His argument that appetite accounts for "the social habit among insects" makes it clear why he seized on Craig. Much of Wheeler's paper examined the satisfaction of a literal appetite: adult ants that feed larvae, he believed, in turn ingest nourishment through physical contact with the developing organisms. Although he invoked emotional wording to describe this "agreeable draught for the nurse," he was happy to substitute a biological explanation for older conjectures that "altruistic" motives drive ant nurture. He imagined appetite to be affective, without being self-conscious.[53]

This exchange of life-sustaining substances among ants was no minor behavior for Wheeler. Inventing the grandiose term "trophallaxis" by combining Greek roots, he was excited to discover "clearly coöperative or mutualistic" relationships in nature itself.[54] His modestly titled talk belonged to a series of theoretical reassessments. "On Instincts" had been Wheeler's topic for a group of Harvard philosophers and psychologists in 1917, eventually published in the *Journal of Abnormal Psychology*.[55] In both lectures, Wheeler weighed in as more than an ant specialist when he attributed psychological qualities such as satisfaction to basic natural processes.

It may seem that Craig and Wheeler made their case for innate drives to counter the idea implicit in American experimental psychology that animals are no more than malleable learners. As students of the late morphologist C. O. Whitman, perhaps the two were staging a conceptual field war. To the extent that disciplinary turf played in, however, it is more accurate to say that biology-oriented words such as "instincts" and "appetites" were now in general demand. Wheeler noted that his paper, "On Instincts," followed a presentation on the same subject, also at the Royce Club, by the Harvard "philosopher and psychologist" Ernest Hocking. Hocking's book, *Human Nature and*

Its Remaking, another publication of the banner year 1918, made much of instincts as "the chorus of innate hungers and impulses."[56] It was Ralph Barton Perry, though, a philosopher and club member, too, who most deliberately synthesized what science and philosophy had to say about natural inclinations.

Perry fully agreed that motives must figure in any discussion of behavior, as he explained in "Docility and Purposiveness" in the January issue of the *Psychological Review*. "The learning animal is in each case being *driven* by something," not just "a specific stimulus-reflex, but" behind that, "a general propensity." Like Craig and Wheeler, Perry broke with centuries-old Western thinking that pictured instincts as utterly fixed. To be sure, "purposive or teleological processes are hereditary"; but the ends and, even more, the means are "highly modifiable," which he termed "docile."[57]

To hear Perry so much in tune with Craig and Wheeler underscores how broad this interest in goal-seeking behavior was. Indeed, Perry went out of his way to cultivate science. Although he never performed experiments himself, his article summarized work by major figures in "animal psychology" such as Thorndike and Watson, as well as human studies by McDougall and Freud. He remained nervous about scientific materialism and dogmatism. Many "human purposes" are intangible, he added here, finding "no place at all in the field of nature as that is defined by the physical sciences." The science of his day also verged on being a "cult," he warned in a book the same year, a "form of religion" with faith in itself that displaced healthy "doubt."[58] He showed no wish to fight science, however, perhaps because he sensed that some experimenters were already relinquishing narrow empiricism. The thought that an organism feels a need to pursue an end seemed to offer these biologists and philosophers common ground.

Other examples of cross-disciplinary discussions of the self-regulating activities of organisms can be readily cited. In *Dynamic Psychology* (1918), for example, the Columbia University psychologist Robert Woodworth described a "drive" as a "mechanism" capable of being "aroused," especially by "other persons acting as stimuli." Woodworth did not work with animals, but he chose this theme for guest lectures at the American Museum of Natural History. His phrasing avoided endorsing autonomous psychic forces, long rejected by experimentalists as mystical entities. A drive nonetheless resembled the "something" internal named by Perry that was distinct from the physical environment. They were "prime movers," Woodworth concluded, accompanied by "the emotion, as a conscious state."[59] This made emotions signals of the elusive drives.

The scientists among these writers may well have wondered all along how an unseen inclination—by whatever name—could be tested. Although Craig and Wheeler believed that they had witnessed appetite, overall there was no clear progression from abstract analysis to either laboratory protocols or fieldwork. The distance between the polemics and experiments of Edward Tolman epitomized the methodological problem.

Tolman produced speculative papers in quick succession as a young psychologist at Berkeley in the early 1920s. They were not notably original. Calling instincts "driving adjustments" in a 1922 essay, he insisted that motives are observable in behavior, as did Craig, Wheeler, and Woodworth.[60] One can guess why Tolman set out to argue that affective traits can be studied empirically. Beyond genuine interest in the question, rooted in his admiration of Perry as a mentor, perhaps he took up a big subject to make his mark, especially if laboratory collaborations with his doctoral students were off to a slow, even troublesome, start. Tolman himself had been assigned a narrow dissertation topic at Harvard around 1914: comparing the effects of pleasant and unpleasant odors on human memory. For years, his laboratory publications remained closely tied to sense and learning. In the meantime, his Berkeley doctoral student, Zing-Yang Kuo, issued a high-profile attack in 1922 on innate instincts in the *Psychological Review*. Intriguingly, Kuo proposed that drives might somehow be acquired—that is, learned—in embryo, opening the way for fetal research. Apparently more concerned with reputation than innovation, however, Tolman apologized publicly for Kuo's zeal, and declined to take up embryology.[61]

Tolman may have been reluctant to devise tests for possible neonatal origins of appetite because he lacked training in biological methods. There was no sure path from any grand idea to a scientific specialty equipped to assess its value. In this case, feeling and will, now outdated terms for a drive, were once believed to inhere in the mind. By 1918, however, intelligence signified problem-solving. When innate impulses began to regain academic stature, experts accustomed to working with bodies more easily claimed them.

Walter Cannon, for one, came back from the war obsessed with the individual organism as an unstable collection of physiological impulses. His *Bodily Changes in Pain, Hunger, Fear and Rage* of 1915, based on animal experiments begun in the 1890s, had positioned him to study wound shock abroad. After the war, his Harvard laboratory was the site of new trials to explain interactions among the nervous, endocrine, circulatory, and digestive systems. His live subject was the "'reduced' animal": most often a cat whose arteries were tied, nerves severed, or brain surgically altered.[62] Cannon's frequent lectures to psychologists and psychiatrists showed off his experimental results, and yet they were also requests for collaboration. Much bodily upset was identical when measured physiologically, he discovered, and hence emotions were presumably distinguished from each other by their "subjective quality." Perhaps he hoped that the psychologists who read this phrase in the *American Journal of Psychology* in 1927 would suggest ways to penetrate a cat's mental world.[63] Still, within the disciplinary bounds of medicine, Cannon amassed substantial evidence during the 1920s on the physical processes connected with affect.

In fieldwork, already gaining popularity among animal behavior experts before wartime theorizing about appetites, one finds the same disjunction between endorsement of emotion as a scientific subject and unevenly collected information. Researchers seemed initially overwhelmed by the unrelenting activity of organisms in natural systems. They commanded facts, and yet, like Cannon, fell short methodologically in investigating the operations of purposive effort.

The pioneering visits of biologists to Barro Colorado Island in the Panama Canal Zone were a case in point. Located on an isolated land mass created artificially by the canal system, the field station that opened in 1923 represented the combined efforts of American scientists, the US government, and private donors. It was "a tropical Noah's Ark," wrote W. C. Allee after his research stay in 1924, "a wonderfully fine zoological garden."[64]

Wheeler, one of the project's prime movers, arrived in 1923 with a trail of recent publications celebrating natural impulse as a concept. Following his papers on instincts and appetites during the war, he told one professional audience in 1920 that a scientist must never give up being a "lover" of nature. In 1922 he denounced college laboratories as teaching mere "dry-rot fungus," instead of "life itself." Once in the Canal Zone for a two-month sojourn, however, taxonomy was Wheeler's preoccupation, as if the jungle's fecundity required his attention to this rudimentary task. His family "went bathing while Zetek & I collected [indecipherable] on the large & poisonous manzanillo trees on the beach & put them in cloth bags for [indecipherable] at the lab." Two weeks later: "Spent the whole day at the laboratory, writing up notes and labeling specimens."[65] Allee, who chose to live more primitively than Wheeler, compared himself to "Robinson Crusoe without any Man Friday." His *Jungle Island* (1925) described the island's curious animals in a narrative for older children.[66] Both men were excited by tropical fieldwork, but also seemed daunted by the profusion of unfamiliar organisms. Field studies of animal emotions were not immediately undertaken.

Whether examination of the appetites and their accompanying emotions took place in the lab or the field, delay in the implementation of a new idea is not surprising. The researchers were eager for concrete data, however, and so it was equally predictable that they began investigations without settling conceptual questions. Exactly what an emotion is, why specific animals should be experimentally favored, and whether emotionalism sustains or threatens life were still up in the air.

Controversy about the nature of emotion after the war focused on whether feelings are immaterial. Although McDougall's writings seemed the obsessive takeoff point for every discussion, his American critics missed his genuinely fresh idea: emotions are signs. As of 1918 there was much agreement that purpose-seeking and sensibility were connected. An empiricist such as Tolman all but threw up his hands rhetorically when describing subjectivity, however, calling the "problem of *motive*" a "problem of desire, emotion, instinct, habit, determining set" in 1922. His cautious inclusiveness did not stop Tolman from applauding and then dismissing McDougall's basic thesis, first stated in 1908 in *Social Psychology*, that emotions are paired with instincts. "McDougall's doctrine" had "overwhelming appeal," Tolman conceded, but it was "merely mentalistic." In 1931 Calvin Stone was distrustful for the same reason: "Recently McDougall has repudiated the physical element of this so-called *psycho-physical* disposition." Even so, Stone's skepticism appeared in an essay entirely devoted to the subject, "William McDougall's Methodological Concept of Instinct."[67] McDougall's unashamed theory drew and repelled at the same time.

Although these American psychologists cannot be faulted for wanting evidence for McDougall's claims, distaste for his philosophizing seemed to deflect them from constructing their own working definition of emotion, or even recognizing the value in McDougall's ongoing revisions. McDougall had seized on the functional approach to communication then gaining currency, for example, epitomized by *The Meaning of Meaning* of 1923 by the Cambridge-educated linguists C. K. Ogden and I. A. Richards. Their book purported to be a foundation, in the words of its subtitle, for "The Science of Symbolism." McDougall incorporated this emphasis into his *Outline of Psychology* the same year. He identified emotions as the means "to recognize our own state, and to regulate" it, as well as signals of "similar emotions in our fellows" to which we respond "with appropriate co-operation or opposition."[68] His American critics ignored McDougall's idea of feelings as monitors, however, and debate remained centered on the emotions' tangibility.

Scholarly excitement about drives also produced less reflection about animals than might be expected. Only Wallace Craig stated clearly that renewed interest in affect reopened basic points of evolution: namely, relationships among organisms and the survival value of feelings. At a time when most American experimenters were laboring over how an individual organism moves from an instinct to a goal, Craig argued in bold historical terms in his 1922 essay, "A Note on Darwin's Work on the Expression of the Emotions in Man and Animals." Dismissing the "absurd intellectualism" of Darwin's premise that an animal devises a conscious intention before baring its teeth or engaging in sexual display, Craig insisted on "the more fundamental fact that *emotion* is important," as well as an experience shared by animals and people: "Emotion is a great part of human life."[69]

In practice, Darwinian assumptions ruled wartime behavior research, just as Craig saw. Belief in an evolutionary ladder permitted experimenters to shift easily from animal to human subjects, and appreciation of natural agency underwrote appetite as a concept. The quiet influence of Darwinism did not lead to systematic review of the first principles of evolution, however, and Craig was not in a position to press the issue. The same spring that his article appeared, he lost his job as a professor of psychology at the University of Maine. Although he lived another three decades, he never secured a similar post.[70] His evolutionary perspective on animal emotions—a solid basis for comparative studies—seemed stillborn.

Indeed, the use of rats in behavior experiments became almost normative during the wartime period, in large measure because they were convenient. In contrast to earlier scientific reports that included tips on the care and handling of birds, fish, raccoons, pigs, dogs, and primates, Karl Lashley declared simply in 1918 that "the rat is the animal most studied and perhaps the best fitted for laboratory study," as if his readers needed no convincing.[71] Donald K. Adams, who completed a doctoral dissertation under Yerkes in 1927, chose cats instead, because his goal was to replicate exactly, and reassess, Thorndike's famous learning trials of the 1890s. He defended the scientific advantage of a cat's psychological "complexity," but also acknowledged "greater difficulty and expense" in keeping them, as compared with "the mass production method commonly employed with rodents." Even McDougall opted for utility at the Harvard laboratory. He wanted "some quickly breeding species" to test inherited traits and judged rats "more intelligent, more hardy, and in every way more satisfactory than mice." Specialization reinforced practicality in narrowing experimentation to a few animals. Although Yerkes never lost his appreciation of zoological variety and spared nothing to set up a primate colony after 1924, the great apes were his sole postwar research subject. Animal studies were bifurcated between "the rat and the chimpanzee," Adams noted, in his case for the middling cat.[72]

The disappearance of a range of animals from laboratory tests did not keep them from reappearing as symbols, still dependent on evolutionary premises for their meaning. Here researchers made no more inquiry into why a rat might frighten a baby than they asked elsewhere about the rodent's psychic typicality. John B. Watson, willing to recognize "fear, rage and love" as inborn emotions, tried to demonstrate that an infant thereafter acquires affective responses by provoking the child "with a white rat, a rabbit, a dog, a monkey, with masks with and without hair, cotton wool, burning newspapers, etc."[73] His choice of objects may seem random, until considered in relation to British wartime therapies of reading animal fables to shell-shocked patients and interpreting the animals in their dreams. McDougall, for one, believed that the soldiers regressed to the point of resembling "animals" or "savage men" who "sometimes die from mental influences of emotional nature."[74] If these treatments based on thoughts of animals aimed to meet the men where they were, emotionally speaking, then perhaps Watson similarly judged a rat and a primitive face to be stimuli adapted to a baby. He gave no reason for his selection, however, and any evolutionary premise went unexamined.

The last question left unopened was whether emotions have value. It was neglected in part because feeling remained a secondary theme in comparison to process: how organisms accomplish purposes. Although words like "appetite" and "drive" connoted affective energy, their emphasis was on actions executed in time. Debate about which psychic cause was more instrumental often neglected how the catalyst felt. "Instinct and Intellect" (1846), by the Harvard philosopher Francis Bowen, phrased the choice almost identically to the young psychologist Leonard Carmichael in his paper, "Heredity and Environment: Are They Antithetical?" (1925).[75] What was innate—and not how it was experienced—was the lead issue.

The main reason for virtual silence about the worth of the emotions, however, was that researchers had already made up their minds. Emotion was at best a necessary evil. Donald Adams voiced a recurring complaint of experimenters when he blamed his cat's "tantrum" at confinement for her inattention to her learning task. Her upset was an annoyance that slowed learning. General theories likewise described emotionalism as a regrettable stage. Although McDougall championed affect in his analytic scheme, he named "self-consciousness" the cornerstone of "true volition" in 1920. Any "action issuing from such a conflict of impulses and desires is not true volition." "Bodily stabilization" was the way Walter Cannon characterized homeostasis, the concept he popularized to sum up his many experiments. His word valorized a placid state, and he lectured readers of *The Wisdom of the Body* in 1932 that "homeostasis is not accidental, but is a result of organized self-government."[76] So uniform a message about controlling emotion was consistent with centuries-old wariness of feelings as unreliable moral guides. Darwin broke with this tradition when he attributed survival advantages to the emotions. But World War I once again displayed irrationality at its most destructive. Scholars of behavior were fascinated by a force of nature that they also distrusted.

It is not surprising that the scientists left questions about emotions and animals less than fully vetted before they set off on postwar projects. Affective experience in the animal kingdom has recurred through the ages as a problem of utmost difficulty. What is noteworthy is how quickly their sights moved from death on a catastrophic scale to the scientific puzzles of reproduction. Generation, nurture, and maturation of organisms in communities were the topics of the 1920s that further opened research about motivations and reactions. As intriguing, the work was often a family enterprise. Scientist-husbands, their wives, and young-adult children collaborated professionally. There was an apparent synergy between domestic subjects and these family teams.

However surprising the next step, World War I had been an intellectual turning point when scientific observers of behavior wondered what impels living things to act at all. Although there was a personal component to the discussion, the scholars were not changed suddenly from rational to emotive investigators. But experiencing the war pushed their varied feelings enough to the surface to make emotion itself an object of inquiry. Before and after, they were emotional men, with private feelings interwoven with their science. "At a few minutes after noon his spirit passed," Robert Yerkes wrote sadly in his diary in July 1924 at the death of Chim, one of his first chimpanzees: "the heart went with him from my anthropoid research." His suffering "will haunt me forever." "It can never be the same again." By early August, however, Yerkes was once more "hard at it," meaning his work, and longed "to reach the farm and settle for a solid rest" before the start of another term.[77]

3
The Family Passion

> The male and female may reverse the customary order of behavior, in which case the female, instead of the male, does the mounting. . . . Thus it is seen that the pattern of behavior is not stereotyped and invariable, but is dynamic and adjustable.
>
> —C. Ray Carpenter, 1933

In the postwar years, sexuality was the first topic to engage American animal researchers now intrigued by untutored appetites and aversions. Although Calvin Stone was skeptical about the fieldwork on primate relationships conducted by C. Ray Carpenter, Stone's former doctoral student at Stanford, the issue was methodology and not procreation as a theme. For his dissertation, Carpenter had faithfully extended his mentor's castration experiments. Whereas Stone tested whether sterilized rats continue to copulate, Carpenter surgically reduced hormone-regulating tissue in male pigeons to measure change in the "reproductive cycle" from courtship to nurture. Stone did not hide his opinion that Carpenter's subsequent trips to study wild primates were a distraction, and Carpenter handled the advice by agreeing: "As you might imagine, I am pretty well tired of dealing with qualitative field observations."[1] Verbal accord aside, Carpenter tracked monkeys and apes in Central America and Asia throughout the 1930s. He watched females "put their arms around each other," both sexes manipulate a "pendant clitoris," and the "piston movements" of heterosexual copulation. It was not all romance: during one summer his research team killed seventy females in search of pregnant subjects.[2] Here nonetheless was sexual bounty on a level to induce Carpenter to ignore Stone's counsel. There was no dispute, though, about the goal: explaining the psychology of reproduction.

There were unsettled intellectual issues from the start, however, and the social setting of investigation was not a simple choice of lab or field. As American animal researchers returned to empirical work in the 1920s after the flurry of wartime theorizing, they half-raised basic questions: what is sex, and is it emotional? Eager for data, they defined sexuality broadly. Carpenter gave voice to standard practice in 1933 when he listed "mating, nest-building, laying, brooding, and rearing" as his pigeons' "*secondary sexual behavior*" beyond copulation.[3]

There was less confidence about whether animal sexuality had an experiential dimension that might qualify as emotion. Rhetoric implied that animals have feelings in addition to instincts: Stone encouraged "sexual aggressiveness" in his rats, and Carpenter witnessed primate "behavior" that "is probably the results [*sic*] of emotional excitement." As always, no animal was able to self-report verbally, however, and the role of sensibility in the reproduction of organisms ranging from protozoans to primates remained elusive. In all, the new research turned toward sexuality, a subject that usefully narrowed, in Edward Tolman's words, "the problem of desire, emotion, instinct, habit, determining set."[4] Scholars nonetheless set out with just a modest idea of what they were looking for.

They did not make this scientific journey alone. After the war, wives and grown children increasingly became collaborators in their capacities as research assistants, ghost writers, coauthors, and publicists. Whether Mrs. Carpenter would be coming to the jungle was a practical question for the site manager in the Canal Zone in 1935. He pressed her husband in order to avoid unnecessarily buying "a big store of provisions" that would strain "our slim income."[5] The consequences of family participation for animal behavior study transcended logistics, however. The scientific activity of these women and their offspring reinforced analysis of the animals' family-like habits and yet muted sexual explicitness. Talented and successful writers, the wives communicated well with lay readers. They produced an articulate defense of the family as an organic entity, a vision that mirrored their own lives. Not only were they educated and ambitious but also married with children, and happily so. Marjorie Hill Allee, Cornelia James Cannon, Margaret Morse Nice, and Ada Watterson Yerkes were the most prolific of the domestic partners.

But here is the catch: scientist-husbands stood side by side intellectually with bright, assertive wives. Pushy and prim women did not single-handedly corral research innovation. Rather, distrust of passion informed sex study from the outset. Aging animal experts, such as Walter Cannon and Robert Yerkes, showed discomfort with sexual emotions in the social theorizing they undertook around 1930. At the same time, they were dispersing funds for sex experiments, and none of the young empiricists, like Stone and Carpenter, objected to their elders' quiet moralism. At home, all the scholars were invested, like their wives, in households that were orderly, productive, and formally patriarchal. The prudishness that spilled both ways across the private-public boundary was a family collaboration.

My emphasis on the interactions of sex research and scientific families does not seek to reduce investigation of animal sexuality to an expression of personal values. Sex science, both experimental and theoretical, and families, high-powered and moralistic, did intersect, but never in a simple way. No matter how much the humans worried about unbridled desire, empirical studies showed the animal kingdom to be saturated with a drive to mate. Social theories devised by these same scholars that downplayed the agency of sexual pleasure sat uncomfortably on a contrary body of evidence. My discussion examines these components: empirical sex research, scientific activities of families, and broad estimates of the significance of sex in animal societies.[6]

Copulation

As the attention of researchers with an interest in affect came to rest empirically on sexuality, previous scientific concepts were not wholly abandoned. The relationships of pleasure and pain, instinct and learning, and random acts and purposiveness were woven into new studies. Sex entered animal science, in other words, in familiar terms. Although William Morton Wheeler refrained from concluding that courting insects seek pleasure, his lecture notes confidently affirmed their "apparent insensibility to pain" as evidence of "intensity of [the] pairing instinct." Because Calvin Stone wondered if male rats learn coitus, he permitted "precastration indulgence" before neutering some, while others were celibate until the operation. Considering sex-related behavior, Leonard Carmichael believed a fetal guinea pig moved away from the experimenter's annoying touch because it possessed a "truly end-seeking and purposeful" capacity.[7] Curiosity about sex thus supplemented ongoing conversations about the causes of behavior. When investigators restated a general concern with motive as a specific inquiry about sexuality, however, they increasingly treated the body—and not the mind—as the locus of emotion.

Sexuality became a favored subject for three reasons: plentiful funding, philosophical relevance, and laboratory practicality. Money flowed through the Committee for Research in Problems of Sex (CRPS), begun in 1922 and chaired by Robert Yerkes. The group was as institutionally elusive as it was financially powerful. When C. Ray Carpenter acknowledged its support in a 1933 article, he described it as an arm of the National Research Council (NRC), a board of scientific advisers the government established during the war. By 1930 the NRC received 100,000 dollars a year from the Rockefeller Foundation, Yerkes informed his committee, a gift that determined the CRPS budget. "As you are well aware," Walter Cannon, one member, nonetheless reminded him, "Mr. Rockefeller has never allowed his name to be associated with the activities of this Committee."[8] This back-room atmosphere did not change the committee's wealth. Jockeying for an institutional home for himself in 1923, Yerkes let Yale know that he then controlled 50,000 dollars annually through the CRPS. "Some progressive university" would surely want him: "If Yale is the right one then I hope that the opportunity may come of it."[9] Even beyond his retirement from Yale years later, he dispensed individual grants of a few hundred or a thousand dollars, until he stepped down from leadership in 1947. Nearly everyone who studied sexuality in the United States during those decades was in the committee's debt.

The intellectual rationale for the program was by no means the scientific explication of emotion, an esoteric subject for hard-core scholars, but rather the role of sex in eugenics first and foremost, and in Freudian theory to a lesser extent. Although eugenic mating and a healthy libido would seem exclusively human concerns, early laboratory successes at explaining the "endocrinological aspects" of sex, in Yerkes's words, favored awards to research on animal proxies, including amphibians, birds, and lower mammals.[10] Solving society's problems biologically was nonetheless the real goal.

This agenda helps explain the serpentine relationship of John D. Rockefeller, Jr., to the CRPS and, more deeply, the discomfort the scientists felt due to their sponsor's expectations. Rockefeller was already known and respected for his support of the Bureau of Social Hygiene (BSH), an organization said to oppose prostitution but in fact promoting eugenic reproduction. If the CRPS was to seem objectively scientific, then the labyrinthine bureaucracy that at one time made Yerkes his committee's liaison to the BSH should be left unexposed. Many CRPS committee members and grant recipients indeed favored eugenics, but still were uneasy with the reformism that paid for their science. Cannon grumbled to Yerkes in 1930 that if the request he received for a laudatory statement about sex research was the "incidental activity of a secretary" at the BSH, he wished "to tell her that I am extremely busy." Yerkes advised compliance, however, and Cannon soon found himself at a "Birth Control Luncheon Conference" at a New York club, paid for, including his travel from Cambridge, by the BSH. Themes for discussion were precirculated: "Will the extension of birth control result in the elimination of the unfit or will it merely reduce the number of the fit?" Although Cannon's wife, Cornelia, published frequently to tout family planning, Walter resisted the strings attached to funded research. In the end, though, he went along.[11]

Psychoanalysis similarly relied on sexuality to anchor a grand vision, although Freudianism otherwise mainly resembled eugenics in how cautiously American researchers endorsed it. The excitement about Freud of William Morton Wheeler, an ant specialist, may seem unlikely, but his ambivalence was typical. He had recently "read some twenty volumes of psychoanalytic literature" when he spoke at the Royce Club in 1917. It was "a veritable cesspool of learning" containing "extravagant statements," yet also "a very considerable residuum of accurate observations . . . of the dream, of the perversions of the nutritive and sexual instincts, of the erotic conflicts and repressions and of the surviving infantilisms."[12] Calvin Stone's mind was just as divided. His wife described him after his death in 1954 as "a close student of Freud's work." When he toured Germany and Austria in 1932, however, Stone visited clinicians such as Magnus Hirschfeld of Berlin's Institut für Sexualwissenschaft, but neither Freud nor his followers. So, too, Karl Lashley made a distinction in 1914 between "the nursing instinct in young mammals" and "the 'pleasure sucking' described by students of human behavior," but he refrained from attaching Freud's sexual meaning to oral stimulation. These university-based scientists were empirically minded and moralistic, to the point that they danced around Freud. Transformed into a popular manifesto of liberated self-expression, psychoanalysis became an American fad after the war.[13] Animal behavior experts kept their distance from this fervor. Even so, Freud's emphasis on sex, viewed as science, lent their projects validation.

Far more practically, sex research using animal subjects went forward because it was suitable to laboratories. The uniform prewar opinion about the difficulty of studying "the affective aspect of consciousness," Yerkes's judgment in 1911, quickly evaporated once the scientific problem of drives came to focus on bodily sexual acts. Postwar researchers in effect followed Lashley's 1918 advice: a little creativity in designing a modest task for a simple species was preferable to training complex animals to manipulate an elaborate apparatus.[14]

What could be easier for a rat than having sex, or cheaper for the investigator? Not only did Stone's baby rats reach puberty in about sixty days but he also recycled adults from learning trials, because he thought sexual performance was psychologically unrelated to mastering a maze. In 1932 he summarized techniques in use across the country for acquiring quantitative measures of "sexual drive": the frequency of copulation following separation from the opposite sex, ablative surgery, hormone injections, the number of crossings of an electrified grill in order to copulate, or the speed reached by a sexually stimulated subject placed in a revolving drum.[15]

If the meaning of sexuality was broadened to include fetal and neonatal responsiveness, good tests were just as elementary. Leonard Carmichael collected salamander eggs from a New Jersey pond and then stored them in either tap water or a liquid sedative, before reviving the anesthetized batch and prodding all with a pin. He examined development of a sense of direction in newborn kittens by dropping them from a height as soon as a day after birth and recording the ages they began righting themselves in the air before landing on a pillow.[16] Scrupulously minimalist in their explanations, both Stone and Carmichael refrained from inferring that copulating rats seek pleasure or falling cats try to avoid pain. Whatever psychological insight the experiments provided, however, they were eminently doable. Animal sex studies promised a continuous research stream because they scrutinized a natural activity that begins early in the life cycle.

Although this burst of laboratory sex testing was in a way logical, on second thought it made less sense. Before the war, the science of animal behavior had been on a path toward increased fieldwork. Because reproduction occurs nearly everywhere, observation in the wild seems an ideal method. In 1940, however, Walter Cannon warned a CRPS applicant that "narrowly biological aspects of the subject" secured most funding, strange criticism of bias coming from a physiologist who had spent nearly two decades dispensing the awards. His personal file on CRPS business confirmed his impression. Communications to the group described "intra-ocular transplantation of endocrine glands," "fractionations of the anterior pituitary," and "injection of urine of pregnant women" into rabbits.[17] Reversing the prewar trend, field studies were left mainly to researchers on the professional margins, including C. Ray Carpenter, underemployed throughout the Depression, and Margaret Morse Nice, with no academic post.

When Carpenter transferred his systematic habits from Stone's laboratory to the Central American jungle in late 1931, he demonstrated that field analysis of sexuality had potential. His primate notebooks adapted concepts and methods recently devised for his dissertation on captive pigeons. Tracking howler monkeys that winter, he differentiated between primary and secondary sexual behavior, then inscribed a subheading at the top of each page—for instance, "duration of infancy," "play," and "vocal expression"—and filled in observations as he made them. To assess maturation in young animals viewed from a distance, he created rough size categories, such as "an infant size no. 1." The following summer, when he fortuitously traveled with anatomists and pathologists who employed native hunters to collect spider monkeys, he was more often able to inspect the reproductive organs of autopsied animals.[18] Throughout, conditions were sufficiently rugged to make scientific rigor difficult. On one trip, his camp was five hours by mule train from the nearest steamboat landing, and on both treks, he spent patient hours locating monkeys and then waiting for copulation. "Brachiating"—propelling themselves through treetops using arms and tails, high above the scholar—"is a rather common form of locomotion." Although Carpenter's subjects further expressed their opinion by showering him with branches and excrement, he got a monograph out of his trials, *A Field Study of the Behavior and Social Relations of Howling Monkeys* (1934), published by Johns Hopkins.[19]

Neither the colorful behavior of Carpenter's primates nor his careful scholarship was enough, however, to get him a steady job. Lab work on animal sex seemed to dwarf other approaches. Exchanging hearsay about openings in colleges and zoos dominated his correspondence with Stone until 1940, when Carpenter began a long career at Penn State, initially in educational psychology. In the meantime, he held a string of soft-money posts in Yale's Laboratories of Comparative Psychobiology, Harvard's Museum of Comparative Zoology, and Columbia's School of Tropical Medicine. He joined field parties to Southeast Asia and Puerto Rico and received grants from, among other sources, the CRPS.[20] Although field investigation of primate sexuality kept him afloat, and held his interest, it was not the key to professional success.

At about the same time, Margaret Nice upgraded the tradition of the backyard naturalist when she launched a multiyear study of song sparrow reproduction on a sixty-acre tract near her Ohio home. Books about familiar flora and fauna, aimed mainly to cultivate sensibility in women and children, had long been available. Nice's detailed log of one daughter's reactions to flowers and animals, "A Child and Nature" (1921), resembled this genre. In contrast, her extended investigation of specific birds and their offspring, beginning in 1928, demonstrated her biological training at Clark University. To track migration, the US government gave birders aluminum bands "to individualize the birds," and it compiled reported sightings. Because Nice's goal was "life-history studies," however, she also marked her subjects with "colored celluloid rings" visible "in the wild with the aid of field glasses." Sixty-nine mated pairs were banded at her site's "height of prosperity" in 1932. For each couple, she minutely observed courting, brooding, and nurturing, as well as responses to routine accidents such as the destruction of eggs by predators or disappearance of a partner between seasons. "Up and down Interpont I tramped in sunshine, rain, and snow," often rising before dawn and sitting for hours on her campstool, notebook in hand, her meals delivered by her husband or a daughter. Not until a year into the project did "I see coition for the first time," and yet by 1933 she was "banding the great-grandchildren of my beloved Uno."[21]

For a wife and mother, Nice achieved remarkable recognition, thanks mainly to her dauntlessness. Her song sparrow work appeared in parallel popular and scholarly forms: the highly readable *Watcher at the Nest* (1939), published by Macmillan, and two dense volumes of *Studies of the Life History of the Song Sparrow* (1937, 1943). By the mid-1930s she was associate editor of two journals sponsored by ornithological societies. More strategically, she befriended European ornithologists—Konrad Lorenz, Ernst Mayr, and Niko Tinbergen, among them—and strenuously publicized their ideas in the United States, for which they gratefully opened doors for her abroad. Closer to home, the ecologist W. C. Allee called Nice "a careful observer of bird habits" in 1931 in his *Animal Aggregations*.[22] To most animal behavior workers, especially laboratory psychologists, however, she remained unknown.

Whether the distance between lab and field was wide or narrow, empirical studies of sexuality strongly accented the possibilities of data gathering in investigation of the emotions. Taken together, manipulating hormones, counting sex acts, and catching a rare glimpse of copulation in nature yielded much information. Even so, the new research did not put an end to overviews of affective experience. When William McDougall updated his *Introduction to Social Psychology* of 1908 as *Outline of Psychology* in 1923, his discussion resembled American textbook psychology. He analyzed the relationship between feeling and instinct, listed primary and secondary emotions, and distinguished near-equivalents like temperament and mood.[23] From the concrete viewpoint of sex research, however, all these words risked sounding like guesswork.

The many measurements and observations derived from both experimental and field studies boiled down to two basic conclusions: the sex drive is impressively strong, and the body largely regulates it. Laboratory scientists constructed problematic situations for animals, forcing them to choose between food and sex, for example, or making the pain of traversing a wired surface the cost of copulation. Although sex did not always prevail, animals seemed amazingly adaptive in its pursuit. In the Stanford lab, Carpenter's "completely castrated" male pigeons, now without "sexual secretions," were

Figure 3.1 Margaret Morse Nice receiving an honorary doctorate at age seventy-one in 1955 from Mount Holyoke College, her undergraduate alma mater. Courtesy of Mount Holyoke Archives and Special Collections.

induced "to coöperate in coition" by "a ready female which postured and engaged in secondary sex behavior and vocalizations." Wild animals likewise sought strategies to procreate. When two mates of 4M, one of Nice's song sparrows, were successively killed one spring and summer, he courted "a third wife."[24]

Of these two scholars, Carpenter was far more surprised by the vigor of compensatory responses. Ablative surgeries, based on a physiological understanding of sexuality, were so standard in laboratories that Carpenter was unprepared for the persuasiveness of the female bird's "social actions." In contrast, Nice assumed from the first that organisms belong to environments, and yet she, too, considered the urge to reproduce to be essentially biological. Neither worried about what the animals felt subjectively, or what mental triggers impelled them. This emphasis on the body seemed to represent a transformation of psychology rather than desertion of the field. Carpenter and Nice warmly recalled their undergraduate teachers—respectively, William McDougall and G. Stanley Hall, both psychologists.[25] They seemed satisfied, however, to equate emotions with bodily acts and to narrow inquiry to the single life process of reproduction.

Significant as this trend was as a way to pin down emotion, sexuality did not altogether displace other behavior subjects. Ironically, Robert Yerkes, the man in charge of the CRPS, remained a generalist who did not let sex study take over his research. His trips to Cuba in 1924 and Africa in 1929, initially to learn how to care for captive primates and later to acquire them, only incidentally involved field observation. Learning experiments were the early focus of his Yale-affiliated Florida primate colony, also established in 1929, although the rural setting facilitated sociability and hence mating. A birth in 1933 was a reason for Ada Yerkes to take a day off from watching a young ape manipulate sticks, blocks, and sand: "Baby born to Palii & therefore no exper. w. [*sic*] Gua or play."[26] Whereas motherhood was respectable enough to merit attention, her husband was uneasy with masturbation. Writing in his journal, Yerkes seemed offended when coworkers pointed out the "frequency of erection" of his favorite, Prince Chim: "I am strongly suspicious that he was taught or encouraged in a bad habit rather than the opposite." Yerkes's final book, *Chimpanzees* (1943), promised "a composite portrait of chimpanzee personality," and this included sexuality. Nonetheless, the text's dramatic opening about sighting a male in the jungle fixed on "his evident independence, self-sufficiency, physical mastery of the situation, and his perfect fitness."[27] The words conveyed masculine power, but not orgasmic excitement. Although Yerkes fathered in the new sex research, the money awards he distributed did not revolutionize his own thinking.

In all, sex science did not advance in a cultural vacuum, and Americans' traditional misgivings about sexual passion shaped the lessons learned from new discoveries. To be sure, the field of animal behavior had its own strong postwar trajectory to renew empirical work and pin down such concepts as appetites and drives. Eugenics, Freudian theory, and the morally liberated spirit of many Americans encouraged sex study for varied reasons, and favored no single conclusion. Animal researchers were most nearly influenced, though, by the scientific collaboration of their spouses and children. This domestic foundation favored a socially conservative conclusion: biology naturally controls sex, and for humans, the agent of restraint is the family.

The Family Business

The gratitude to his wife expressed by Leonard Carmichael in his presidential address to the American Psychological Association in 1941 was suggestive of the family effort behind reproductive research. "I have carried out a prolonged series of studies" of fetal guinea pigs, he acknowledged, "assisted by a number of colleagues and especially by my wife."[28] This cursory reference to Pearl, unnamed in his speech, opens the promising and yet tricky topic of how scholarly families shaped sex science between the wars. Households of animal behavior experimenters were nearly all of a single type: headed domestically by a highly educated wife, each was a working unit devoted to the husband's career. The immense literacy of family members resulted in an abundance of documents detailing habits of communal labor. The challenge is to discern paths of intellectual influence, because in practice collaboration was not simply patriarchal. At the same time, there was no disagreement about values: although natural principles should be free to guide animal societies, including human ones, sexual emotions were unimportant, or even dangerous.

Wives were the lynchpin of family scientific production, and although providing husbands professional support seemed to be their common motivation, there was a range to their strategies. At one end of the scale were successful authors whose themes dovetailed with their spouses' science. Cornelia Cannon earned sufficient royalties as a novelist who explored eugenic descent to pay for a three-month family car tour of Italy in 1929 in the Cadillac she purchased. For Marjorie Hill Allee, it was true enough that her fourteen novels were "concerned with Friendly ideals," as a 1945 obituary published in a Quaker journal explained. The fictional settings of her cooperative message were oftentimes based on the communities of biologists she knew well, however, located at the Marine Biological Laboratory at Woods Hole and the University of Chicago.[29]

Private activities of other wives could impact their husbands' science more directly. By 1929, when Ada Yerkes's name first appeared as Robert's coauthor of a major study, *The Great Apes*, she had already been translating scientific reports, summarizing the literature, and ghostwriting for him for two decades. Margaret Nice turned similar collaboration to her advantage in 1924 when she published her first book, *The Birds of Oklahoma*, with her husband Blaine as second author. Although his university post as a physiologist was the reason she cataloged this state's birds, the avian "passion"—to borrow from her autobiography's title, *Research Is a Passion with Me* (1979)—was her own.[30]

It was these four scientific wives and mothers whose work connected most strongly with the takeoff of sex research in the 1920s. All began postsecondary education at small colleges: Cannon at Radcliffe, class of 1895; Yerkes at Barnard, 1898; Nice at Mount Holyoke, 1906; and Allee at Earlham, leaving without a degree in 1910. Marrying between 1901 and 1912, they were near the end of childbearing at the end of World War I, eager to resume intellectual labor.[31] In comparison with other American women

Figure 3.2 Yerkes family, 1924, including Ada, Robert, and children David and Roberta. Courtesy of Robert Mearns Yerkes Papers (MS569), Manuscripts and Archives, Yale University Library.

involved in animal study, before and after, they were unusual. An older generation tended to remain single, and a younger group acted mainly as research helpers to husbands, not intellectual partners.

The unwillingness of these accomplished mothers to bypass family for science implicitly censored the lifestyle of earlier unmarried scholars as too sterile. Although by now male peers were ceremoniously honoring the aging spinster Margaret Washburn, electing her president of the American Psychological Association in 1921 and soon feting her with a dinner to mark her twenty-five years at Vassar, these wives did not follow her example. William Morton Wheeler's enthusiastic citations of the ant studies of Adele Fielde, a former China missionary he met at Woods Hole, also unmarried, were respectful, too. His admiration did not make them equals, however, nor did Fielde leave a legacy of women choosing her lonely path.[32]

Althea Sherman explicitly counseled Margaret Nice against spinsterhood, as well as amateurism. An esteemed ornithologist living in rural Iowa, at home to care for her parents, Sherman praised Nice in 1921 for "your work done," but "I do not for a minute forget the greatest task, the rearing a family of children." At the same time "too many women are dabblers": "With a family of growing children it is fine to know every thing [*sic*], to be interested in every thing [*sic*], but we can excel in but a very limited number." Because Nice bore her fifth child in 1922, Sherman's endorsement of parenthood was apparently unneeded. Nice heeded her friendly urging to specialize, though, when she suspended her writing on child development in favor of the birds that thereafter consumed her. In 1952, when Nice recalled their "inspiring friendship," she did not add that she had taken a different personal path, with Sherman's blessing.[33]

A younger cohort of spouses of animal behavior scientists did not achieve the public stature of these four. The simple reason was that they were recently married and just beginning families. Between 1915 and 1932, Kathleen Tolman, Minnie Stone, Pearl Carmichael, and Marianna Carpenter wed rising men who had an academic interest in animal sexuality. Merely sporadic evidence that the wives helped out in the lab or went along to the field may not mean that they were intellectually unengaged. Perhaps couples discussed work privately. Now, however, it was more likely a childless wife who excelled in a famous husband's shadow. For example, Lucile Quarry Mann, married in 1926 to William Mann, entomologist and one-time student of William Morton Wheeler, subsequently traveled the world with her husband and worked in his office at the National Zoo, where he was director. Her only children were the orphaned zoo animals she cared for in their Washington apartment.[34]

The exceptional scientific impact of the middle group—mothers in command of households by the war's end—underscores the unusual promise of the 1920s for American women, and especially educated wives with a bit of free time. On the one side, the triumph of women's suffrage culminated decades of middle-class volunteerism. It was a heady moment. On the other, the glorification of domesticity had not begun, according to Betty Friedan's chronology. "Mine was the first college generation," entering just before World War II, "to run head-on into the new mystique of feminine fulfillment."[35] In the meantime, Marjorie Allee, Cornelia Cannon, Margaret Nice, and Ada Yerkes promoted twin values: family and science.

Their attachment to family in principle was rooted in personal sentiment. In a diary Ada Yerkes kept in 1926, as she oversaw construction of a new house in New Haven, she wrote a history of where the couple had lived since their marriage in 1905, followed by a poem titled "My Prayer." "Careers are naught to me / Nor fame, nor management of men / I want a house" for "endless happy service to the ones I love / And there to meet the world with courage and / A cheerful heart." Ada and her peers had lived long enough not to be altogether starry-eyed. These were academic families on limited budgets. Cornelia Cannon and her husband, sharing space with five children and two maiden aunts, never could afford to buy a house in Cambridge. For the same reason, Ada was "glad and sad" at the "launch" in the New Haven dwelling: "Robert feels it a heavy responsibility for our finances, & his health and other duties. This takes the joy out of it for me."[36]

Two couples were grieved more acutely by the death of a child. After a street accident in 1923 took the life of their ten-year-old son, the Allees left Chicago for fieldwork in the Canal Zone, soon followed by Marjorie's last pregnancy. Pneumonia caused the death of nine-year-old Eleanor Nice in 1927. In her memory, her parents donated 150 children's books to a hospital and a library. Despite hardships, or perhaps because of them, home and the family in it were precious. Ada concluded her reflections about number 4 St. Ronan Terrace by naming "such a lovely place—my dream come true."[37]

This evident emotionalism about family was counterbalanced by the women's dedication to household order. They displayed a rationality perhaps predictable among wives of scientists, expressed in family planning, management of servants, and use of technology. Although the couples' number of children ranged from two to five, nearly all the parents articulated eugenic ideas that presumably influenced them privately. During the war, Marjorie Allee compiled a typescript on "Alcohol and Racial Degeneration," where she linked drinking to progressive "sterility" over generations. Cornelia Cannon championed birth control for decades before her death in 1969, certain, as she declared in the *Atlantic Monthly* in 1922, that "innate qualities" produce "the inferior-minded" who are "a handicap and a menace," as well as her own superior offspring.[38]

The servant problem was also a constant theme, in part because maids helped care for children. Robert Yerkes, father of two, noted in his diary in 1918 that Ada "is too much tied down by lack of domestic service." The year before, Cornelia Cannon complained to her husband about "Florence's domineering ways" with the children, which "makes them naughty": "I shall have to keep her hands off the older two." Ada Yerkes favored technological solutions for scarce help. Discovering bungalow architecture in California in 1915, she marveled that the housewife finds "in compact space and of easy access, the baby's crib, water faucets for dirty little fingers, the tradesman's door and the telephone," all redressing the deficient "supply of maids or men." Although Ada was late learning to drive among these wives, still without a license in 1924, on the whole they aimed for efficient domesticity by means of devices such as automobiles, telephones, and typewriters.[39]

The goal was not simply to modernize "our own small kingdoms," in Ada's words, but also to organize families to permit women "an ever increasing part in the world's housekeeping." For these wives of researchers, welcoming inventiveness in their households was matched by promoting science in public. Their enthusiasm represented more than spousal loyalty, because they were reasonably well trained themselves. Although Allee and Nice chose a typical girl's path by majoring in English, Nice subsequently earned a masters degree in biology. Yerkes completed a degree in botany, and Cannon, though feeling free like many college women to be a generalist, studied zoology under the young Harvard instructor Charles Davenport.[40]

None doubted that science expands knowledge and serves society. The high point of Allee's novel, *The Great Tradition* (1937), occurs when her heroine, Merritt, a graduate student in animal science at the University of Chicago, is invited by her mentor to accompany his family to Barro Colorado Island, precisely the kind of site " 'in the tropics that has stimulated biologists from Darwin on,' " he assures her. " 'You yourself may be in the great tradition.' " Beyond the personal doors being opened for Merritt, research unites her fellow female students, coming from various American regions and international homelands, as well as different races. "These were her friends, loyal and devoted," Merritt reflects, with "a real affection and understanding for her and each other."[41] Although the word "friends" alluded to the peaceable ideal of Allee's Quakerism, she made the day-to-day practice of science the means to the end.

There was a problematic inverse relationship, however, between praise of investigation and the wives' firsthand acquaintance with it. The simple social applications articulated by women who had little role in experimentation could sound charming or shrill, but in either case lacked intellectual caution. The Allees' coauthored book for children about Canal Zone fauna, *Jungle Island* (1925), explained how "the long struggle between white men and yellow fever" made possible "the big wonderful work of building the Canal." The grand words were in all likelihood Marjorie's, the partner who did the writing but left no record of handling data. For Cornelia Cannon, the "psychologists" who conducted IQ testing among wartime troops were "searchers for truth"—"members of our race who have led us up from barbarism and keep us from sinking back into it." Bright as she was, she knew Robert Yerkes mainly as a summertime neighbor, and yet she boldly distilled his massive evidence into a call for reform to align democracy with natural inequalities.[42] By contrast, the extreme intellectual reserve of Ada Yerkes stands out sharply. Warnings about "fragmentary evidences" and how "little is at present known with certainty" fill *The Great Apes*, a literature review of more than six hundred pages compiled mostly by Ada, who had the language skill to synthesize scholarship in Spanish, Italian, German, Dutch, and Latin. So, too, notes of her observations of Robert's experimental animals spanned two decades, long enough apparently to instill scientific modesty.[43]

The wives' perception of their task as disseminating knowledge could similarly lead them to romanticize animals. Free from professional pressure to appear hardheaded, they permitted themselves some emotion that, at best, became empathy. For Cornelia Cannon, who focused on biology's human applications, the countryside and its animals were happily distant from "the tenement" and its "germs of physical diseases . . . and moral degeneracy." On a camping trip with her sister in 1917 in their native Minnesota, "every barnyard we pass" had its hens, pigs, and horses, not to mention "an old shepherd dog" that made their sojourn seem almost pastoral.[44]

A self-consciously literary author, Cornelia perhaps sensed the silliness of her idyllic tone. The emotion of Margaret Nice, in contrast, was inseparable from her science. "My affection for these particular birds had grown so strong that my attitude had changed into one of fierce guardianship," she confessed in *The Watcher at the Nest*. She fatally turned her "shotgun" on "the enemy"—a squirrel—when it threatened her fledglings. Although this comic image of Margaret as a menacing protector did not make her research any less solid, her willingness to caricature herself in print distinguished her from male peers. In truth, she was passionate about nature. "The world would be a far lovelier and more wonderful place to live in," she wrote with unashamed feeling, "if we left some space for the wild creatures, some forests for the beasts and birds, some swamps for the water fowl, some prairie for the wild flowers."[45]

No matter what spin wives put on animal behavior, though, their husbands were intimately connected with the women's ambition. Just as wives willingly tagged along, husbands depended on them. One crucial outcome was that their ideas cannot be fully separated. Whereas Robert Yerkes was Ada's public face, she was his private voice. From 1916 until 1935 Robert intervened with editors to place his wife's writing. Often, but not always, her subject was his research, so when Harper and Brothers replied to him that Mrs. Yerkes's essay succeeded only "to elaborate the difficulties incident in animal experimentation rather than throw new light on animal behavior," he missed a chance at popularization. To get Ada's papers in print, the couple sometimes added his name as lead author, or else approached a friendly editor of a scholarly journal. Most confusing, her manuscript prose appeared in his publications. In a typescript titled "Animal Individuality and Genius" and identified as Ada's, one primate they studied was "slow, erratic, lacking in ingenuity and resource," and another, who was "our favorite," had a "gentle disposition." The same two animals appeared by name in *The Mental Life of Monkeys and Apes* (1916), sole-authored by Robert, as, respectively, "slow, inattentive, and erratic" and "gentle, docile, and friendly."[46] How this blended wording came about is a mystery: did he use her notes, or did she draft his pages? Either way, indistinguishable voices of wives and husbands were a logical result of family production.

The same composite literary identity appeared in research papers coauthored with grown children. In an intriguing case, William McDougall, another firm eugenicist and a father of five, explained in 1931, "I have written this article in the first person," although "my son, Kenneth, has done the larger and more difficult share of the observational work." The situation was to an extent as exploitative as it seems, because the father used his son's tests of insight in monkeys, raccoons, rats, and wasps to promote his own unpopular argument for "teleological causation in the natural world." The favor also ran the other way, however. Although McDougall had at best a modest professional interest in either animals or North Carolina, where his family now lived and the experiments took place, Kenneth went on to write a dissertation on the state's marine vertebrates for a PhD at Duke in 1943.[47]

This symbiosis of family interests was one sign of the children's integration into their parents' science from a young age. For more than a decade after 1906, McDougall took extensive notes, never published, on his children's development. Margaret Nice recruited her daughters as early to be her assistants. It may have seemed a short step from the day in 1920, when she "boosted four-year-old Barbara up to a dove nest," to the car trip from Chicago to Oklahoma with twenty-six-year-old Constance in 1937 to survey the state's birds. When Robert Yerkes felt lost linguistically in Dakar in 1929—"I am speechless in too many languages"—he marveled at the mastery of French of his daughter, Roberta, who "chattered away" with a local scientist. Translating for him was the reason she had come, because her ear for languages resembled her mother's. Yerkes was likely more comfortable the year before when he coauthored a paper on memory in chimpanzees with his son David.[48]

The children's loyalty to their fathers' careers had limits, though. David Yerkes became an architect, and Angus McDougall, the only McDougall son to survive World War II, was a sculptor. Marriage remained the life goal for daughters, and almost none wed scientists.[49] The family economy of the animal behaviorists nonetheless peaked when the children came of age between the wars.

Their privately held values added up to a self-referential social ideal: family unified by a common task and organized by a sexual division of labor. They left no evidence of domestic discord, and Robert Yerkes, for one, effusively expressed his love in poems, jokes, and testimonies. It was his habit to mark the couple's wedding anniversary with a fake check made out to Ada for the amount of her worth, "one billion dollars and 00/100" in 1949. Although he could be sentimentally silly, writing as "Robert, Bobs, Grandbobs, as the case may be," more often he chose words like "workable, profitable and happy" for their "decades of coöperative living." He recognized that their marital experiences were not identical. If we had had six children, instead of two, "it might have been more fun for me, but harder for you," he mused in 1934.[50] In this gender gap, one senses frustration in the midst of the wives' unrelenting helpfulness. Was it Ada who filed away her rejection letters from publishers? Did Marjorie, an established author, feel that she "providentially" stepped in—as her husband's colleague put it—to become family breadwinner in 1930, when Warder Allee developed crippling back tumors? What satisfaction did Cornelia reap in 1931 when a Harvard friend jokingly named Walter Cannon " 'the husband of the famous novelist?' "[51] Despite the sincerity, even fervor, of the couples' embrace of family, there was a personal cost for accomplished wives.

These family practices of postwar animal researchers did not have a predetermined intellectual outcome. The scholars appreciated the difficulty of understanding reproduction and the emotions that might accompany it among animal species. It was clear that a procreative impulse was not always a sex drive, for example, and sexual pleasure was distinct from parental satisfaction. What the scholars knew personally in closely managed households nonetheless affected their theorizing: they downplayed sex, highlighted gender, and distrusted unsupervised emotion.

Sexless Societies

The empirical studies of sexuality sponsored by the Committee for Research in Problems of Sex between the wars were not matched by a theoretical emphasis on sex in conclusions about animal societies. In fact, nearly the opposite was true: arguments stated or implied that sexual activity was at best incidental or, worse, disruptive. In 1931 Warder Clyde Allee ended his book *Animal Aggregations* with a pointed refutation of the idea that reproduction is the basis of biological communities. If society's evolutionary origin resembled a "gang"—a "consociation of adult individuals for nonsexual, co-operative purposes"—then "sex may have evolved from the mutual stimulation."[52] Mating was not the cause of group life, but instead a byproduct. One reason for this disjunction between experimentation on procreative behavior and such cool general assessments was perhaps the inevitable time lag in circulating new data. As likely, moral disapproval of sexual emotions, quietly influential in the field from the first, was more visible in broad theorizing. Families philosophically committed to self-restraint were part of the situation. The result was a scientific tug of war between preoccupation with sex and skittishness about it.

Beginning in the mid-1920s, the experts on animal behavior expressed their views in a new wave of synthesizing declarations in one way or another about evolution. In addition to Allee's theory of crowding as sociability's primitive root, Walter Cannon popularized homeostasis, William Morton Wheeler toyed with emergent evolution, and William McDougall expanded on teleology. Although Robert Yerkes, tenaciously empirical, was simultaneously narrowing his focus to the great apes, he ironically gave voice to the men's common motive: to explicate human beings by comparing species. Thus in 1925 he described his anthropoid research as "an indirect approach to certain problems in human psychology."[53] The grand sweep of their ambitious works seemed to reflect the wish of aging scientists to reach conclusions. By now nearly all were beginning to experience disabilities due to cancer, tumors, deafness, or emotional collapse. Writing at this personal turning point, and as committed family men, they had little positive to say about either sexuality or emotion.

When Allee and Cannon held forth in a way that their science resembled philosophy, they treated emotions as unruly. "We need to pool our ignorance and attempt positive steps toward training emotion" to reach "emotional stability," Allee urged in the journal *Science* during World War II. His plea was most noteworthy for its unusual bluntness: previously, he barely spoke about felt drives or reactions, as if affect was an unneeded scientific concept. His principle of aggregation meant that nature programs organisms to mass together for survival, with dominance and gender providing supplemental cohesion. Among protozoans, "mass spermatozoa retain fertilizing power longer," he explained in a 1937 lecture, just as "massed eggs divide more rapidly than if isolated." By the same logic, mating was " 'the co-operation of two individuals.' "[54]

Peck order was another "one of the fundamental principles of biology" that further organizes groups, and it made sense to Allee in 1932 to separate male and female chickens for parallel studies of their respective hierarchies, as if sex identities were discrete by nature. Although he acknowledged that "certain behavior traits" correlate with specific social patterns, he avoided words denoting emotion and seemed pleased that "many modern psychologists have discarded the concept of consciousness," making it easier for him to treat sociality as "automatic." A rare reference to animal temperament not only appeared in a personal anecdote, but also implied that affective evidence is misleading. Driving his family's cows to pasture as a boy, Allee had watched the "half-Jersey" Leah move up from "the most hooked and butted member" to "seniority," even though she was "still small, and usually inoffensive."[55] Structural advantage—which for this herd meant following an experienced leader—was the sole social determinant for Allee. Any personality trait or transient feeling had no clear meaning in his scheme.

A year after the publication of Allee's pioneering book on animal populations, Walter Cannon presented homeostasis to "the general reader" in *The Wisdom of the Body* (1932). Older than Allee by more than a decade, Cannon had devoted a long career to the emotional physiology of animals and humans, most memorably the catastrophic disturbance of life functions caused by wounds on the Western Front. This deep acquaintance reinforced his initial equation of emotion with upset. His warning to dyspepsia sufferers around 1900 to avoid meats that "excite animal passions" was consistent with his current praise of homeostasis as "bodily stabilization." Although "appetites" are essential "goads" to seek satiation, rest is the healthy state, and control through "self-government" the means.[56]

The solitude of the body in Cannon's vision contrasts sharply with the fundamentally sociable organisms of Allee. Cannon's perspective makes sense for a man trained as a doctor, but the Depression now induced him to add a social application: "the nearest equivalent to the fluid matrix of animal organisms" is "the system of distribution" of a nation. He was optimistic that "specially organized control" by government during the economic crisis would not restrict individuals in pursuit of "the highest activities" of "adventure and achievement."[57] Sex was presumably a less worthy pastime, and Cannon's book did not dwell on it.

The virtual absence of comment on sexual emotion in either theory did not mean that Allee and Cannon were unfeeling men or uninterested in sex research. Privately, both began to notice their own disease symptoms in 1930, soon requiring successive surgeries and tipping the balance in their marriages, as wives became nurses.[58] Professionally, Allee's laboratory was largely funded by the CRPS, and although philanthropic politics later disrupted the arrangement, in 1949 he was still beholden to the Rockefeller Foundation to the point that its "wish is a command." Cannon, a member of the CRPS for two decades, was so invested in the grants that he scolded a journal editor in 1924 for publishing castration experiments performed on animals without anesthesia, regrettably sponsored by his board. Anyone "who has been careless in operating on animals, who has inflicted unnecessary pain," Cannon wrote insistently, should "be informed that that sort of works [*sic*] is reprehensible in any case, and furthermore, is likely to jeopardize seriously freedom of animal experimentation in this country."[59] Much in character, he ignored sex, the project's ostensible subject, in his tirade against wanton cruelty. Whereas it is hard to tell what Cannon expected from sexual experimentation, it was clearly not to enhance pleasure.

William Morton Wheeler and William McDougall resembled Allee and Cannon socially, and they, too, expansively theorized. Although their wives—Dora and Anne, respectively—were not scientific collaborators, the women embodied rationalized housekeeping. Dora was a stalwart Republican who crossed paths with Cornelia Cannon in Boston-area causes. Anne shared her husband's ideas about positive eugenics to the extent that she bore five children. Both men by this time were also confronting private trials. In 1926 Wheeler was hospitalized for two months in the hundred-year-old institution originally named the Bloomingdale Insane Asylum. Although a colleague politely chose the words "upset nervously," Wheeler left treatment convinced that even the behavior of "scientific men . . . is determined to a very considerable extent by their desires."[60] Memoirs of McDougall invariably mentioned his isolating deafness. Jerome Bruner, later a Harvard psychologist, recalled having "to raise one's voice to be heard," which led Bruner and his fellow Duke students "to formulate our points more concisely": "Foolishness in a loud voice somehow seemed the more foolish."[61] Neither Wheeler nor McDougall was connected with the CRPS, but both read Freud with unusual interest. Whatever might be expected of such complex men, however, they did not approve of sexual emotions.

Wheeler lectured and wrote with explosive energy after his release from Bloomingdale, and his ideas about emotion were a jumble, albeit a consistent one: passions are regressive, and yet sex differences are the root of progress. Whether explicating ant behavior or the sweep of organic growth, Wheeler seemed prone to philosophize. Emergent evolution was already much discussed by scholars when he weighed in pessimistically in September 1926. How can change be improvement, Wheeler asked, in view of evidence of "regressive development" such as "higher emotivity, increasing insanity, criminality and mob-psychology" in civilized nations?[62] Just that summer he had slammed psychology for ignoring the body, the root of organic dysfunction. Reviewing John B. Watson's *Behaviorism*, Wheeler dismissed this "queer science" of practitioners "who subtly vegetate on cloud-banks of ratiocination and pure asexual love." Wheeler fully accepted that "instincts, emotions or appetites" include "uglier propensities," and in lectures in Paris in 1925, he all but traced degeneracy to uncontrolled sex. Because of "the antagonism between sexual and intellectual activities," societies "approach the social insect type when they incline too much toward reproduction." Yet the final twist was that humans must nonetheless imitate these insects. Reproduction, "the raison d'être of insect societies," recurs through animal species precisely because it unites females, "the social sex par excellence," and males, "the antisocial sex." The benefit is not simply community survival, but "division of labor," too, needed in the long run for high culture.[63] This made ants the perfect model for people: a gendered society where reproduction occurs without lust.

On the surface, McDougall's late-life insistence that nature is so vast as to include undetected paranormal psychic forces greatly distanced him from Wheeler's microscopic scrutiny of ants. They were companions in theorizing, however, and in their wish to see passion sublimated. From the early 1920s on, at the same time Edward Tolman was arguing that single organisms have purposes, McDougall boldly declared humans "purposive" when "they conform to teleological laws." Attacking "intellectualist views" (meaning, defining reality as rational) and also "the mechanistic account" (reducing all to matter), McDougall identified affect as life's essence.[64] This metaphysically conceived emotion was a far cry from raw sex, and yet he imagined the life force to be part of nature, although he sometimes called it "spirit." Thus he praised the book, *Extra-Sensory Perception* (1934), by his protégé and colleague J. B. Rhine, as "the first fruit of the policy of naturalization of 'psychical research' within the universities." When a correspondent in 1935 understandably mistook McDougall for a Christian, he replied that his only tie "is a loose one with the Unitarians." Instead, he stood loyal to "a science of imponderables," and animals other than humans were too lowly on the scales of insight and will to figure in McDougall's final works.[65] Although parapsychologists in the 1960s tested the effects of ESP or psychokinesis on mice, plants, and fungi, McDougall's pathbreaking early work on instinct and emotion had little direct influence on this legacy.[66] He expected sexual impulses to be progressively left behind.

Neither American leaders in the field of evolutionary behavior, such as these four, nor empirical sex researchers, like Stone and Carpenter, produced a sharp analysis of animal sexuality that approached *Social Life of Monkeys and Apes* (1932) by the British zoologist Solomon Zuckerman. Although European experts on animal societies were by no means fixated on sexuality, the intellectual climate abroad was sufficiently unrestrained to permit Zuckerman's imaginative work.[67]

The crux of Zuckerman's proposal was that monkeys and apes use sex instrumentally to manipulate social relationships. Precisely because sex for them is so protean, the outcome is "liberation of these responses in asexual situations," meaning, these animals deploy sex nearly symbolically. Zuckerman reported scientific observations of both males and females, "always to some degree in a sexually excitable condition," where the animals enlisted "kittens, puppies, foxes, and even snakes as sexual objects," as well as masturbated with "fruit, glass and sticks." This "diffuse sexuality" is integrated into "a system of dominance": not only do powerful members command sex but normally submissive individuals also learn to acquire favors, such as food or protection, by offering themselves sexually. Zuckerman's phrase "'prostitution' or modification," making the words equivalents, toned down any implication that the animals were behaving immorally when they traded sex.[68] Based on fieldwork in his native Africa, as well as experiments and historical records at the London Zoo, where he now worked, Zuckerman treated sexuality across the evolutionary order as a social fact with far-reaching consequences.

His citations speak volumes about his acquaintance with American research, and his failure to be inspired by it. He relied on observations of primate sexual activity compiled by G. V. Hamilton and Harold Bingham, both Yerkes's protégés, and by Ada Yerkes for *The Great Apes*. For how animals think, he read Edward Tolman and Donald K. Adams, also Yerkes's students. These choices in the literature followed the traditional strengths of American animal science: data-collection as a method and cognition as a subject. Zuckerman was also familiar with recent experiments in sexual biochemistry, mainly by Stone, and yet he mentioned no general American works on animal emotions, mating, or nurture.[69] Rather, his thesis extended the argument of *Sex and Repression in Savage Society* (1927) by the anthropologist Bronislaw Malinowski: that there is "plasticity of instincts under culture." Although Zuckerman did not credit primates with the foresight or memory required for human inventiveness, he saw these animals as intelligent enough to separate sex from its biological function and use it to mediate social power. However much the men's ideas on the malleability of sex already converged when they met in London in 1929, Zuckerman acknowledged Malinowski in his book. To be sure, Malinowski strenuously rejected Freud's claim that the Oedipus complex drives all societies, but he fully examined sexual emotion on his own terms.[70] American scholars offered Zuckerman no comparable analysis.

These Americans, intrigued and repelled by sexuality, remained a presence in the nation's science until World War II. Like William Morton Wheeler, who collapsed and died on a subway platform on his way home from work in 1937, they left satisfying academic routines reluctantly. William McDougall, summering each year at his new English country house and wintering at Duke, died of cancer in North Carolina in 1938. In 1945 Walter Cannon succumbed to complications of the same disease, traceable to x-ray exposure incurred during his animal experiments at the turn of the century. W. C. Allee, confined to a wheelchair by the late 1930s, pictured himself as "a paraplegic doing active work as a professor of zoology." Disgusted when administrators "put me on the retired list" in 1950, he left Chicago for the University of Florida, dying five years later.[71] With the exception of Marjorie Allee, all the men's wives outlived them. So did Margaret Nice, who with characteristic energy organized relief for European scientists at the end of the war. Her correspondence for the cause showed that there was something left of the circle of students of animal behavior who took so mixed an interest in reproduction. When she nagged Allee in 1949 to help secure funds from the Rockefeller Foundation for her revered Konrad Lorenz, Allee promised to give "my recommendation of aid for a Nazi," despite "differences of opinion on political matters." This said, "I do want to be less politically naïve than Dr. Lorenz is reported to have been."[72]

Allee's response epitomized his generation's conflicting views about how social values figure in science: research is disinterested, but be careful to be right. In sex studies, whether the subjects were animals or humans, this meant to be wary of passion. In 1933 the CRPS was distressed when some grant recipients tried to sell their discoveries to manufacturers developing hormone-based products. When the committee learned that Eli Lilly and Company planned to adapt "the gonad-stimulating substance from the blood of pregnant mares" for "commercial production and sale," an NRC official called in the scientist who made the contract and rescinded his grant. "The interview was painful," the administrator informed Yerkes, particularly because this scholar was not alone in thinking "the public interest should be protected and science can be advanced by a patent." On the surface, the objection was to private gain via "royalties under patents and contracts." Even so, the thought that sex research might promote pleasure-seeking likely occurred to the sponsors.[73] Indeed, no similar protest accompanied a 1939 proposal connected with the involuntary sterilization of "feeble-minded young women." The applicant wrote: "The investigation would involve a study of daily vaginal smears, morning rectal temperatures, and possibly, also, an examination of the urine," followed by the surgery, and "if a recently ruptured follicle is found, it and the entire tube on that side would be removed and prepared for histological study."[74] Whereas commercialization of hormones raised tempers, eugenic control brought no special comment.

This pattern of response makes it unsurprising that Yerkes, in charge of the grants, described primates moralistically when he revised Zuckerman's idea of the social adaptability of sex acts for his last book, *Chimpanzees* (1943). The meaning of prostitution for Zuckerman had been nearly Marxian: primates invest sex with value as a traded object. Yerkes traced the same behavior to character traits. Although he, too, identified "sexual allure and varied forms of solicitation" as components of power plays, he saw these gestures as specifically feminine: "As for females, wiles, trickery, or deceitful cunning, which are conspicuous by their absence in the male list, are favorite resources."[75] There was something unpleasant for Yerkes in the way his female chimpanzees appeared to detach copulation from its reproductive function to gain an advantage. This was not the personality of his beloved Ada, who in later years was often found in her New Haven garden tending prize-winning flowers.[76] If, to Yerkes, Ada epitomized the natural state—devotion to family as the anchor of a husband's labors—then the diverse sexual practices of the rest of the animal kingdom suffered by comparison. Similar partner-marriages of many of Yerkes's colleagues meant that they brought comparable private experiences to sex research.

Calvin Stone, both younger and less philosophical than these social theorists, progressed from performing ablative sex surgeries on rats to supervising experiments on human puberty and intelligence, all supported by the CRPS. How Stone's investigations of hormones then led to an intense interest in electric shock—as it influenced perception, learning, and emotion—seemed to mystify Yerkes. "I dare not comment on your reports because I am relatively ignorant of the effects of electro-shock," Yerkes wrote guardedly to Stone in 1953. This was not quite true. Yerkes's shock experiments on mice in 1908 demonstrated that while mild or extreme pain impeded learning, moderate discomfort hastened mastery.[77] The method was nonetheless anomalous for Yerkes, who for five decades relied mostly on food rewards. Perhaps Yerkes's hesitancy grew instead from Stone's fascination with neurology, particularly when the induced experience was pain. In taking up the subject of suffering, however, Stone was not alone.

4
The Rediscovery of Pain

> Every investigator working with rats must surely have verified our observation that these animals defecate and urinate during rough handling or during the first days in a maze, a discrimination box or the like.
>
> —Calvin S. Hall, 1934

Beginning in an earlier era, when psychology was indistinguishable from philosophy, pleasure and pain were understood to be opposites, and so it seems logical that scientific studies of animal sexuality were joined in the 1930s by investigations of suffering. World events pushed forward tests of the susceptibility of organisms to harm. In 1939, the first year of declared war in Europe, the New York publisher Harper and Brothers issued a short account of experiments designed to probe the psychological causes of "the neurotic pattern" in rats. The animals displayed convulsions, tics, clenched paws, rigid postures, and self-isolation in response to testing. The catalyst was a combination of mental and physical discomfort. First taught to choose an opening marked by a simple visual pattern that led to food, the rat hit a wall when the experimenter instead placed the cue in front of the obstruction. Electric shock delivered through the starting platform induced the rat to continue jumping, at the same time that surgical lesions of the cortex disrupted learning further. For the study's author, Norman R. F. Maier of the University of Michigan, the "conflict set up" between past training and new dangers and between instinctive self-protection and forced action explained the dysfunctional behavior.[1] In bitter ensuing controversy, he defended this psychological argument about emotional imbalance against a colleague who claimed that simple sensory shock—a loud noise, for example—was just as unnerving. The men nonetheless agreed that the hurt organisms incur in environments diminishes survival ability. In the scientific language of the time, they acknowledged animal pain.[2]

Although attention to impairment in animals was not new, there was a marked increase between the wars in the kinds of disability studied and the methods devised to manipulate an organism's defenses. Robert Yerkes's conclusion that a moderate dose of electric shock improved rat learning was unusual in 1908 amid commonplace use of food rewards in training. He now shifted his focus to psychological injury, along with fellow scholars. In a 1934 collection of papers on "mental disorder," Yerkes and his wife Ada attributed pathological behavior in primates to "an overexacting environment." Two years later they measured "avoidance (fear) response in chimpanzee," when the experimenter showed the animals various objects ranging from a badminton shuttlecock to snake-like lizard.[3]

These tests, where creating discomfort was the express purpose, advanced side by side with laboratory practices that produced pain as a side effect of good intentions. New surgeries that experimentally altered bodily systems once again had earlier origins, and from the first, scientists stressed beneficial outcomes instead of the more obvious infliction of harm. In 1902 Shepherd Franz claimed "greater friendliness" in cats after he severed connections in their frontal lobes to observe the consequences for learning: "A few that tried repeatedly to escape from me before the operation, kept close to me afterward." After World War I, surgical interventions to diminish normal brain, reproductive, and endocrine functions gained currency and progressed in the same affirmative sprit. Whereas "it was necessary to drag the animal" to a difficult task before her lobotomy, one team reported in 1935, she was so placid afterward that she seemed to have "joined the 'happiness cult of Elder Micheaux,' and had placed its burdens on the Lord!"[4] Whether or not we consider such treatment of healthy animals inherently cruel, researchers made provision for their subjects' pain when they budgeted recovery time from surgery in their protocols. In at least this way, the experimenters recognized the animals' reactions to invasive surgeries, although producing distress was not the stated research goal.

Interest in pain in the laboratory somehow mirrored the substantial global problems of the 1930s: economic hardship, totalitarian rule, and colonial expansion. These human trials did not keep American investigators of animal behavior from increasingly traveling abroad, and at times their writing conveyed their sense of the vulnerability of organisms. Circling the globe in 1935, with stops in Beijing to teach physiology and Leningrad to lecture, Walter Cannon witnessed life outside Harvard. In 1940, as outgoing president of the American Association for the Advancement of Science, he expressed foreboding. "The body physiologic and the body politic" alike faced "harmful" forces requiring anticipatory "control" to secure "margins of safety."[5]

Yet far more often these American specialists in animal sensibility came home sounding unexpectedly triumphant. Grim experiments and sobering journeys ended in redoubled confidence in the inherent orderliness of nature and the capacity of science to cure its practical imperfections. "Perhaps our examination of the physiological antagonism which can develop between the cortex and the thalamic area," Cannon assured an audience of doctors in 1936, will explain "how a return of internal peace may bring miraculously a return to health and happiness."[6] From a practitioner's standpoint, logic supported this conclusion: pain scrutinized in laboratories could be an intellectual key to remediation outside. Professionalism further guarded scientific travelers from disillusionment. Received in foreign lands as respected authorities, they caught glimpses of suffering in the background of research projects, cordial colleagues, and orchestrated events. They were perceptive and principled enough to be disturbed by sights of poverty, militarism, and fear. Even so, they held onto their conviction that understanding pain in test animals served the public good. Beneficent nature was on their side.

Human beings were the object of their altruism, however, and animals were just the means. Only rarely did researchers of animal behavior think about their subjects' adverse responses in ways unrelated to the efficiency of experiments. When Yerkes spoke philosophically in his obituary of Cannon, he implied an ethical trade-off: antivivisectionists proceed—wrongly in his view—as if "other organisms are more worthy of considerate treatment and preservation from accident or disease than is man."[7] Along with the preeminence of human interests, Yerkes's statement endorsed the probability that animals suffer. Indeed, interwar studies of inflicted mental and bodily damage made little sense unless his peers similarly assumed that animals are more than machines.

Doing Harm and Detecting Pain

In historical perspective, the 1930s research on induced distress was notably self-conscious about emotion. Calvin Hall coined the term "emotionality," meaning an "upset or excited condition," in his doctoral dissertation completed at Berkeley in 1933. Based on the commonplace laboratory observation that rats are fearful in a "strange enclosure," particularly an open space with nowhere to hide, Hall placed his subjects in an eight-foot-wide circular field. He then measured the rats' frequency of defecation as an index of anxiety, in relation to their willingness to eat, a sign of adjustment. He discovered that his rats were so susceptible to stress that one died of starvation.[8] What was new here was how deliberately Hall provoked unpleasant sensations. Learning had been the intended subject of American studies at the turn of the century that entailed the experimental impairment of animals. Although these earlier tests clearly caused injury, and revealed the scholars' beliefs about animal sensibility, the scientific focus on suffering emerged only gradually.

In the summer of 1917, as American troops began to go to war, the lead article in the new journal *Psychobiology* was a potent demonstration of the utility of surgical injury to investigate the neurology of intelligence and, as it turned out, sensation and emotion as well. The coauthors were Shepherd Franz, by now an old hand at operating on animals, and Karl Lashley, at the time a postdoctoral fellow starting a career-long quest to understand the brain. Although their collaboration may seem intellectually fated, they connected somewhat by chance. Franz was working with human patients at St. Elizabeths Hospital, the massive government-run facility for the mentally ill in Washington, DC. Lashley was reasonably close at Johns Hopkins. Their experimental question—do rats subjected to frontal lobe lesions retain acquired skills?—was clear in their unglamorous title: "The Retention of Habits by the Rat after Destruction of the Frontal Portion of the Cerebrum." Each man made a contribution. Franz had performed the same procedure on cats and monkeys around 1900. Lashley had a supply of rats that already knew the maze.[9]

The interest of the paper from the viewpoint of animal sentience lay in the details. Lashley especially had a knack for uncovering original findings in routine projects—for example, crediting rats with curiosity in his essay, "A Simple Maze" (1918). In "The Retention of Habits," the authors first disposed of their planned inquiry, affirming that postoperative rats have memory, and then moved on to "disturbances of sensibility" that caught their attention. They confidently reported testable changes, specifically, loss of smell and "tactile sensitivity." To be thorough, they also recorded "incidental observations" of "apathetic pauses" on the path and "abnormal reactions" at the food dish.[10] Seemingly intrigued by how injury affects drive and appetite—both fresh psychological concepts of the war era—they mentioned pain only as a problem of method. They used ether to anesthetize the rats for surgery. Afterward, they delayed retesting for about twenty-four hours. Even so, descriptions of postoperative working rats as still "very weak" and fiddling with "head bandages" indicate the animals' physical discomfort. In all, the researchers were satisfied that prefrontal surgery did not erase learning, and they were equally pleased that their work opened new ground. Most obviously, if "the habit reaction is not necessarily cortical," then what sustains it? Conversely, how did the cerebrum influence the rats' senses, and possibly their moods and pain?[11]

Lashley was the dominant partner both in this experiment and soon in the expanding allied areas of ablative techniques and pain studies. His intellectual intensity, accompanied by qualities suggesting his narrowness and callousness, were consequently influential. Not only did the 1917 paper flatly reverse Franz's prior conclusion, but its protocol's clean design omitted the older man's tests of animal affect, using a control group for comparison as well. Because Franz subscribed before 1910 to the then-popular view of learning as "simple associations," he was prepared to find loss of habit due to surgery. So he had argued in papers on cats in 1902 and monkeys in 1906. At the same time, he took steps to confirm his impression of emotional change. Wondering whether "surgical shock" alone disrupted his cats' performance, he created a control group subjected to brain surgery without ablation. To estimate normalcy of affect, he placed the same cats at different stages of treatment in front of caged mice.[12]

All these marks of Franz's early experience now disappeared. The coauthors' thesis that efficient maze running survived injury mirrored a holistic approach to brain functions now gaining credence. The rats' failures to demonstrate sniffing, respond to touch, or show excitement, though presented as meaningful, were merely anecdotal. There was no formal controlled testing. Although Franz, too, had abandoned an associative theory of intelligence by the 1910s, the embrace of neurological complexity and adoption of a streamlined method represented a style identifiable as Lashley's. He took up tantalizingly elusive problems in manageable ways.[13]

This combination of esotericism and pragmatism was a magnet for graduate students. Although laboratory procedures aiming to alter animals surgically or induce dysfunction did not always overlap, Lashley trained a startling number of practitioners devoted to one or both. At the University of Minnesota, where Lashley became an assistant professor in 1920, Calvin Perry Stone and Carlyle Jacobsen began early careers focused, respectively, on castrations and lobotomies. When Lashley moved to the University of Chicago in 1929, he taught doctoral candidates and postdoctoral fellows including Frank Beach, Donald Hebb, Norman Maier, T. C. Schneirla, and Joseph Yoshioka, who studied questions ranging from reproduction to neurosis, often with surgical components. In 1935, when Lashley left for Harvard, Beach and Hebb followed him. Clifford T. Morgan, a recent PhD of Leonard Carmichael at Rochester, soon joined the Harvard group.[14]

An undergraduate recalled advice from Morgan to choose "a 'Lashley-type' problem," which for this young man meant measuring distance perception in rats subjected to cortical lesions. Lashley's mentoring and his iconic reputation tell the same story: he was a presence in the experimental modification of animals. Predictably, there was a ripple effect. For example, Jacobsen, as well as Stone's student, C. R. Carpenter, both worked with Yerkes's primates at the Yale Laboratories of Comparative Psychobiology in the early 1930s.[15]

In a memoir of Lashley following his death in 1958, Frank Beach inserted the 1920 letter written by Lashley that recruited Calvin Stone to Minnesota. It bespoke Lashley's creativity and ambition, and his low estimate of animal pain. A long list of possible dissertations for Stone included the "effect of drugs, diet, starvation, hormones, etc. upon the learning process," "production of abnormal development of brain in fish and amphibia," and tests of rats' senses "by strychnine and extirpation methods." Lashley's distinction between topics that do "not seem to attract much attention" and projects "pretty sure to make a man in psychology" in effect counseled Stone that the laboratory was no ivory tower. As practically, because "some surgical technique" was needed for work in "functional neurology," Lashley advised Stone to "pick it up this summer" by finding "a third year medic to help you out." This offhand encouragement to gain rudimentary skill was the logical outcome of Lashley's opinion of rats, his favored subjects, as having "relatively simple and homogeneous cerebral physiology."[16] Whether he believed, as a corollary, that they felt no pain, the animals' comfort was not his priority.

Although operations that were deliberately injurious and casually crude easily rivet our attention, two distinguishable research streams mixed together in Lashley's letter: experimentally altering bodies to infer normalcy by the contrast, and inducing psychopathology by unpleasant external or internal stimuli. The animals were harmed irrespective of the human intention. Chronologically, physical reduction or enhancement, typically by surgery or injection, began in the 1920s and preceded purposeful efforts to generate psychic distress.

In these anatomical and physiological trials, discussions of palliative care of the animals increasingly disappeared from published reports. Sometimes it seems a researcher had something to hide. Calvin Stone made no mention of anesthesia in studies begun under Lashley of sex drive in castrated male rats. Private correspondence nonetheless identified Stone as the object of Walter Cannon's tirade in 1924 against a colleague who has omitted anesthesia and "inflicted unnecessary pain."[17] In another case, exactly how Leonard Carmichael handled affect in pregnant cats and their fetuses is a mystery, but his accounts are chilling. In contrast to the mid-1920s, when Carmichael dulled responsiveness in amphibian eggs by immersing them in chemical solution, mammals were his subjects a decade later. But the technique was nearly identical: he observed the movements of fetal kittens surgically removed "just before birth" and placed in "saline solution held at blood temperature by an outer water bath."[18] This was the extent of his comment about the treatment of mother and offspring. Neither Stone nor Carmichael set out to cause suffering, but neither did they show concern for the reactions of sentient creatures to invasive procedures.

Especially when animal subjects were complex in evolutionary terms, a planned psychological disturbance could be added to physical intervention to gauge departure from normal functioning. Because Lashley focused increasingly on neurology, his students consulted other authorities on how to disorient organisms by modifying their environments. In the mid-1930s Carlyle Jacobsen, for one, drew on Ivan Pavlov's experience with " 'experimental neurosis' " to enhance otherwise standard prefrontal surgeries on two of Yerkes's chimpanzees, Becky and Lucy. In light of the intricacy of primate brains, Jacobsen recruited Dr. John Fulton of Yale to perform successive ablations that removed as much as thirty-seven grams of tissue. The purpose was to assess postoperative memory of established skills, a long-standing question. During her training, Becky unexpectedly developed "temper tantrums" resembling the acquired abnormalities Pavlov had observed in his laboratory. Unable easily to recall where the experimenter hid objects, the chimp "rolled on the floor, defecated and urinated" when a lowered screen blocked her view of the field. Once she associated the obstruction with her frustration, seeing it triggered recurrent collapse. Needless to say, Jacobsen was intrigued by Becky's contrasting indifference after surgery, when she "continued to play or to pick over its fur" in response to the barrier and her erroneous choices.[19] Initially, such manufactured distress seemed an accidental byproduct of the animals' prolonged labor. Yet it is not surprising that researchers subsequently manipulated moods to investigate disorders of temperament.

In Edward Tolman's laboratory at Berkeley, the deliberate disturbance of rats' instincts did not cause obvious pain, because there was no surgery. But for a rat, "a stereotropic or wall-following animal" that prefers darkness, it was stressful to be confined in a well-lit circular field. Indeed, precisely because Tolman did not perform operations, his students demonstrated the trauma of disrupting drives by changes in environment. The reason for open-field trials was to examine the influence of emotion on learning, an issue about which Calvin Hall, the most systematic of Tolman's protégés, changed his mind. Publishing his first results in 1934, Hall judged the rats' "emotionality," measured by decline in appetite and increase in elimination, "a barrier to the satisfaction of needs." By 1936 his conclusion was more nuanced: inherently fearful, rats cower when threatened, a disadvantage when activity is useful and yet an asset that may "facilitate adjustment" at other times.[20] The rat's timidity deterred learning, in other words, but caution served it in danger.

Hall's modest experiments were in truth path-breaking. He isolated rat emotion for analysis by equating upset with observable behavior. He questioned the truism of Western thinking that feelings are a weakness. He showed the psychic toll of an unsuitable environment. If Hall's subjects had been something other than rats, his insights might have made his career.

The social dislocations of the Great Depression were simultaneously focusing other psychologists' attention on human suffering, often conceived as abnormality, and the study of animal anguish did advance as a related theme. Robert and Ada Yerkes contributed the only paper on animal "psychopathology" to a hefty 1934 collection on "mental disorder" sponsored by the Carnegie Foundation. Whereas they withheld judgment on animal deviance, they eagerly promoted "neglected opportunities for fruitful research."[21] Good science was needed to address an alarming situation. In the editors' view, psychic distress in human beings worked "mischief" by sustaining "healers, magicians, cult leaders, quacks, astrologers, radio advisers, and the like." Some of the afflicted went on to experience the "defective self-direction" and "unbalanced performance" of "the distinctly dangerous classes of society." For troubled primates, there was no "systematic review or digest of the literature," but the Yerkeses had personally witnessed either unusual lethargy or "emotional explosion."[22] Perhaps the couple's high standard for scientific reporting kept them from mentioning Jacobsen's work-in-progress with Becky in the Yale laboratory or the rat tests sponsored by Tolman, Robert's former student. Yerkes's prominence all but guaranteed, however, that his endorsement of research on disorders of temperament stepped up experiments on animal stress.

In this context, Norman Maier's choice of the word "neurotic" in the subtitle of his *Studies of Abnormal Behavior in the Rat* (1939) hardly seemed sufficient for the animals' seizures and catatonic states. Illustrated by close-up photographs of rats in postures of "convulsion" or "semi-rigor" with "clenched fists," the book combined virtually every feature of existing protocols: disrupted training, painful stimuli, surgical lesions, and an exhausting schedule. Maier presented selected "case histories," as if the numbered rats were patients, but there was a common core. Once the rats learned to find food, the same signals were changed arbitrarily to lead them to punishment. Blocked escape routes, air blasts, and electric shock prevented avoidance or inaction. Some of the animals were subjected to frontal lobe surgery, but all by necessity repeated dysfunctional behavior to a wearying extent. Between ten and twenty trials a day for several weeks was the average, making the rat that ran randomly in circles and collapsed after 161 tests a typical example. Miraculously, a diminishing number of "neurotic attacks" occurred in a few animals that adjusted to the pressure and, on their own, achieved a "cure."[23] Maier's goal, however, was to drive the rats crazy, ostensibly to aid suffering humans who felt similarly harried.

These varied techniques designed to disturb adaptive responses in animals carried over from laboratories into fieldwork. Although the intention was not to reduce an organism's chance of survival, the researchers did not seem to reflect on potentially hurtful consequences of transferring methods outdoors. Much as Maier tampered with his rats' habitual path to food, for example, T. C. Schneirla moved leaves bearing scents that helped guide "raiding" army ants (*Eciton*) to their prey in the jungles of Panama. Schneirla's background was animal learning, specifically, a 1929 dissertation at the University of Michigan that cleverly substituted ants for mammals in a maze to observe the insects orient themselves spatially to acquire food. On Barro Colorado Island soon thereafter, he used military terms—"bivouac," "booty," "pillaged regions"—to describe the aggressiveness of local ant species. Beyond noting temperament, however, Schneirla did not apparently think that he might harmfully disrupt the ants' sensibility by meddling with their trail. His account was strictly factual: "hesitating ants gather on the two sides of the 'break'" he created. For Schneirla, relocating objects along "a *chemical trail*" was simply a component of "standard experiments," now repeated in the wild.[24]

In the end, it is difficult to determine whether animal scientists had second thoughts about causing pain, because they wrote in disinterested terms. C. R. Carpenter, who compiled volumes of meticulous field journals during the 1930s on world travels to study primates, largely refrained from passing judgment on his heterogeneous activities. On the one hand, he worked to preserve natural habitats. In Sumatra in 1937, where he advised Dutch administrators on endangered orangutans, he expressed disappointment about local soldiers who were "poorly informed about the regulations and have had developed [*sic*] little knowledge or interest in these protected animals." On the other hand, he endorsed surgical, hormonal, and chemical alteration of semifree subjects. His 1938 notes for planned experiments on hundreds of gibbons and macaques, transported from India for a colony in Puerto Rico, included castration and lobotomy. "Produce experimental homosexuality" and "produce drug addiction" were also on the list.[25] Having castrated pigeons in Calvin Stone's laboratory for his dissertation, Carpenter now seemed prepared to take further steps. Not the lead investigator, he may also have had little choice. In 1940, his observations in Puerto Rico, just before he left for a permanent job, ended up being exclusively about behavior. Perhaps the experiments on the primates' bodies never occurred. Still, nowhere in his private log did he doubt that they should.[26]

Carpenter's comments nonetheless pinpoint the overarching reason for the wide boundaries set by animal experimenters for their own behavior: the good of science. His defense of Sumatran primates was based on the expectation that "American scientists" would undertake "a thorough field study of the behavior and ecology of orang outans [*sic*]" in 1941. His engaging confidence masked questionable assumptions. Well-intended specialists who entered the orangs' domain, Carpenter implied, could not harm them. By the same logic, physical or psychological damage imposed on laboratory animals for the sake of advancing knowledge was somehow good. If faith in science was one blind spot, naïve underestimate of world politics was another. Japanese aggression across the Pacific was sufficiently obvious by 1937, when Carpenter wrote in anticipation of future American research, to make his hope seem shortsighted.[27] The human world of the 1930s was filled with pain. Animal laboratories mirrored it. In these controlled spaces, and in society at large, experimenters viewed the dysfunction of organisms through the lens of their science.

Traveling a Disordered World

American specialists in animal behavior typically mixed work and leisure when they went abroad during the 1930s. C. R. Carpenter was largely truthful in 1938 when he told a Rotary club in Malaysia that he "liked occasionally to flee from the routine of academy duty and live in more primitive surroundings." To be sure, acquisition of knowledge about "animals below man" was his serious purpose, and distant travel offered insight into human nature as well.[28] Indeed, there was a political reason for his quick return to Asia following fieldwork in Sumatra the year before. With American laboratories importing 20,000 monkeys annually, mostly for medical experiments, India was initiating a partial "embargo" on the trade. Carpenter was on his way there to transport hundreds of future breeders to Puerto Rico. Articles clipped by Carpenter gave plentiful evidence of protests in English-language Indian newspapers by humanitarians distressed at the monkeys' future "long-drawn-out tortures in vivisectional laboratories." The single American report in the *Boston Post*, picturing a smiling keeper befriending a caged monkey, treated the animals' stateside arrival as a charming distraction.[29] If there were lessons in these events for the conduct of international science, Americans seemed the least concerned party, at least as measured by publicity. In a similar mood, there was a limit to how much American scientists absorbed the world's troubles that they witnessed abroad.

Not every animal researcher left a record of foreign travel. Older scholars invited to speak and younger fieldworkers seeking material were more apt to go abroad and write up their impressions than rising men turning out publications. There was sufficient overlap between experimentalists and travelers, as well as consistency in the Americans' reactions, however, to ask whether international exposure affected research, and particularly understanding of suffering.[30]

In the decade or so before the war, American analysts of animal sensibility visited Europe, Asia, Africa, and Latin America. They gave lectures, collected specimens, conducted fieldwork, and toured for pleasure. Scientific goals and acquaintances afforded them a kind of protective buffer as they moved about, and yet professionalism could not altogether keep them from feeling vulnerable. Birding in the Hungarian countryside in 1938, Margaret Morse Nice joked that her "Anglo-Magyar phrase book"—offering translations of " 'Are you sure you love me very much? What is the way to the marriage bureau?' "—did not help her order breakfast.[31] Amusing as her inconvenience

Figure 4.1 C. Ray Carpenter, on right in the back row, and colleagues on Asiatic Primate Expedition to then-Siam, 1937. Courtesy of Eberly Family Special Collections Library, Penn State University Libraries.

was, the world of the 1930s remained vast, technologically and culturally, and foreign places could seem alien. Discomfort was part of the travel experience.

Curiosity, though, most often scientific in nature, was the chief motive for going abroad. At age forty Calvin Perry Stone had never before left the United States when he traveled for seven months in Europe in 1932. Growing up poor on an Indiana farm, he spent World War I supervising Army mental tests in Arkansas. A paper presentation at the International Congress of Psychology in Copenhagen was now his official reason for leaving Palo Alto, but in truth Stone undertook a psychologist's grand tour. In England, Scotland, Germany, and Austria he faithfully visited universities, laboratories, zoos, and asylums, meeting and dining with nearly every psychologist of note. Letters sent home to his wife and children were a minute record of his observations. Occasionally impressed—for example, by a man "doing remarkable things with apes" at the Dresden Zoo—Stone more often thought European practices deficient. "How can Drever" in Edinburgh "give a stimulating course in animal psychology if he has never performed an animal experiment himself?"[32]

Whether praising or censuring, however, his interest never flagged. Because the English "are not accustomed to modern plumbing," Oxford students wore bathrobes outdoors to reach the few bathtubs, still "better than standing in a tub." Invited to "a small lunch" by Wolfgang Köhler in Berlin, Stone was unprepared for a spread of ham, cheese, eggs, bread, rolls, and coffee. Insulting private remarks did not diminish his zesty cheerfulness. The intellectual heir of Wilhelm Wundt at Leipzig was "a sprucy old boy." Wundt's daughter, around age fifty, was "an old german [*sic*] girl." Nothing measured up to American standards. Glasgow's mental hospital resembled "a poor class of state institution in America."[33] Stone's sense of national advantage seemed to enhance his enjoyment.

For American scientists who ventured beyond Europe, journeys begun for practical purposes unexpectedly, and pleasantly, offered dreamy escapes from obligations in locations so exotic as to seem fantastic. Steamships were the typical transportation, and for these intellectuals, reading and writing were the main pastimes. Most boarded with "a traveling library, containing books for the voyage," in the words of the ornithologist Frank Chapman, en route to Panama from New York. Good intentions of using the time at sea for work evaporated. Bound for Yokohama from Vancouver in 1937, C. R. Carpenter found "it impossible to read heavy serious papers," though a colleague "plays at doing so." Instead, he picked up *Gone with the Wind*. Even the intense Robert Yerkes indulged himself by drafting an autobiographical essay off the coast of Africa. "I've never seen father look so like the contented cow," his daughter marveled, "eating & sleeping enormously, & placidly reviewing himself autobiographically to his great satisfaction." The travelers enjoyed this mental release—despite seasickness, cramped cabins, annoying fellow passengers, and anticipated dangers. Yerkes did remember to dose himself with quinine before landing in Dakar.[34] Still, there was a seductive imaginative dimension to journeys aimed officially at field study and specimen collection.

Upon arrival, unfamiliar landscapes sustained the pleasure of leaving everyday cares behind. Carpenter was charmed as he traversed Japan and China in 1937, just months before the Japanese army occupied a wide region from Beijing to Shanghai and Nanjing. Noting "beautiful" scenes as soon as he passed through Oregon's Cascade Range to meet his ship, he described Japan as a land of "striking topography" and "marvelous beauty." From a distance, the inhabitants belonged to the panoramas. "Truly Chinese landscape" was the sight from his train window on the way from Hong Kong to Canton: "Coolies—porters—vegetable, flower & rice gardens." Having walked through the " 'Chinese city' " in the port of Shanghai, he felt China close up to be "a hodge podge of smelly, crowded humanity." Nor was he unaware on shipboard to Singapore, traveling first class with a handful of others, that "2000 deck passengers" endured "terrible" conditions "stowed away like freight on three decks."[35] But he continued to be drawn deeply to Asia. "The girls were extremely pretty," he wrote in his journal in the Thai countryside, "with clear dark olive skin, cheerful smiles and dark slanting opened eyes." Even after getting down to the serious work of tracking gibbons, Carpenter remained a tourist. In search of a campsite, he took time to notice priests "clad in bright yellow flowing robes" as "a colorful group as they kneeled on the polished but dark teak floor, bowing deeply and chanting to the dimly lit bronze Buddah."[36]

Fieldworkers had long written up their expeditions as travel literature, and although Carpenter did not publish these impressions, his companion, Harold J. Coolidge, a Harvard zoologist, was adept at communicating the adventures of science. Plans to court future readers reinforced the inclination of American scientists to savor the exotic. In 1933 Coolidge had filled his book, *Three Kingdoms of Indo-China*, with seaborne pirates, hundred-pony caravans, and white-water rivers. Collecting fauna was the trip's purpose, but the book meanwhile evoked the " 'Never Never' lands of Indo-China." More tamely, Frank Chapman in Panama described *Life in an Air Castle*, spent scanning the treetop through binoculars, as an "Assisian dream."[37] The ever-literary Cornelia Cannon, crossing Russia by train from Vladivostok in 1935, wrote movingly of her visit to the final residence of Nicholas II, complete with its walled "garden where the doomed family had been allowed to exercise." That the Soviet secret police detained her for

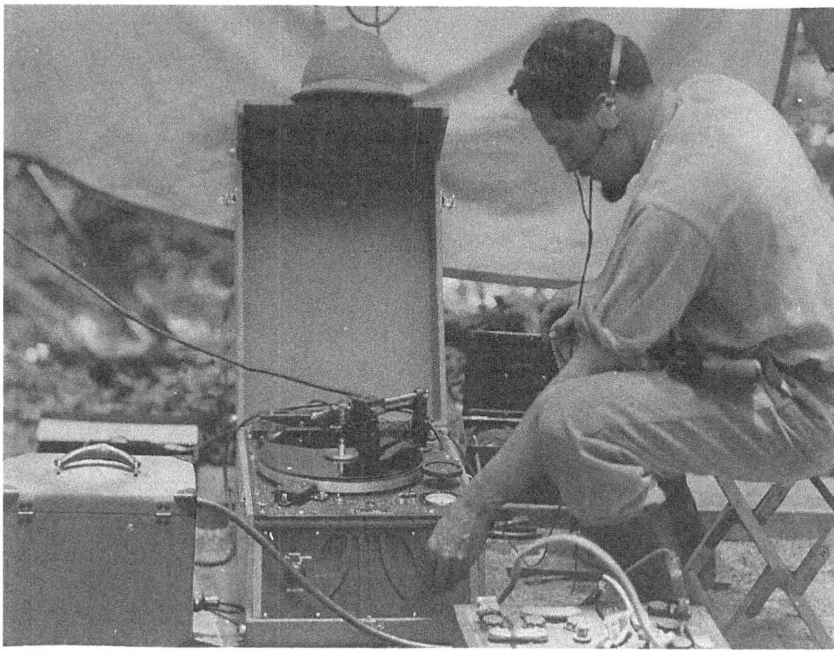

Figure 4.2 C. R. Carpenter assembling equipment to record gibbon calls, Siam, 1937. Courtesy of Eberly Family Special Collections Library, Penn State University Libraries.

Figure 4.3 Carpenter at the jungle's edge. Courtesy of Eberly Family Special Collections Library, Penn State University Libraries.

taking photographs—not recognizing the wife of Ivan Pavlov's personal guest—added to the romance. No doubt disappointed when Harper and Brothers refused to publish her story, Cornelia had done her best to capture the pathos of the Romanovs' fall.[38]

The thrill of experiences akin to fantasy did not prevent these highly educated travelers from also comprehending real dangers. Disease-bearing insects, unsanitary accommodations, and unpredictable human situations were regular components of scientific journeys. Although the Cannons on the Trans-Siberian Railway did not face the tropical fevers and headhunters reported elsewhere, Cornelia worked methodically to limit risks to her husband, in fragile health due to exposure to x-rays in his laboratory years before. As, in her words, an "ingenious American," devoted to "defending Father," she rigged up a curtain to trap cinders, "spend most of my time killing flies," and arranged with their inefficient waiter for "two meals a day for Walter, and one for me." Although she brushed off her encounter with the government, she watched "political prisoners" laboring beneath "wooden towers with a man with a gun on the top of each." "How many broken hearts," she reflected, "those places represent."[39]

Indeed, just as the gratifications of touring were in good measure imaginative, so the burdens were less bodily discomforts than disturbed emotions. "Father is making notes on these matters to present to the proper authorities," Cornelia laughed gently in her letter home, "as an evidence of our appreciation of their generous hospitality to us." Here was the pain of travel—feeling out of place and out of control—and "Father, like a man," determined "to grapple with the situation" head on.[40]

Much like the laboratory rats unbalanced by novel situations, the Americans were most thrown off by mundane details. Whereas scientific tasks remained an anchor, drinking and speaking became vexing problems. Alcohol consumption abroad was often a shock. Walter Cannon, who in 1917 pledged abstinence on his way to France, was sufficiently worldly by 1935 to know that drinking beer in Siberia was preferable to tap water. Calvin Stone, in contrast, seemed dogged by "the inevitable problem of drinking." "Being myself"—his reason for declining liquor at lunch with the psychologist C. S. Myers at one of London's "most exclusive clubs"—risked causing social offense. Having chosen a "Temperance Hotel" in York, without the bar that made other lodgings "distasteful," Stone later accepted wine "called Vermuth" from German hosts, though "it tasted a little like cough syrup." Not a convert, he soon disparaged Heidelberg in a letter home as "the seat of learning or beer drinking."[41]

More indulgently, Carpenter in Asia, neither under the eye of distinguished peers nor moralizing to his children, admitted in his journal to enjoying champagne and sake. But excess was another matter: he disapproved of "two young English bucks" who passed the time at sea "drinking continuously from 10:00 a.m. until mid-night." Even Margaret Nice, who otherwise found Europe liberating, did not drink freely. "Every serious student in America must acquaint himself with what is being done in Europe," she passionately lectured fellow birders. In Hungary she turned down tea spiked with rum, however, preferring mineral water and, when good manners required, a glass of wine.[42]

The discomfort posed by language barriers also centered more on giving offense than simple misunderstanding. Nice, who like most American scholars commanded decent German, felt "ignorant and rude" in Hungary, because "I could not even say 'Good day' or 'Thank you.'"[43] But for intellectuals who liked to study, language deficiency was more easily remedied than a habit of temperance. "Father sits beside me studying

German," Cornelia Cannon noted three days west of Vladivostok. Soon "our porters," two "blonde and pleasant" men, were "teaching Father Russian." So ambitious a journey could not help but intensify the Cannons' awareness of linguistic diversity. They met "a sweet young Polish woman" who "speaks perfect English" because she was "educated in Shanghai." Their "little Japanese fellow traveler" was on his way to Lisbon as a diplomat "after learning Portuguese" in Brazil.[44]

If these remarks seemed merely Cornelia's chatter about curiosities, Walter fully appreciated the politics of speech. In Paris in 1929, preparing for lectures that he had written in French, he "employed a charming young Frenchwoman to come and soften his American accent and teach him conversational French to relieve his agonizing over the language," his daughter recalled. Not always able to be so fastidious, the Americans felt the awkwardness of being visitors. "We chatted a bit, he in Deutsch and I in English," Stone wrote casually about a conversation in Leipzig, but soon worried whether "most of what I said was understood" when he delivered two scientific papers in English.[45]

Science, however, was the great unifier, and colleagues abroad were a welcoming presence who lessened the anxieties of dislocation. Walter Cannon in Leningrad almost reverently cited "the world-wide uniformity of scientific phenomena" in his plenary speech to the International Physiological Congress, the practical reason for his long trip. But for Cornelia, it was the sight at the station of "dear old-young P.P."—Professor Pavlov, a "beautiful old saint" at age eighty-six—that told her why she had come. In 1923 the Cannons had hosted Pavlov in the United States following the men's years of correspondence about digestive experiments.[46]

Exchanges of letters about research were often the foundation for international visits. For Robert Yerkes, finally meeting "Dr. Kohts" in Moscow in 1929 helps explain why "it seems more thrilling to be going to Russia than to Africa." A fellow student of chimpanzees, "Mrs. Ladygin-Kohts," he declared in print in 1925, "has published in Russian a book" with "some of the best pictorial and verbal descriptions of the animal ever presented." Foreign-born scholars living in the United States wrote letters of introduction for less well-connected American acquaintances. Although Calvin Stone had recently critiqued William McDougall's idea of instinct, McDougall, at Duke, opened doors for him in England. "An Englishman, it seems," Stone remarked in London, a bit surprised, "is bounden to do something for anyone who is properly introduced." Just so, Margaret Nice gratefully recognized Ernest Mayr, a new emigré from Berlin to the American Museum of Natural History, "to whom I am largely indebted for my introduction to European ornithology and ornithologists."[47] All these personal contacts gave substance to the theoretical universalism of science and made distant places less foreign.

Racial concepts were a different kind of byproduct of biological science that, sad to say, similarly aided acclimation. Whereas the American travelers achieved reasonable intimacy with their peers, they drew hasty conclusions about broader populations based on physical qualities. In French Indochina, Harold Coolidge almost obsessively used color to describe himself: he "was the fourth white man" in the group of naturalists that composed their ship's sole "white passengers" and later camped in a village where "no white man had ever slept." A frontier post had a tiny "white population" to oversee "a prison for nearly four hundred desperate criminals." Boston-born, Harvard-educated, and just twenty-four, Coolidge to all appearances managed his disorientation amid a mass of dark faces by repeating his difference.[48] C. R. Carpenter, Coolidge's companion in Southeast Asia on a later trip, was more relaxed about race, praising a settler of European descent who "knew more about the habits of the orang outans than any person that I have met"—also the father of a "1/2 Dutch 1/2 Japanese" son.[49] Miscegenation was presumably familiar to Carpenter, a native of North Carolina.

Whether human diversity was distressing or curious, current science instilled confidence that physiognomy reveals character. "We all know it when we see it," Cornelia Cannon wrote of "that unmistakable look of the man of culture," absent in Siberia, she believed, because "they have certainly liquidated that group successfully." Although she toyed with the idea that a refined "face" is an acquired trait—by "diet or leisure or the process of learning or thinking in abstractions"—parentage was never far from this eugenic advocate's mind. The "Jews from New York" at her Moscow hotel came to see the Soviet system, she assumed, because of their "racial interest in social problems."[50] In these ways Americans steeped in biology used race theory to interpret strange surroundings.

They were equally prepared to look down on religions they considered superstitious. To a point, rituals were quaint, but it did not take much to dismiss the irrationality of Buddhist monks at prayer "like so many sparrows on a limb, dressed in filthy yellow robes."[51] In a crisis, pious beliefs were an easy target for blame. Robert Yerkes remained publicly gracious to Madame Rosalía Abreu for her "hospitality" in 1924 on her Havana estate, where his team studied her many primates. After Chim, his own chimpanzee, became ill there, however, his journal harped on her Catholicism. With Chim in "an alarming condition" from fever, "she offered [him] the bottle of water of 'the Lady of Lourd [*sic*].'" Once he died, Yerkes saw his hostess in a different light: "Whims and superstitions are difficult to separate from her rationalized observations." Although this was the uncharitable voice of despondency that left Yerkes "weak and listless," the experience made visible his hierarchy of cultures: rational science atop senseless faiths.[52]

The scientists' training offered fewer direct lessons about political systems, but professional lives devoted to orderly experiments made the prevailing colonialism of the 1930s generally a comfort for these traveling Americans, or at least not an object of protest. From the safety of Cambridge in 1923, Walter Cannon disparaged "disorderly China," subject to the whims of warlord armies. Once the Rockefeller Foundation successfully launched its oasis of Western medicine, Peking Union Medical College, however, he visited in 1935, honored by its Chinese physiologists as "our old Master."[53]

The American government's administration of territories in ways profiting animal study elicited even more praise. The Panama Canal's completion in 1914 displaced residents and animals from millions of flooded acres, but the ornithologist Frank Chapman nonetheless cited the National Research Council as the "guardian of wildlife" when it designated the newly formed Barro Colorado Island a field station protected from hunting and deforestation. None of the scholars who thereafter made the comparatively short trip from the United States to observe Panama's insects, birds, and mammals questioned their country's right to the Canal Zone. When colonial administrators were instead Europeans, American fieldworkers dealt with them matter-of-factly, usually to secure permits or information. Carpenter circulated a questionnaire to Dutch-led military patrols in Sumatra asking if orangs typically harmed crops or humans, and whether natives hunted them in turn. In the Laotian uplands policed by the French,

Harold Coolidge "dug out white shirts and trousers, and went up to play tennis with the Commandant."[54] All judged colonial regimes by a government's attentiveness to science and seemed to take their own sponsorship by the Western powers for granted.

European politics caused the Americans far more distress, because the evidence favored no clear conclusion. In Cannon's Leningrad speech, three of the four states he applauded for nurturing science—Soviet Russia, Italy, and Japan—also restricted their citizens' freedom. By the same token, "soviet energy, new building, new projects," in Cornelia's words, including her favorite birth control clinics, were exhilarating for researchers who expected their own discoveries to aid rational planning.[55]

But censorship of the couple's letters, the sight of naked prisoners heading east locked in boxcars, and "the unobtrusive slink of the Russian," next to "the fearless mien of the Americans," understandably gave the Cannons pause. "Father is feeling the queer atmosphere of oppression," beginning "with my arrest and he said later, 'Don't you feel as if you would like to get out of this country?' " Gratefully aware that they toured as "privileged persons," Cornelia recognized "a queer kind of horror that seizes you now and then," when "the bear shows his paws."[56] Simple vulnerability was a part of the unease, but so was their sense that individual liberty in a science-minded state meant so little.

Germany was as confusing as the Soviet Union, less because of a façade as a grand experiment than due to the country's past as a scholarly hub. Dismissively calling mounted birds in a Paris museum "a beautiful picture," Margaret Nice predictably gushed over exhibitions prepared by Germans as displaying "their abhorrence of superficiality, their determination to get down to reality, their passion for truth." American professionals felt a personal stake in this shimmering reputation. "Here a great many of the American psychologists came for a part of their early training," Stone explained to his wife after visiting Leipzig. Mainly touring scientific institutions where his acquaintances had studied, Stone rarely commented on politics, and then as a neophyte. "Political developments are hard to fathom here in Germany," he confessed in June 1932, months before the appointment of Adolf Hitler as chancellor. He blamed the "demagoguery" of German, British, and American newspapers for obscuring understanding, and yet he seemed unreflective when he brushed shoulders with trouble. He was unable to meet the Gestalt psychologist—and Jew—Kurt Lewin, for example, "who has left for Stanford."[57]

In 1949 Nice was still arguing with Warder Allee about what scientists should have done under the Nazis. "As to Dr. Lorenz, it seems to me rather a pity to spend time on his one-time favorable attitude to certain aspects of Nazism," an opinion Allee did not share. She had not wondered in Austria in the summer of 1938, then Lorenz's guest in a country formally allied with Germany, if she had a political responsibility. Captivated by the European varieties of Redstarts and Serins she tracked with Lorenz for a coauthored paper, she seemed oblivious to how ominously Central Europe was changing.[58]

Once the war riveted the attention of American animal scientists on the world's human distress, they responded much as they had as travelers: as intelligent specialists. Largely untrained to think systematically about societies, they reached out individually to fellow professionals with empathy and a dash of condescension. Although Margaret Nice offered in 1946 to coordinate American relief for European ornithologists—"tea, coffee, chocolate, powdered milk," as well as "reprints, books, and journals"—her preferred approach was more direct: " 'Adopt' an ornithologist abroad whom you know. Send him or his family packages as often as you can."[59] In the same modest way, researchers who traveled the globe before the war caught glimpses of insecurity and fear. Occasionally they themselves felt anxious in strange lands. Preoccupied by the scientific purposes of their travel and welcomed as notables, however, they understood just so much of the decade's unstable mix of colonialism and totalitarian politics.

Nonetheless, they were also intellectuals who believed in their social obligation. Although their international experiences produced no revelations or conversions—no mass release of rats from painful protocols—these animal experts wrote more than laboratory reports. The universal mission of science to enhance life was not a new idea in the 1930s, but it gained energy as American practitioners saw more of the world. Put simply, knowledge would overcome suffering. The moral dilemma was how much pain should be required of animals for this happy outcome.

In Praise of Nature

The general drift of scientific thinking among American animal researchers during the 1930s was straightforward: they believed that the natural endowment of organisms tends toward order, and science enhances the process. In a few cases, there was an equally simple link between international exposure to human trials and redoubled efforts to relieve hardship. At home after circumnavigating the globe, Walter Cannon, for one, led the Soviet-American Medical Society, American Bureau for Medical Aid to China, and Medical Bureau to Aid Spanish Democracy.[60]

For Cannon's peers, however, the connections of personal experience, experimentation, and scientific conclusions were often less clear. Although all agreed that the goal of science was discerning natural laws for sake of human betterment, animal researchers typically thought about problems in the nature-nurture terms of theoretical dispute. Their explicit question was whether behavior springs from biological endowment or learned conditioning. They said less about what pain is and why induced suffering in animals profits humans, the issues embedded in their experiments. The writings of animal scientists suggest answers. As they embraced genetic and physiological causes of behavior, pain became a negative quality: the disruption of otherwise beneficent nature. Animals, moreover, fit neatly in a progressive vision: they were instruments of remediation in the hands of scientists. Strangely but logically, optimism about nature and science grew in inverse relation to a sense of human distress.

As war spread step by step across the globe in the 1930s, these intellectuals broadcast what they imagined to be a new perspective: biology causes ongoing, goal-seeking activity. In his presidential speech to the American Psychological Association in 1940, Leonard Carmichael confessed to having been "under the domination of a universal conditioned reflex theory." Now arguing that adaptation is "a result of maturation"—and hence "natural"—he made learning-oriented behaviorism his foil. The body was still a "machine," but the environment did not impose one lesson at a time; rather, the source of "timing and interplay of these mechanisms" was internal.[61]

Carmichael overstated his conversion from nurture to nature. One of his first articles—"Heredity and Environment: Are They Antithetical?" (1925)—had advocated compromise. Even so, studies of genetics and growth were indeed taking off. By 1932 Calvin Stone had repeated Yerkes's 1913 comparison of temperaments in wild and laboratory rat populations. Calvin Hall titled the concluding paper of his series of field tests of rat fear, "The Inheritance of Emotionality."[62] Edward Tolman rejected atomized scrutiny of "stimulus-response connections" in favor of holistic appraisal of "biological organisms immersed in environments." The common conviction was that unimpeded nature follows lawful patterns of maturation. As Calvin Stone marveled in 1934, "orderly development in young animals has always held in the opinion of man a high place among the wonders of nature."[63]

Even if these scholars' affirmations of natural agency were less radical than they thought, their excitement set off a wave of rediscovery of kindred thinkers. Young men who accounted themselves psychologists, albeit routinely modifying bodily functions in their laboratories, reached out to unlikely mentors. Calvin Stone made Charles Darwin the centerpiece of his 1942 presidential address to the American Psychological Association. Although Stone began by acknowledging the centenary of the birth of his American forebear, William James, he titled his talk, "Multiply, Vary, Let the Strongest Live and the Weakest Die—Charles Darwin." Similarly, Calvin Hall drew on Walter Cannon's forty-year study of emotional physiology for his own perspective on biology and temperament. He did not cite Cannon's work in his first article on rat anxiety in 1934, but four years later the senior investigator figured prominently Hall's summary paper.[64]

With greater risk, Stone devoted a full essay in 1931 to William McDougall. On the surface, the paper was a standard rejection of McDougall's unverified claim that emotion is a " 'psycho-physical disposition' " that manifests instinct. Read more closely, it was a vehicle for Stone's quiet agreement with McDougall on the influence of inborn dispositions. Sentence after sentence of Stone's prose flatly summarized McDougall: for example, "the instinctive disposition" is the "most permanent and unchanged aspect" of character. Although criticism was Stone's stated purpose, McDougall seemed little more than a mask for Stone's intellectual migration toward natural determinism.[65]

More poignantly, Leonard Carmichael, leaving a faculty post at Brown in 1936 for a deanship elsewhere, renewed contact with Wallace Craig. He wondered why Craig, already a vagabond scholar in the early 1920s when he was Carmichael's teacher, never published his ideas from "the seminar at Harvard which I remember so well." They "were too vague and uncertain and theoretical," Craig replied from England, and besides, "I discovered that many of my students, even my undergraduate students, had better minds than my own." Although Carmichael chided Craig for being "too modest," the correspondence was really about Carmichael's mounting interest in inborn appetites. Enthusiasm for "the biological aspects of psychology"—words that he chose in a letter to Yerkes the same year—triggered Carmichael's insistent search for precedents.[66]

Because their day-to-day work consisted of minutely tracking responses set off by specific organs or biomolecules, researchers often jumped rhetorically from reporting their tests to grandly declaring natural order. The few who tried to describe the process, however, were surprisingly optimistic about the capacity of organisms for harmony, considering the extent of animal distress in laboratories and human difficulties outside. By 1937 Robert Yerkes's friendly interaction with his captive chimpanzees encouraged his conclusion that "coöperative behavior is exhibited throughout the animal kingdom."[67] The ecologist Warder Allee crusaded for the same proposition. In an outpouring of speeches and essays around 1940, Allee argued that animals ranging from protozoans to primates demonstrate "unconscious cooperation" because of its " 'survival value.' " "Personally, I was well and thoroughly trained in this orthodox biological doctrine": the "egotistic emphasis" of Charles Darwin and the grim cycles of overpopulation and mortality of Thomas Malthus. It was in the laboratory that the benefits of aggregation became clear to him, once he observed that having "neither too few nor too many present" favored growth and learning.[68]

Both men were sufficiently self-aware also to realize that they were preaching: if all living things have the seed for good, then humans should embrace this better nature, with scientists to guide them. When Yerkes confessed in the same paper to his initial error of treating primates with "restraint and compulsion," he did not hide his disapproval of "the usual let-it-alone, cut-and-dried, standardized procedures which we use with such laboratory animals as mouse, rat, rabbit, or guinea pig." In effect Yerkes moralized that "fairness, justice, patience, friendliness and sympathetic interest" were not only decent but efficient, too, to avoid animal "contrariness."[69] Exactly how disturbed he was by his colleagues' callousness, and how much he expected to change them, is hard to say. Yet in the end, his cheerful message about cultivating chimp sociability was more than reform-mindedness. Almost in defiance of current events, Yerkes believed evolution was progressive.

The same researchers who celebrated life's promise could be hesitant, however, about the ability of current science to unlock and apply nature's mysteries. These were the self-critical reservations of successful practitioners who otherwise proceeded with confidence. They nonetheless voiced doubts about the truthfulness, universality, and applicability of scientific knowledge.

Edward Tolman, for one, entered philosophical dispute as to whether empirical methods directly apprehend nature at all. In 1932 he called psychology "a map-account" of the mind, translating "real, rich, immediate experience" into a shorthand of signs "in the interests of prediction and control." "The physicist or the chemist" does the same. Two decades later T. C. Schneirla aptly termed Tolman's stance scientific "phenomenalism" and acknowledged it as a healthy brake on overconfidence. Abstract concern about the reliability of facts did little to deter American data-gathering, however. Tolman and Schneirla, both avid experimenters, likely aimed to make their findings trustworthy by limiting their claims for what they measured.[70]

More informally, international travel pushed American scientists to reflect on the truism that science is universal. Was there uniformity in nature and science across diverse landscapes and cultures? Feeling lost among unfamiliar birds in Europe in 1932 compelled Margaret Nice to consider how ornithology could be transatlantic. "Handicapped" by regional guidebooks "of uneven value" and sporadic help from local experts, "I puzzled out the birds as best I could by myself." She nonetheless recognized maturation as a common process across species and believed complete study entailed attention to method first, and only then geographic variety. "Life history problems are the same everywhere," she lectured American readers of the journal, *Bird-Banding*, "and very important work is being done in this field across the ocean."[71]

In Leningrad three years later, Walter Cannon delivered the same message. Praising governments that funded science within their borders, he equally censured nations that restrained communication. Having witnessed how fragile liberty was in Soviet Russia, he defended "free inquiry and free intellectual intercourse without respect to national boundaries" in his plenary speech to the world's physiologists gathered there. Although he heartily agreed with Pavlov, who years before declared "'the omnipotent scientific method, will deliver man from the existing darkness,'" he added social substance to the generalization: international researchers must collaborate, with the current conference as a model. More clearly than Nice, Cannon saw the decade as "troubled times," and yet both had a concrete sense that internationalism, with its discomfiting uniqueness of places, complicated science's aspiration to achieve single truths.[72]

There were second thoughts, too, about the justice of animal-human comparisons, the premise behind experimentation. Even sunny-tempered Robert Yerkes expressed uncertainty. Caution was essential in transferring expert understanding of affective behavior in animals to human situations. After lauding cooperativeness in primates in 1937, Yerkes held back from predicting that methods devised "to cultivate unselfishness in this animal" would solve "our most baffling human problems." He stood by his first-hand observations of chimpanzees' capacity for "mute assistance" and his faith of "forty years" in reasoned empiricism. The path forward was "patient objective inquiry" into "the facts, conditions, principles, relations, and ways of controlling animal behavior." Against the backdrop of the New Deal's rush to reform, however, he warned that it was tricky to generate constructive emotions in humans: are "selfishness and coöperation" truly "antithetical," and could scientifically engineered generosity be too much "of an obviously good thing?"[73]

Walter Cannon's essay, "'Voodoo' Death," synthesized the scattered notes of skepticism about nature and science in a sober response to the general triumphalism. Although he did his share of prodding audiences to order "the body politic" in imitation of the healthful balance of organisms, his paper was about collapse: "whether an ominous and persistent state of fear can end the life of a man." In contrast to his colleagues, who treated emotions casually as offshoots of growth, anxiety causing bodily catastrophe was Cannon's main subject. The editor of the *American Anthropologist* accepted the article two days after the journal received it in October 1941. At this ominous historical moment, Cannon's distrust of emotion found an audience.[74]

Once again, he took up the central question of his life's work: how body and mind interact in the destabilizing emotions of "pain, hunger, fear and rage." Were Australian Aborigines such "bewildered strangers in a hostile world" that their belief that "a bone being pointed at them" was lethal actually made it so? To find the answer, in 1934 Cannon began contacting Western-born governors, doctors, scientists, and missionaries worldwide. His commitment to the idea of an evolutionary hierarchy structured his data-gathering. Non-Western peoples were primitives to Cannon, much as death-row prisoners were similarly humans-in-extremis, reported by a Sing Sing psychiatrist to experience "deflation of personality." The many stories Cannon collected did not change his theory. "Powerful emotions"—in this case, fear—"associated with ingrained instincts" can fatally disrupt "the 'internal environment' of our living parts." But he declined to dogmatize. Whereas "ethnologists" tend to affirm emotion's psychosomatic potential, "physiologists and physicians" are likely to call the mind-body transfer "impossible."[75] His tone was existential, admitting science's failure to explain, or impede, death itself.

Intellectually appealing as "'Voodoo' Death" was for its assertion of nature's limited solace, Cannon's view was in no simple sense realistic but rather shaped by his conviction that strong emotions upset the body's balance. In a 1940 letter, the anthropologist Lauriston Sharp of Cornell, an expert on Aboriginal society, shrewdly identified and then rejected Cannon's premise that feelings lie at the core of human experience. Never doubting that scientifically mysterious deaths occur, Sharp explained them as rational responses to cultural beliefs. When he witnessed the demise of individuals who had violated taboos, "I was impressed from start to finish by the typical absence of apparent signs of fear on the part of the patient," as the person took "what I called an unemotional fatalistic view of their impending death."[76]

Sharp did recognize affective components. The victim of magic felt "anxiety concerning the benevolence of one's fellows and probably very often, too a sense of guilt."[77] Even so, the scholars' exchange represented an honest disagreement about the place of emotion in evolution. With man as his subject, Sharp had nothing to say about whether animals are emotional creatures. But he objected to a vertical arrangement of peoples that equated unscientific thinking with irrationality. Ironically, both men valued reasonableness as the defining trait of human beings. Sharp went so far as to question the existence of felt pain in a ritualized setting. Together, they reviewed key points in current debate about emotion, at the same time that they displayed the scientist's typical wariness of the passions.

In fact, the detachment of Cannon's anthropological inquiry from his private experience was striking. Cannon's personal enemy was pain of soul and body: lifelong episodes of debilitating "melancholy" and now advancing cancer. The consummate scientist, he said nothing about himself when describing organic decline under duress.

With similar lack of connection, he left animals out of his curiosity about suffering that informed "'Voodoo' Death." Although he faithfully promoted anesthesia and merciful euthanasia in laboratories, empathy never shook principle. The "benevolent intent of the investigators" must prevail against antivivisectionists who were "meddlesome zealots," he repeated to the physiologists gathered in Leningrad. For four decades, Cannon experimentally manipulated animals' "pain, hunger, fear and rage."[78] His late ethnographic work on fatal emotionalism made clear, however, that serving humanity was his goal all along. He apportioned his caring in hierarchical fashion. Being a man, as well as an evolutionist, Cannon favored his own.

Cannon's forthright pessimism about the costs of emotionalism helps clarify the perspective of his peers by means of comparisons and contrasts. All agreed that the seeds of learning—now conceived as development—were somehow organic. Animal behavior experiments were a convenient lens on normal growth, with human beings the real object of interest. Unlike Cannon, however, who imagined passions to be instrumental, others tended to equate emotion with deficiency. Prepared to admire the elegance of developing instincts, they disrupted comfortable adaptation by surgical, hormonal, and situational methods. Making things go wrong—one organ, substance, or circumstance at a time—presumably identified the causes of healthy maturation. Cannon saw pain, however, as more than the absence of good. Because emotions were immediate, and sometimes frightful, they were unlikely to fade away.

Together, however, they saw pain as nature's failure and believed in science to restore health across the organic world. This view mediated their experiences outside laboratories. The expansive global apparatus of animal behavior study—ranging from specimen collection to far-flung colleagues—filtered the suffering these American travelers encountered abroad. As international war advanced and the United States entered it, American professionals participated, as they had in 1917, as scientists.

5

Animal Emotions in the Shadows

World War II

> The soldier who takes a pot shot at a duck on Guam Island may think he is firing on one of 50 million Holarctic mallards. Actually he is killing one of the 50 (or fewer) remaining individuals of the Marianas mallards (*Anas oustaleti* Salvadori).
>
> —Ernst Mayr, 1945

On the eve of World War II, animals as conceived by American researchers were far less the blank slates and passive learners once imagined and instead goal-seekers driven by pleasure and pain. Now, however, events in science and politics threatened to displace the individual—the integrated organism capable of feeling—as the object of study. Evolutionary genetics and international conflict strangely conspired to encourage panoramic viewpoints and bypass the humble single animal. The American animal experimenters who increasingly accounted the emotions to be part of behavior responded to these developments as both scientists and citizens.

On the surface, population genetics presented a familiar celebration of nature. Lecturing at Columbia University in 1936, the geneticist Theodosius Dobzhansky marveled that Darwin paradoxically traced the destructive culling mechanism of natural selection to "the appallingly great reproductive powers of living beings." Dobzhansky was similarly awed by nature's fecundity. Although his hosts at the Universidade de São Paolo in 1948 were horrified that he trekked in the jungle to collect specimens, here he saw "the meaning of organic evolution": "The gene [has] transformed air, water, fire, and earth into exuberance of living matter."[1] Darwin's reflections on life processes had led him to consider the emotions in his classic book, *The Expression of the Emotions in Man and Animals* (1872). In contrast, Dobzhansky was unconcerned with affective experience. His principal subjects, *Drosophila* or fruit flies, displayed none of the tenderness, fear, or anger cataloged by Darwin. Scientific inquiry into emotion was challenged by the geneticist's split focus: in Dobzhansky's words, on "micro-evolutionary phenomena," at one extreme, and "the genetic structure of populations," at the other. An irrepressible naturalist in the field, Dobzhansky studied fragments and aggregations in his laboratory. Investigating neither the mind nor body as traditionally conceived, he aimed to explain nature from angles where the emotions could barely be found.[2]

Animal experimenters sensed the gap between their own observational studies of individual behavior and inferences about populations based in microanalysis, statistics, and evolutionary theory. The psychologist Daniel Lehrman wrote ominously in 1953

Figure 5.1 T. G. Dobzhansky collecting flies in Brazil, 1965. Courtesy of the American Philosophical Society.

about uncertainty in assessing visible evidence: "Modern physiological and biochemical genetics is fast destroying the conception of a straight-line relationship between gene and somatic characteristic." Crowded as Lehrman apparently felt by molecular study as a "burgeoning science," as one scholar put it, he did not abandon behavioral tests of higher animals. Microbiology did not steamroll established macroscopic viewpoints because many committed American researchers continued to perform ablations, modify hormones, and manipulate habits. Nor did they unthinkingly follow their own precedents. Using the same techniques, they moved forward from questions of individual development to social adjustment and, in the process, reemphasized evolution to explain dynamic relationships. Indeed, new scales of analysis in the life sciences—either too small or large to be detectable by the human senses—invigorated useful debate. The sweeping evolutionary ant research of Edward O. Wilson, for one, unexpectedly highlighted the uniqueness of "primate individuality," a socially dysfunctional trait, in Wilson's estimate, that nonetheless "brought us to the threshold of civilization in the first place."[3] How personality works continued to command scientific attention precisely because evolutionary genetics gave new twists to old assumptions. For integrated organisms, talking about the emotions still made sense.

It is tempting to imagine a connection between the disorienting proportions of genetics and the unfathomable scope of World War II. Leaving an estimated fifty million soldiers and civilians dead by war technologies, disease, and famine, international aggression transformed the value of the self—the centerpiece of traditional philosophies—into a tragic question. Perspective—an abstract-sounding but everyday experience—changed between the 1930s and 1950s. Millions of sufferers, like millions of molecules, decentered the observer.[4]

There was no simple link between world politics and organic science, however. Physicists, chemists, and engineers memorably upgraded weaponry, and psychotherapists responded to intensified trauma. In contrast, because animal experimenters now had routinized ways of testing basic questions, they could remain distant from the war effort if they chose. Whereas a pioneering cohort had much to learn and gain in 1917, the field had achieved its own momentum by 1941. Time and again professionalism overrode political loyalties. Research cohorts had differing views of wartime obligations, for example: older American scientists who remembered World War I publicly discussed the new crisis, whereas younger men tried to continue their work uninterrupted. Similarly, disciplinary affinity pulled together investigators otherwise divided. European ethologists and American comparative psychologists, all macro-level analysts, memorably feuded about nature-nurture after the war, but they also met and argued. A sturdy professional framework stood between animal behavior scientists and current events.[5]

This American impulse to pursue basic science amid crisis, on the surface a mark of isolationism and privilege, was not altogether exceptional. Although unimpeded research at home helped shift the worldwide balance of scientific power toward the United States, colleagues in war-torn countries also tried to work undisturbed. In Munich, Karl von Frisch went to lengths to compromise with the Nazi state in an effort to protect his bee experiments. His biographer writes "it is difficult to shake the image of a scientist who escaped the horrors that surrounded him by burying himself in his work." In Austria, Konrad Lorenz "avoided all politics because I was absorbed with my own concerns," he recalled later. This was a half-truth that obscured Lorenz's collaboration with Nazism, but his rationale genuinely bespoke his scholarly passion, as well as the utility of the rhetoric of scientific devotion to justify past actions. Set in comparative national perspective, the civic detachment of many American animal researchers was not unique. It was one expression of scientific engagement that cut across warring nations.[6]

Locating American assessments of the emotions of animals in both dynamic science and tumultuous history is a matter of finding the right angle. Almost in spite of themselves, investigators who worked with animals after the turn of the century acquired insight into how species reproduce, secure nourishment, and endure pain. Although the no-nonsense empiricism of American practitioners curtailed theorizing about what an emotion might be, feelings figured ever more prominently in experimental designs. There was a working premise that animals have sensibility. Here I examine three successive influences on American animal behavior study: interwar genetics, wartime service of animal behaviorists, and early Cold War debate about animal motivations.

Genes and Organisms

In 1933 Thomas Hunt Morgan of Caltech became the first American to receive the Nobel Prize in Physiology or Medicine, honored, as the citation read, for "discoveries concerning the role played by chromosomes in heredity." Although a number of peers had also studied the transmission of biological information, Morgan determined that "genes lie in a serial order in the chromosomes" and may be plotted on "a chart or map." As his acceptance lecture stressed, understanding the "mechanism" of descent promised to give scientists predictive ability. His concern with cellular transactions meant that convenience could dictate his choice of experimental subjects. Before his pioneering use of fruit fly populations around 1910, Morgan wrote about sea spiders, earthworms, and aphids, as his substantive interest shifted away from embryo growth toward cell reproduction. Morgan's award was one sign of public expectancy in the 1930s that the cell held life's secrets. Whereas virtually every investigator now felt compelled to acknowledge genetics, however, many remained physiologists, psychologists, or ecologists. Microbiology nonetheless made their explanations—based on organs, consciousness, or habits—seem somewhat old-fashioned.[7]

Despite ongoing discussions of eugenics begun by Francis Galton in the 1860s, genetics was a new field opened up after 1900 by the rediscovery of plant experiments by Gregor Mendel. Particularly after 1930, when the Rockefeller Foundation heavily endowed research in cell biology, the area strongly pulled anyone interested in evolutionary patterns. John Thomas Patterson of the University of Texas, once a student of the biologist C. O. Whitman, transitioned from embryonic studies of armadillos to *Drosophila* genetics. At Columbia University, Leslie Clarence Dunn and Dobzhansky, more focused on populations from the first, left their respective specialties in poultry and beetle heredity to develop fruit fly colonies. Ernst Mayr, in contrast, remaining loyal to the classification method of systematics as curator of Pacific birds at the American Museum of Natural History, sent long letters to Dobzhansky about genetics. In 1939 Mayr marveled at the relevance of Dobzhansky's laboratory work to the annual return of wild birds to specific breeding sites, causing, Mayr speculated, "genetic differences between these neighboring populations": "So much about birds[,] but who would think that such a thing was occurring in *Drosophila*?"[8] The genetics of the 1930s seemed to draw scholars by its elegant combination of detail and breadth.

Experts who were firmly invested in other perspectives on organisms reacted with more hesitation. For men focused on the physiology of animal behavior, conference agendas included clusters of like-minded colleagues and lineages of teachers and students, all adhering to macroscopic methods. One aim of a "Symposium on Heredity and Environment" at Princeton in 1947 was to promote research on "the genetics of behavior patterns," and yet its five speakers were mainly known for manipulating bodily systems, including embryos, organs, hormones, and senses. The senior three—Karl Lashley, Calvin Stone, and Leonard Carmichael—had been innovators in the 1920s, and the junior two—Frank Beach and Clifford Morgan—were Lashley's students. Whereas Stone tried to sound current by endorsing "crossing closely related interfertile echotypes" to observe "recombination of their genes," Lashley and Carmichael instead chose to demonstrate mastery by reviewing their own work on brain functions and fetal responsiveness, respectively.[9]

Figure 5.2 Dobzhansky in his lab with fly jars, 1959. Courtesy of the American Philosophical Society.

Seven years later at Cornell, a meeting on "Group Processes" showed the persistence of holistic study, despite the absence of towering figures. Only Frank Beach from the earlier gathering gave a paper. Stone died that year, Lashley was gravely ill, and Carmichael was otherwise busy as Secretary of the Smithsonian Institution. Theodore Schneirla of the American Museum of Natural History, who also trained with Lashley in the 1930s, in turn mentored Daniel Lehrman, who was now on the program. So were the formidable Europeans Konrad Lorenz and Niko Tinbergen. In contrast to the physiological emphasis of the older American cohort, Beach and Lehrman championed a model of biosocial development, against the Europeans' claims for innate behavior structures. All, however, examined the visible interactions of individual animals, although published transcripts of the sessions contained allusions to genetics.[10]

In light of the evident adaptability of macro-analysis, it seems odd that Beach bemoaned its weakness in "The Snark Was a Boojum" (1950), unless one reads his essay as a jeremiad, true to an extent but so dire in tone as to inspire redoubled effort. He scolded fellow psychologists for ever more obsessive use of the Norway rat as a proxy animal, causing an intellectual wild goose chase akin to Lewis Carroll's satiric hunt for the mythical snark: "Are we building a general science of behavior or merely a science of rat learning?" Perhaps carried away by his clever rhetoric, Beach mistook research stability for stagnation. Among his copanelists of 1947 and 1954, for example, Clifford Morgan did recommend rats for tests of possessiveness ("the rat hoards just to hoard"); but Carmichael promoted amphibians, Lehrman specialized in pigeons, and Lashley compared amphibians, cats, ponies, and primates, along with comment on the postmortem of "the brain of Chauncey Wright, a distinguished mathematician."[11] Genetics did not bring an end to scrutinizing organisms as systems, and the subject of animal emotions advanced by means of a new element in deprivation methods. Scientists who had withheld physical elements of normal growth now withdrew the social nourishment fostering companionship as well.

Organ-altering surgeries were nearly the signature American innovation between the wars, and although experimentally modifying animal brains and reproductive tissues thereafter declined, the premise that bodily states determine behavior remained firm. A rudimentary pleasure-pain dichotomy continued to be influential. Sexuality commanded research attention, as it had since the 1920s, and along with anatomical ablations and hormone infusions, experimenters manipulated environments. In one 1949 study of ovulation, rabbits' "vaginas were each stimulated 10 minutes with a glass rod." Whatever agreeable sensations the rabbits experienced were counterbalanced by the likely pain of ferrets whose "eyes were removed" to test if disrupted day-night light patterns changed "the gonadal cycle." Frequent inclusion of electric shock in protocols, moreover, reflected belief that suffering is formative. Chimpanzees reared for months in darkness received "shock severe enough to bring vocal protests" in an effort to motivate their subsequent recovery of visual acuity. In his 1949 book, *Frustration*, Norman R. F. Maier reported applying air blasts, manual prods, or shock to rats to force "abortive responses." With dark humor, Harry Harlow ruled "the double-compartment grill box" to be "the most efficient torture chamber which is still legal."[12] All these practices followed two assumptions: a pleasure-pain sensory dualism represents the simplest emotions, and body and mind are well integrated.

The popularity of experiments involving affective physiology still did not produce a clear concept of emotion. Although arguably wrong, William McDougall had at least asserted that our varied feelings, which he listed, are mental signals that activate instincts. In an old refrain, Frank Beach now promised to sweep away such remnants of "metaphysical dualism" by means of the "purely objective terms" of animal research: "There is no necessity for postulating a 'psyche' which acts upon the animal's 'soma.'" The crux of Beach's 1950 lecture at Cornell University was how emotions work, however, and not what they are. "Functions," used as a normative noun to suggest the body's healthful operations, was one key word, and "psychosomatic," the term Beach chose for its dysfunctions, was another. His emphasis on process was the legacy of the enduring behavioral method of describing observable situations. Willy-nilly, and somewhat embarrassed, he also alluded to subjective states: "a 'neurotic' male dog" or "intense fright" in mammals. Seen positively, Beach's paper belonged to an American tradition of cataloging "emotional factors."[13] As typically, he distrusted philosophy and aimed for a science free of abstraction.

The difficulty of disinterestedly assessing facts, however, left researchers vulnerable to uncritical applications of their basic assumptions. One strange outcome of the conviction that body sites produce behavior was the wartime search for an organ that enables animal homing. A summary article, "The Bird Navigation Controversy" (1948), pointed to dissimilar catalysts behind the research: strategic bombing and relativity theory. After the military initiated bird studies to help pilots locate targets and return safely to base, scientists took up the relation of internal and external motion. Possibly "birds (and other animals) might orient themselves by detecting the Coriolis force in the semicircular canals or other organs whenever the organ was subjected to velocity relative to the earth." The mysterious organic basis of directional sensitivity opened the debate to parapsychologists. Although extrasensory perception (ESP) was not quite a biological capacity, their arguments made it the hidden locus of remarkable behavior. In 1951 J. B. Rhine, protégé of the late William McDougall at Duke, called intuitional sensibility "unconscious in man," thereby leading one to "*expect* to find ESP in lower orders of living things as a presensory mode of orientation." Writing on pigeons, his colleague J. G. Pratt named ESP as "some energetic basis that transcends sensory functioning."[14] These inferences exploited the unknowns of conventional research, while duplicating its ideas. ESP was not the property of a cell; rather, it was a sense awaiting empirical confirmation. Occult science ironically attested to the persuasiveness of a macroscopic perspective.

To be sure, even geneticists of the World War II era, with high-resolution technology still evolving, to an extent inferred microscopic processes by calculating distributions of traits in observable populations. Their interest in aggregations did not generally include the organisms' social interdependency, however. In contrast, organ-based research, at first glance limited to individual homeostasis, unexpectedly elicited work on animal communities and the emotions at play in them.

In the United States, experimentally isolating animals began in response to questions about sensory experience and, specifically, how much growth is innate versus learned. Leonard Carmichael memorably suspended amphibian embryos in solution in the 1920s, reporting that their reactions resembled his control group once he washed off the eggs and gently prodded them. The forced delay, restricting learning, did not change a presumably inborn capacity. More drastic methods retarded sense development by 1950: not only were some primates reared in darkness, but other trials "involved bandaging or encasing an infant chimpanzee's lower arms and legs in cardboard tubes and thus preventing the normal manipulative use of hands and feet."[15]

All along, investigators recognized that physical separation was emotionally charged. Laboratory anecdotes described animal mothers resisting experimenters who tried to take away their offspring. Isolation protocols now gradually added tests of socialization to measures of sensory growth. In 1942 Frank Beach examined whether sexual performance is instinctive, when he compared male rats raised alone with those brought up in the company of spayed females or fellow males. It was the solitary animals that reacted most vigorously to estrous females, and yet they were also abnormally prone to "fights" that precluded copulation, due to "intense heat behavior on the part of the female and increased excitement on the part of the male."[16]

Such surprising discoveries led to bolder conclusions about the psychology of social deprivation. The *Journal of Genetic Psychology*, Beach's outlet, issued three articles in 1960 by Zing-Yang Kuo, where Kuo linked forced solitude with "animal fighting." Ironically, the ideology of communist China, which Kuo had by now fled for Hong Kong, inflected his claim that communal training of various species in his laboratory portended a "tranquil mammalian society not only without fighting, but also with the dominance orders or social ranks eliminated." Kuo's work had an audience in the West, however, because isolation methods were well respected there. Indeed, it was Leonard Carmichael who arranged for Kuo's publications. The simple practice of separating single organisms from their normal social environments was the historical connection between Carmichael's humble biology and Kuo's nurture philosophy.[17]

The Americans' odd combination of organ ablation, social segregation, and conclusions about affect was most visible in the career of Harry Harlow. By the late 1950s, when Harlow's "affectional or love variable" caught public attention, he had taken the standard path from somatic to social experiments. A 1989 memorial sketch downplayed his brain-lesion studies as yielding "little of major significance," and yet primate neurosurgeries account for thirty-three articles by Harlow and his students between 1939 and 1973. In step with doubts still widely voiced by American colleagues about natural endowment, in 1951 he ridiculed the notion that internal "tissue-tension" triggers instincts. At the same time, he endorsed inborn drives as the underpinning of "motivation aroused by external stimuli." This intellectual sleight of hand made it logical for physiological psychology to take account of social catalysts. Predictably, Harlow devised isolation methods: infant monkeys reared alone, with inanimate mother-surrogates for emotional shelter. Photo-illustrated articles conveyed the pathos of babies clinging to wire or cloth frames topped by severe or sweet wood faces. Yet for all its success, Harlow's work was in essence a reduction experiment, where the animal did not lose brain capacity but maternal care instead.[18]

As imposed deficits shifted in purpose from explaining sensory growth to exploring personality, the nature of the group became the real subject, albeit unstated. Beach, Kuo, and Harlow elicited abnormal development for the sake of social health, and their corporate interest ironically made them resemble geneticists. Dobzhansky's point in *Genetics and the Origin of Species* (1937) was that a species is a mass genetically bounded entity, "an objective phenomenon rather than a projection on nature of some concepts created by the investigator."[19] Although he cited reproductive activity as the agent of speciation, his cohort of geneticists emphasized the transmission of physical traits, and not the affective character of either the sexual mechanism or temperamental outcome.

Not so the scholars of tangible animal societies delimited by common habits. These experimenters spoke eagerly—and facilely—about emotions as social glue. Harlow's choice of words for an infant monkey's "affection or love for the cloth mother" exposed an intellectual dilemma apparent to any evolutionist: the comparability of species.[20] Beginning in the 1890s, Harlow's predecessors warned against anthropomorphizing and the related danger of imagining other creatures with human-like awareness. Decades of body-based experiments all but abolished subjectivity, but they also sidelined evolution. Explaining behavior with sole reference to a single life had preoccupied the American scientists who studied animals holistically between the wars. Now Harlow's cross-species comparison, tossed out rhetorically, made him seem out of touch with the classic question: which living things, if any, have feelings like ours?

In Harlow's defense, whatever intellectual guidance now remained about the emotional similarity of animals was problematic. Like most specialists, early geneticists—the rising champions of evolution in American science—left the study of sensibility to others. American psychologists who still worried about comparisons were apt to repeat the old rule: don't project yourself onto your subject. In 1942 T. C. Schneirla belittled the "anthropomorphic fancies" of a colleague who described one ant type as displaying a "'pattern of human cruelty and sadism.'" A decade later, he more politely wondered if bees communicate intelligently, the claim of a notable European, and yet repeated, "let us not leap to extreme readjustments on a basis of analogies and impressions." Concurrently, Konrad Lorenz vigorously concluded that subjectivity is universal: "it is without doubt the existence of the *experiencing* subject, the presence of an 'ego,' that represents the sharpest distinction between the living and the nonliving." With these contradictory positions as the wide poles of discussion, the chance of an ordinary empiricist like Harlow devising a reasonable conception of animal emotion was small indeed.[21]

Yet if one takes a step back from the parties at odds over animal subjectivity, points of consensus reappear among the scientists who conducted experiments on living systems: the integrity of organisms, influence of genes, and presence of affect. While geneticists were isolating cell parts and surveying populations, other life scientists seemed to plod along testing individuals. In fact, in light of the fragments and masses of genetics, treating the body and mind as a dynamic unit could seem nearly innovative. Lorenz gave no sign of feeling deficient in the mid-1940s when he described "a *holistic entity* as a closed system of structures and functions," where the components bear a mutual "*causal relationship*" aimed at "the preservation of the system." His American colleagues in animal behavior, less theory-minded, tacitly agreed. All, too, conceded the value of molecular science, particularly the Americans who seemed impressed by its technical precision and controlled conditions. Their reports on the testable activities of experimental animals typically contained asides about genetic lineage. Thus in 1951 Calvin Hall's mice were "the offspring of crosses between two pure lines of house mice, C-57 and *dba*," as he searched for infant roots of "*emotional instability*."[22] Observing live mice of a known strain, he expected to explain adult temperament.

World War II exploded in the midst of this scientific conversation already in progress. No two researchers experienced quite the same war. For Americans in search of the causes of animal behavior, however, the immediate impact was modest.

Scientists at a Safe Distance

Frustration most likely drove the criticism of the Army's scientific ignorance made by Ernst Mayr in 1945. "The soldier who takes a pot shot at a duck on Guam Island . . . is killing one of the 50 (or fewer) remaining individuals of the Marianas mallard (*Anas oustaleti* Salvatori)."[23] Mayr's plea for evolutionary ecology was at once brave and quiet. Under government watch as a German citizen working in New York, Mayr preached to the already-receptive readers of *Audubon Magazine*. His professional devotion to Pacific island birds left no doubt that his chastisement was heartfelt, but this defense would not reach much beyond his audience of naturalists. Public engagement by other American animal scientists was similarly restrained. Overall, the human tragedy pushed questions about animal behavior into the background.

World War II provoked no major American reappraisal of the balance of intellect and emotion that resembled the dense cluster of writings on instinct of 1918. Perhaps, logically speaking, hard recognition of the limits of learning did not need to be repeated. Interwar scholarship in fact soberly—even obsessively—acknowledged affective patterns imposed by physical endowment. The sheer size of the pool of researchers by 1941, moreover, permitted an unbroken series of animal experiments. For every investigator such as Leonard Carmichael, whose post as national director of recruitment of scientists for defense work consumed his time, there was another like Karl Lashley, squarely fixed on behavioral neurology. In between, C. Ray Carpenter, at last securely employed as professor of psychology at Penn State, published his remaining prewar fieldwork on primates and then volunteered for the Army in 1943. He was just as eager to demobilize in 1945, explaining to his department head, "my employment at the College can be made of more service to the Nation than my service as an officer as I am now currently assigned." Size again—of the US population and economy—generally gave Americans choices. "So great was the strength of the United States that the country never was completely mobilized for war," writes one World War II historian.[24] Animal research had its own momentum and was able to keep it.

Two assumptions ran through this spectrum of personal options: science does have a political dimension, and human needs come first in a time of crisis. First, the pressure of events on the academy was welcomed as a chance for service by some and resisted as an intrusion by others, but it now seemed a fact of scholarly life. Second, the character that counted in war was human nature. Old and young animal researchers, as well as native-born and émigré scholars, reacted differently to these commonplace views. Outside the headlines, however, animal behavior experiments continued at a healthy pace.

The Americans who remembered World War I were most eager to apply their science to current events. Comfortable as elder statesmen, they advocated reasoned planning to defend American traditions, a message consistent with the vision of engineered social order developed in the 1920s in partnership with their wives. As before, they saw emotions as disruptive. At one extreme, in April 1941 Robert Yerkes proposed euthanasia for the common good. Speaking at a "symposium on recent advances in psychology" hosted by the American Philosophical Society, he expressed admiration for Germany, where psychologists were "specialist servants of the state," and his corresponding impatience with the United States, where "individualism" hampered planning. For anyone deemed "biologically unfit," "life should be gently ended," without hindrance of "sentimentality."[25] To be fair, this was one of Yerkes's lingering pet ideas slipped into a hawkish call for preparedness. It nonetheless struck notes often heard in his circle: self-restraint, cooperation, and rationality.

Leonard Carmichael favored the same values in his critique of cultural modernism, in the process making science seem reliably old-fashioned. As president of Tufts College and also director of the National Roster of Scientific and Specialized Personnel, he had a wide audience. On his campus in 1940, Carmichael charged "internationalism," psychoanalysis, and quantum physics with "destruction of the moorings of mankind" and instead urged "true neo-patriotism." There was only a superficial paradox to the notion that "the monstrous irrationality of war" required "an emotional patriotic response": feelings harnessed to "the great common need" were compatible with "human engineering." Carmichael's mastery of American anecdote lent charm to the formula of voluntary sacrifice for the sake of planning. "Franklin's true Yankee statement that we must all hang together or else all hang separately," he intoned at a denominational meeting of Maine Universalists in 1943, "has the sanction of all that is practical and pragmatic in individual as well as national policy." His successive published reports about the thousands of chemists, engineers, economists, and translators registered for service were the logical correlate of what he saw as the nation's first principles.[26] Silently, psychology—and Carmichael personally—stood at the apex of the system of human resource management. In this light, the scientific process of experimentation, analysis, and application was a cultural rudder.

Other aging specialists of animal behavior repeated these thoughts elsewhere. Entirely true to both his physiological theory and personal philosophy, Walter Cannon, veteran of the Western Front and president in 1940 of the American Association for the Advancement of Science, stressed "organized control" as the answer "at this bewildering time, when ancient customs are not respected." The "body politic," like the "body physiologic," suffered "harmful consequences of unsteadiness" when "predatory nations" upset domestic homeostasis. "More control is tolerable if it results in greater human freedom." By contrast, Edward Tolman, a pacifist in 1918 and still dovish, seemed easygoing in *Drives toward War* (1942) when he imagined the "psychologically adjusted man," with all instincts satisfied, to be key to the "abolition of war." Yet even then, affective gratification resembled the triumph of reason. The condition of an organism out of sync with its environment was "*irrational* or *traumatic*," Tolman explained at the same symposium addressed by Yerkes.[27] Happiness and intelligence, both forms of adaptation imagined to produce harmony, turned out to be nearly identical.

Behind these social arguments were the same men's routine experiments on rats, rabbits, or primates. They spoke as if there was no intellectual leap from conducting controlled tests to formulating laws of organic nature. They took their authority as public intellectuals as a given, and their pronouncements exposed the intimacy of science and political philosophy. The impulse to ally research and patriotism also met resistance, however.

Two odd cases that underline nervousness about the wartime political climate for science involved the Nobel-laureate T. H. Morgan and Stanford psychologist Calvin Stone. What is noteworthy about Morgan's 1940 defense of Soviet genetics, published in a communist journal, was not its deviance from his peers' patriotism but instead his use of a pen name: Lucretius Smith. Morgan apparently feared backlash. Tired of the "vituperation and scorn" aimed at "the alleged intolerance of Soviet philosopher-scientists," Morgan praised "the liberating force of Marxist philosophy which expands the field of science." This eminent man assumed a disguise to refute, albeit naïvely, Stalinism's "menacing" temper. A letter from Calvin Perry Stone to Herbert Hoover in October 1941 was also covertly political. All Stanford professors had received a questionnaire from this famous alumnus asking their personal views of US policies. While others signed a joint statement protesting the inquiry, Stone sent Hoover a social-science critique of the questions, labeling them "*Ambiguous*," "*A bird-shot question*," and a "*Mandate query*." Although he kept the tone light by confessing his "somewhat rebellant [*sic*] mood," the episode bore witness to political pressures on the academy, Stone's discomfort with them, and his preference for resisting indirectly.[28]

So complicated were reactions to the proximity of ideology and science that some scholars both embraced and rejected value-laden research, depending on the situation. Warder Clyde Allee, still a devout Quaker, welcomed the chance in 1943 to counsel readers of *Science* to end wars by taming the "competitive, egoistic drives among all animals, ourselves included" with "the even more fundamental cooperative, altruistic forces of human nature." He was nonetheless privately irritated that his distaste for the politics of Konrad Lorenz, a one-time Nazi, threatened his own high standard of professional disinterest. Asked by Margaret Nice in 1949 to approach the Rockefeller

Foundation about support for Lorenz, Allee championed scientific nonpartisanship: "I have no wish to allow differences of opinion on political matters to stand between my recommendation of aid for a Nazi and his receiving it any more than I would want to have such considerations stand between an eminent biologist who was also a Communist and his receiving aid." Lorenz's past did affect Allee, however, so much that he delayed his appeal to Rockefeller by several months.[29] Happy to take a public stance for his favorite message, Allee bristled when political conflict came closer to research.

Unlike their elders, a younger cohort of American animal specialists, born after 1900, preferred academic business as usual. Whereas conviction and careerism pushed an earlier generation toward public service or pacifism during World War I, the new conflict seemed neither to inspire sacrifice nor offer opportunity. The professional disruption experienced by those who entered the military was one likely reason others avoided it.

On the surface, C. Ray Carpenter's three and a half years as an Air Force captain handsomely extended his civilian career in psychology, with specialties in fieldwork and filmmaking. His major assignment was as "technical advisor" on a motion picture to teach downed pilots jungle survival. It was filmed in the Canal Zone, where a decade before he conducted his first observations of howler monkeys. The base newsletter, *The Runway*, gave him the honorific title, "recognized simialogist [*sic*]." He was equally suited for later projects on night vision in pilots and the release of homing pigeons from aircraft.[30]

Frustrations seemed endless, however. Each trip from the unit's Florida home camp to military superiors in New York or film technicians in Hollywood required a travel pass and patience with delays. Every movie detail—the foreign-language words taught, knives pictured, and "polynesian-like [*sic*] natives" recruited—needed clearance, provoking Carpenter in Panama to complain to New York, "It has been impractical for this officer to inform his organization of these changes and get approval."[31] In June 1945, a month after the German surrender, he received a disappointing order to prepare for "Overseas Shipment" to Europe, a directive that Penn State administrators, at Carpenter's behest, did their best to block. Finally discharged in September 1946, he had a four-year gap in his academic vitae. A new series of articles about educating former soldiers suggested the war's professional profit; but the postwar collapse of British, French, and Dutch colonial empires ended his field studies of Asian primates until the 1960s. The world order would have been transformed with or without Carpenter. At no point, however, did his mood as a participant depart from dogged obligation.[32]

After Pearl Harbor, Congress directed all American men between ages eighteen and sixty-five to register with the Selective Service, with those forty-five and younger liable for military call-up. Perhaps Carpenter, who volunteered at age thirty-seven, enlisted to avoid the draft. Harry Harlow, just one month older, continued his brain-deficit experiments without pause, however. It was common for middle-aged animal scientists to keep their distance from the crisis.[33]

The American Museum of Natural History in New York, now a rising research site, was an especially productive harbor for Frank Beach and Theodore Schneirla. The two men arguably defined American comparative psychology in the 1950s, and their wartime collaboration in the museum's Department of Animal Behavior was the catalyst of their future dominance. By 1960 their papers ruled on fundamentals such as the comparability of organisms and nature-nurture. When Beach, department chair in 1943, invited Schneirla to become his associate curator, their ages were thirty-two and forty-one, respectively. Their research not only continued during the war but, far more, remarkably flowered. Their training under Karl Lashley in the 1930s was their academic connection, and yet neither devoted himself solely to brain ablations. Beach's publications hopscotched from sexual maturation in rats in 1942 to the biosociology of animal play three years later. A 1944 Guggenheim Fellowship enabled Schneirla, whose expertise was army ants (*Eciton*), to spend six months doing fieldwork in Mexico. Hardly narrow, Schneirla issued a paper, "Problems in the Biopsychology of Social Organizations," soon after his return. Indeed, he was sufficiently broad to be listed as a suspected communist in the 1947 report of the House Un-American Activities Committee. The same year, he advanced to the rank of curator at the museum. Beach had left recently for Yale. Neither man missed out on professional development due to the World War.[34]

The museum served less fully as a haven for scholars of immigrant backgrounds. Animal science no longer wholly belonged to native-born practitioners like Beach and Schneirla, who grew up, respectively, in Kansas and Michigan. These two made the institution work for them in a way that Schneirla's protégé, Daniel Lehrman, could not match. Lehrman's German language proficiency, presumably first nurtured at home in Queens by Yiddish-speaking parents, led to four years as an Army cryptographer. His military value, at least in this niche, far exceeded the usefulness of Beach, whose foreign languages were so poor that Harvard delayed awarding his PhD for this reason for three years after he finished his dissertation. Even so, Lehrman's war service slowed his education. In 1940 he was already coauthor of a field study of gull nesting with Beach's former boss, the biologist G. K. Noble. He was twenty-seven years old when he completed his BS at CCNY in 1946 and thirty-five in 1954 when he received his doctorate at NYU, where Schneirla was his adviser. All this does not mean that Lehrman was an unwilling soldier, or that the war by itself obstructed his progress. He was the prototypical New York Jew of the era: anti-Nazi by conviction and limited professionally by the academy's widespread anti-Semitism. At the same time, neither his military assignments nor well-placed friends furthered Lehrman's scientific ambitions during the war.[35]

The situation of Ernst Mayr, fellow ornithologist and a mentor of Lehrman at the museum, was more complicated. Scientifically, Mayr was a rising star, moving to an Ivy school, like Beach, when Mayr accepted an invitation from Harvard in 1953. He was also a citizen of Germany until 1950.

Mayr's museum colleagues were among the fifty American friends who supplied affidavits of good character for his final, successful petition for US citizenship, a process he began in 1939. He also kept his job, despite an FBI search of his home in 1942, mandatory supervision by police and immigration officials, and restricted travel outside New Jersey, where he lived. Indeed, the war years were immensely productive. Mayr made his mark as a theorist when his Jesup lectures at Columbia in 1941—integrating taxonomy, genetics, and behavior analysis—appeared as *Systematics and the Origins of Species* (1942). Day by day, his curatorial work involved classifying the thousands of Pacific bird specimens he helped collect in the late 1920s. His informational articles introducing new species or questioning earlier groupings steadily appeared. Although his *Birds of the Southwest Pacific*, described as a "field guide," was not commissioned by the military, its publication in 1945 seemed a patriotic adjunct to American fighting on "Espiritu Santo, Ponapé, Bougainville—or Guadalcanal, which," the museum's chief ornithologist wrote in his preface, "our armed forces have now turned into an immortal name." At the same time, Mayr bravely criticized the Allies in his *Audubon* article. There should be no "needless" tree destruction, brush fires, introduction of cats or goats, or "indiscriminate shooting or killing." But he included a caveat: "War is war, and it would be silly to suggest that an artillery barrage be eliminated in order not to hurt some rare birds."[36] Mayr walked a tightrope as a noncitizen naturalist.

Political insecurity was one reason why émigré scientists cultivated private friendships that, when their fields were close, were also intellectually nourishing. "A lot of your tenses sound wrong to me," Mayr commented on a draft by the geneticist T. G. Dobzhansky in 1947: "But, who am I to judge?" Encouraging Dobzhansky to "get some born American" to go over the prose, Mayr implied that he found colleagues reasonably welcoming.[37]

Like Mayr, though, Dobzhansky knew he was an outsider. His lack of citizenship affected a job prospect at the University of Texas in 1936. Although he had arrived in 1927 from the Soviet Union, his citizenship application was recent, and it was not clear whether Texas law would allow him to receive a state-funded salary. Before breaking off negotiations, he further volunteered "that I have no letters after my name—no Ph.D., no M.S., not even B.S.," because "Russian universities were giving no degrees of any sort." His words seem testy, as if his talent did not need legitimation by academic or government credentials. Whatever his mood, he was used to uncomfortable expectations. In 1962 he recalled that in the eyes of his then-Caltech supervisor, T. H. Morgan, "Russians to him were of two varieties, the 'black' and the 'white.'" Where "the old man" got the idea of this "dimorphism" he could not say. "Thank God, my wife and myself were classified among the 'white!'"[38]

Despite these irritations, Dobzhansky was grateful to be safe and productive in the United States. Days after atomic bombs destroyed Japanese cities, Dobzhansky sent Mayr news of a Ukrainian entomologist who "has lost everything": his specimens, "the museum itself," and "his family, as well as all his earthly possessions." Letters from other European colleagues sadly confirmed Dobzhansky's sense that this was one "case of human woe, probably one of millions." Writing from Holland in June 1945, Niko Tinbergen informed Margaret Morse Nice that he was now "free again" after the "depressing, terrible nightmare of 5 years," two of them spent in a German "hostage camp." Nice rallied American scientists to aid their friends abroad with necessities such as food and toothbrushes, along with scientific publications for locations "where the normal exchange of literature has been cut off for years and many libraries have been completely destroyed." Asian science was as devastated. One of Edward Tolman's former doctoral students, Joseph Yoshioka, held temporary positions in the United States until some point into the war, but thereafter disappeared for good from official American and Japanese records. Zing-Yang Kuo, another past student of Tolman, took refuge in successive Chinese cities during the fighting and then spent his remaining quarter century exiled in Hong Kong.[39]

We are "unbelievably fortunate fellows," Dobzhansky confided to Mayr. Indeed, in 1944–1945 the two men issued their collaborative experiments on the genetics of geographic and sexual isolation in fruit flies in the *Proceedings of the National Academy of*

Sciences.[40] The steady flow of papers by American colleagues who had comparatively few international distractions confirmed the imperviousness of basic science. One extraordinary controversy between the psychologist Norman Maier and his critics—where the issue was whether rat seizures are caused by cognitive dysfunction or auditory shock—seemed all-consuming and also myopic. On his side alone, Maier published twenty-four numbered papers on the subject during the 1940s.[41]

For Americans who devoted careers to examining animal behavior, World War II induced little new soul-searching about human bestiality or, more generally, the power of emotion in living things. The nature of this war and the state of their science are among the reasons. Technological innovations in weaponry based on the physical sciences were so crucial to victory that animal laboratories may have seemed of distant importance. By far the largest number of specialists registered as ready for service on Carmichael's National Roster were chemists and engineers. Ever-improved warships, submarines, transports, tanks, and planes delivered weapons of staggering destructive capacity that reversed the proportion of civilian and military casualties vis-à-vis World War I: noncombatants were now the majority of victims. The devastation of Coventry, Hamburg, or Tokyo perhaps seemed more impersonal than the slaughter on the Western Front.[42] But for whatever reason, these professionals who studied motivations were able to turn away, and they generally did.

Their scientific vocabulary was by now poorly equipped in any case to express the trauma of war. By these intellectuals' own hand, instinct and similar words denoting internal states had evaporated as an American language of emotional analysis. The influence of "heredity in the determination of behavior" was the shared idea of the psychologists invited to Princeton in 1947 for a "heredity-environment symposium." Between the wars, the moderator added, there was a "violent reaction against the presentation of the topic of instinct in William McDougall's (1908) *Introduction to Social Psychology*."[43] "Heredity" reinvented "instinct" in scientifically valid terms. But the change also blurred the traditional link between instincts and emotions. The relationship of heredity to behavior was mechanical cause-and-effect. Sensibility was not considered.

What goes around comes around, however. The concept of evolution, although never far from view in twentieth-century life science, wholly dominated postwar animal studies. In this synthesizing atmosphere, the comparability of species stirred academic debate and, with it, marvel at the stunning variety of the animal world and the intricacy of individual creatures, which included, for at least some organisms, their emotions.

In the Kingdom of the Beasts

For scholars of genetics and behavior alike, the idea of biological descent had increasingly moved to the forefront between the wars. British scientists around 1930 were the first to seize on evolution's potential as an integrative concept. In the wake of the war's human catastrophe, nature's regenerative capacity could seem positively consoling. Konrad Lorenz expressed awe at inexorable life forces. "The mysterious apparatus for transmitting and receiving sign stimuli is age-old," he observed in 1949, "far older than mankind itself." His American counterparts were worlds away from

Lorenz's nearly four years as a German prisoner of war in Russia before his return to Austria in 1948.[44] Evolutionary thinking achieved general dominance, however: its promise of uniform and perpetual laws seemed to captivate life scientists across the West, who at the same time managed to disregard its impersonality and materialism. It set molecules, individuals, and populations in global perspective, boosting awareness of animal variety and encouraging comparisons among species. Behavioral scientists did not lose sight of the individual animal under pressure from population genetics. Rather, they borrowed selectively from the new field, thereby broadening their analysis.

Optimism about nature and adaptability did not prevent intellectual conflict. American researchers solidified their own idiom of animal behavior in rough-edged exchanges with Europeans. The tone of their backhanded internationalism was critical to the point of being derogatory, hardly the best seedbed for conceptualizing emotion. Recognition of feelings in animals nonetheless suffused the Americans' work.

Their prewar preference for animal behavior trials meant they were not the first American scientists to recognize the philosophical promise of evolution. The Columbia University geneticists L. C. Dunn and T. G. Dobzhansky, attuned to the fate of insect populations, readily transformed their science into readable sociology in 1946 in a short volume, *Heredity, Race, and Society*. On the surface a simple primer in genetics and its social uses, the book aligned reproductive biology with political freedom. It had a persuasive breadth that was absent in experimental reports.

The authors counseled strenuous defense, purged of nationalisms, of man's "most precious patrimony—his own hereditary endowment." Not just a matter of counting fruit fly mutations, the biology of genetic combination taught "the absolute uniqueness of every human individual," which "translated into metaphysical and political terms is fundamental for both ethics and democracy." This approachable creed skirted ideological minefields. The scientists were boldly liberal on the concept of human races, understood to be fluid "populations which differ in the relative commonness of some of their genes." Their chapter on eugenics, "Controlling Man's Heredity," was more strained. "It stands to reason that an intelligent mother is better qualified to bring up children than a dull one," making genetic information indispensable. But "sterilization programs become less and less effective," because damaging recessive genes are hard to detect. False values compounded the technical difficulty: "Unfortunately, eugenics has often been perverted to a pseudoscientific justification of social inequalities and oppression." Throughout, the relevance of *Drosophila* inheritance to human affairs was taken for granted. "Understanding man and his place in the universe is perhaps the central problem of all science." In grand terms, Dunn and Dobzhansky made natural selection the foundation of progress. The book's sixteen editions issued by 1964 suggest that Americans welcomed the message.[45]

Frank Beach, trained in rigorous experimentalism by Karl Lashley in the 1930s, was one student of animal behavior who eagerly responded to evolutionary discussion. His laboratory experience gave him practical evidence of the relationship of species. Specializing in the balance of physiology and socialization behind sexual activity, he switched focus from rats to dogs when he arrived at Yale in 1947. At first housed in a hospital basement, the dogs showed Beach that a canine was not "a rat that weighed 25 pounds and barked."[46] Modern evolutionary theory corroborated this empirical lesson.

"An evolutionary approach has proven fruitful in advancing our knowledge of many phases of human behavior," Beach had written in 1945. The occasion was an article about play that, in its ambitious scope, anticipated his best later essays. Beach drew on studies done over the past half century of fish, dogs, cats, horses, and chimpanzees. He reviewed theories ranging from the pleasure-based "joie-de-vivre hypothesis" to learning-centered claims that apparently useless actions teach spatial orientation and cooperation. In 1952 Beach applied the same synthesizing formula to psychosomatic illness, induced experimentally in ferrets, birds, and humans by darkness, cold, or shock. In 1954 he examined how "early experience" in various species affects mature behavior. Perhaps deliberately, he did not harp on evolution, thereby distinguishing himself from geneticists and systematists. Rather, he described his context as "the rapidly growing interest in this area of comparative psychology," a phrase that likely mixed his hopes for a rising field with fact.[47] Evolution was the silent partner in Beach's issue-oriented, cross-species surveys.

As he waded into the panoramic science of dynamic living systems, Beach wrote not only bravely but also sharply. His wit was indicative of the habit of deflating opponents used generally by American experimentalists in postwar interdisciplinary and international discussions. Beach himself had a sense of humor. A female beagle that was "undiscriminating when it came to accepting males in copulation," he quipped in the 1960s, was practicing the mantra, " 'Make love, not war.' "[48] Still, these Americans were more comfortable being critics than theorists when they moved beyond empirical reporting. This made their analysis of animal behavior often sound harping instead of constructive.

From the first, Beach corrected errors with one hand, while he opened inquiry with the other. In 1947 he dismissed the credulous "anecdotal method" in the American Museum of Natural History's popular magazine, *Natural History*, soon after he left his job there. If finger-wagging about anthropomorphism, launched against an article signed by a "veteran guide," seems beneath a Yale professor, the real issue was the comparability of species.[49] Genetics had reignited debate about historical continuities and differences among organisms—the enduring contest between lumpers and splitters—and Beach, a cautious specialist, defended splitting. He had as little patience with narrowness, however. He took fellow psychologists to task in 1950 for producing books whose titles promised explanations of "behavior" when the sole factual basis was experiments on rats. Even his conciliatory efforts had a scolding tone. "The Descent of Instinct" (1955), a title that echoed Darwin's *Descent of Man* (1871), aimed to end the decades-long "revolt" against natural inheritance by "radical Behaviorists," because "unlearned patterns" by now were "generally accepted." Peacemaking, though, depended on presenting his predecessors as bullheaded: "Had psychologists always recognized this fact, much of the fruitless debate about unlearned behavior could have been avoided."[50]

It was evolutionary thinking that fostered Beach's exuberant intellectual ambitions and encouraged him to clear a path by chiding opponents. The arrival of animal science as an international enterprise added to the sense of vast potential and, understandably, also sharpened conflict. At the war's end, the reserve of American animal researchers toward events abroad evaporated. They behaved as consummate scientists: a foreign war had not fully commanded their attention, but new professional contacts were enticing. Niko Tinbergen, Karl von Frisch, and Konrad Lorenz resembled celebrities when they toured the United States giving lectures in 1947, 1949, and 1954, respectively. In return, their American hosts attended conferences in France in 1954, for example, and Germany in 1955.[51]

Along with a degree of civility, however, went a competitive tendency to categorize European-versus-American schools of thought. Daniel Lehrman, arguably the lead American controversialist following publication of his "Critique of Lorenz's Theory of Instinctive Behavior" (1953), looked back regretfully in 1970 at his readiness to "wrangle bitterly." What he once believed was "an analysis of a scientific problem" now read as "an assault upon a theoretical viewpoint."[52] Lehrman did not pioneer the us-them mindset. Tinbergen explained his 1947 lectures at the American Museum of Natural History as alerting "Anglo-American workers to research done on the European continent," following a basic division of labor: Old World ethologists "concentrate on innate behaviour," while American behaviorists "specialize in the higher types of behaviour," especially those of "man." Lehrman repeated the contrast in his sizzling article. The Europeans "are zoologists, physiologists, zoo curators or naturalists, unlike the psychologists who constitute the majority of American students of animal behavior." Even if these characterizations began as honest summaries, the labels were liable to pejorative use. When Tinbergen at Oxford introduced Lehrman, trained as a psychologist, as a zoologist, Lehrman quipped, " 'I took this as a compliment.' "[53]

The breaking point for the Americans was not subject, field, or argument, however, but instead method and, specifically, empiricism. Because the Europeans were content to speculate, verification of their hunches was haphazard. "The assumption that pecking grows *as* a pecking *pattern*," Lehrman wrote damningly, "discourages examination of the embryological processes leading to pecking." Philosophizing meanwhile produced sophomoric conceptual errors, Lehrman added, including "teleology" and "anthropomorphism." Lorenz has "a rather neat, positivist theoretical system" built up from an "a priori emphasis upon the 'native' aspects of behavior," Schneirla concurred in 1952. A decade later, Schneirla again aimed at Lorenz's "non-experimental anecdotal" procedures and literary "obfuscations," habits that separated Lorenz from "responsible scientists."[54] In their charges, the Americans elevated technique to the level of orthodoxy. Species-by-species, behavior-by-behavior tests were the sole route to scientific truth.

It is easy to miss substantial transatlantic consensus in this crossfire: basically, that animal behavior study rests on observation of organisms in environments. For all the ink spilled over the insufficiency of simply identifying a behavior as instinctive, the Americans fully recognized biological endowment. Time and again one heard versions of Schneirla's 1958 rejection of "the barren issue of 'nature vs. nurture.' " None would disagree that each creature is "a natural unit," as Lorenz stated at Cornell University in 1954, or that interaction with its environment triggers behavioral changes. As Beach explained at the same conference, "Living systems engage in metabolic exchange with the surroundings and exhibit some degree of biochemical stability." There was even agreement on method. No matter how often Schneirla censored Lorenz for being naïvely "observational, qualitative, and relatively simple," he equally opposed "rigid use of quantitative methods" and excessive trust in "still and motion-picture photography, sound recorders, activity recorders, and the like." Importantly, "the investigator becomes a dynamic perceiver of his phenomenon."[55]

A distinctive American emphasis on organic development in environments remained. Inborn temperaments, surroundings, and growth were the core determinants of animal lives. This formula yielded insights in specialized studies.

Schneirla had been researching army ants for a quarter century by 1957 when he published, "A Comparison of Species and Genera in the Ant Subfamily Dorylinae." The paper explained the reproductive cycle in evolutionary perspective. The central event was the queen's production of thousands of eggs; but the cause was "excitement" generated by the maturing "*callow brood*," as it raised the "level of social stimulation" by means of "innumerable tactual and olfactory encounters between brood-and-workers and workers-and-workers." This buzz of activity set off a sequence of migration and foraging that acquired "food in increasing quantities" for the queen. She engaged in "voracious feeding" induced by the workers' "licking and other social-stimulative responses," until she lay the next brood. For the several New World *Eciton* species, raiders adhered to a "diurnal light rhythm" during the "nomadic phase," leaving the bivouac at dawn and returning at dusk. The African subspecies *Anomma*, in contrast, consisted of eyeless, underground nesters, also demonstrating a cyclical pattern of colony stasis and mobility, but day-to-day swarming from the nest at random times. Schneirla attributed the capacity of *Eciton* to live above ground to the "continuity of heavier forest cover in the New World tropics." More subtly, the army ants represented an evolutionary advance: "The adaptation of *Eciton* to a terrestrial life has brought a much more complex and intricately organized functional patterns than that of *Anomma*, with its essentially subterranean adjustment."[56]

Schneirla's personal observations of these tiny creatures in Central America expanded here into an inferential analysis of an intricate, unlearned survival strategy that varied ecologically. Laboratory dissection of a queen's reproductive organs and measurement of the workers' light sensitivity supplemented what he learned by careful watching. So did firsthand reports on African ants issued in French and German by European researchers.[57] He got close enough to colony life to appreciate its dynamics—including recurring change in affective tone—and yet stepped back to a point where evolutionary triggers might be imagined. It was a masterful validation, if one was needed, of the American concept of natural development.

When Frank Beach pointed to multiple possible outcomes of a common instinct, he, too, in effect argued against the notion of fixed species behavior attributed by the Americans to European ethology. His 1969 paper, "Locks and Beagles," celebrated animal personality and, more subtly, their degree of freedom. Like Schneirla's analysis, it acknowledged the interplay of biology and circumstance, left room for animal agency in its experimental method, and reported the considerable influence of something resembling emotion. At the same time, the difference between Beach's work with lovesick hounds and Schneirla's jungle treks attests to space for creativity in the American field.

That reproduction was also Beach's subject suggests the influence on psychology the Americans attributed to natural processes. After his move to the University of California at Berkeley in 1958, he devised clever means to measure the responsiveness of beagles to potential mates. In one set of outdoor "visiting tests," a free-roaming dog, either male or female, could approach a row of cages housing members of the opposite sex. Where and how long an estrous female, say, paused en route indicated her sexual interest, as did her reaction to male indifference, either barking at him or withdrawing for a nap. In a second arrangement, a dog was tethered to a long rope, and the researcher noted the time that a courting canine spent in range. Without the cage between them, "Each female was fitted with a specially designed chastity belt."[58]

Beach explained the genesis of the project in a way that underscored how his research had blossomed. In the late 1940s, he reared dogs in isolation before introducing one sex to the other, as he had done earlier with rats. The new free-form method better accommodated beagles' sociability, to the point that the experimental protocol seemed barely needed. Simply logging the choices of Peggy, Eddy, or Ken made it clear that each one had "a system of preference" that induced Peggy, for example, "to accept some males and at least partially reject others." All the females passed judgment on the male personality, moreover, when they uniformly favored "the least dominant of the five males." Once Beach discovered the dogs' unique characters, he did not trace the cause to dissimilar nurture. True to the neurological orientation of his teacher Lashley, he speculated that the cortex in carnivores determines sex drive.[59] What he did show was that some mix of genes, brain, and environment distinguished each beagle's tastes.

It was not the American comparative psychologists, however, but rather Lorenz, Tinbergen, and von Frisch who together won the Nobel Prize in Physiology or Medicine in 1973. Members of the Swedish Academy echoed their fellow Europeans' own emphasis on inherited structures of action when the citation praised "discoveries concerning organization and elicitation of individual and social behaviour patterns."[60] This constituted a general theory of animal development.

Lorenz gave the principle of social triggers of innate capacities an initial formulation in his work with bird flocks between the wars. Members of the same species are responsive to each other's signals, he argued, setting off expression of shared adaptations. After the war, Lorenz's personal "Noah's Ark" grew to include shrews, dogs, and fish, as well as ever more birds. His home, laboratory, and field studies blended together in what seemed an ongoing celebration of animal wisdom. A succession of readable books such as *King Solomon's Ring* (1950) and *On Aggression* (1963) amplified his ideas. Not only does a species have inborn signs to mediate interactions but entire rituals—"a hereditarily fixed copy of a behavior pattern"—that substitute for destructive impulses and, as "ceremony," function "in keeping the group members together." Stepping back, the theory's frame of reference was unashamedly us: human beings. Unlike the Americans, Lorenz did not aspire to experimental objectivity. The "general longing for a bond with Nature" finds a simple outlet when people adopt pets. His own sophisticated observations, anchoring his texts as first-person accounts, were not entirely different from sightings by weekend birders. By the same token, animals teach a morality tale: we are also beasts, and must recognize our aggression and devise rituals to preserve peace through sports and the arts.[61] In all, the idea of organic continuity, seen alike in Lorenz's intimacy with his animal subjects and his unwavering certainty that biology drives behavior, flourished in a scientific environment much taken with evolution.

Whether right or wrong, Lorenz's work had an elegance the Americans lacked. The Americans' empirically based comparisons of species overlooked useful meanings. Daniel Lehrman aimed in his doctoral dissertation to explain the feeding of young doves by parents "more parsimoniously" than by citing a mystic instinct. His answer—that the babies' touch gradually taught the adults to reduce their own "crop tension" by regurgitating into the little ones' beaks—may have been true, but it also missed the forest in the trees. Explaining the sensible benefit to the single caregiver said little about the origin or function of its habit. Rigorous, even narrow, professionally, the mid-century Americans could also be scientifically curious and impassioned. When Lehrman and Lorenz met in person, they discovered common ground as naturalists who were similarly fascinated by birds.[62] As scholars, though, the Americans made a virtue of declining to connect the dots without all the facts.

The elements of successful animal behavior study were nonetheless in place in the United States. For decades, animal research subjects had seemed mere sites to investigate, first, learning, and later, pleasure and pain. An evolutionary perspective did much to restore a sense of the variety and uniqueness of species, or, said more grandly, the intrinsic importance of all animals. Reproduction—the pivotal moment of biological descent—became the compelling research focus. In the context of the life cycle, the brain could no longer be viewed only as a learning device, or the body as a simple hormone manufacturer.

The subject of animal emotion thereby gained strength as the disciplines of psychology and biology converged. The migratory excitement of Schneirla's ants, amorous preferences of Beach's beagles, and symbiotic parenting of Lehrman's doves were affective states, if not exactly resembling human emotions. This appreciation of animal sensibility was by no means altogether new. At least since World War I, experiments on the nervous and endocrine systems underscored the body's role as an agent of behavior. In them, the organism was conceived to be individualistic and static. Evolutionary analysis was sociological and dynamic by contrast, but it also depended intellectually on the prior work.

Science was poised for this transition before World War II. Although the wartime experience of American animal behavior specialists was far from uniform, as a group they took solace in natural processes seemingly immune to human interference. They identified individual emotions as part of this organic world. More unexpectedly, the Americans began to speak about animal consciousness—the long-discarded concept—as a component of evolution. This thought opened the door wider to animal emotions.

6
The Animal Mind Reinvented

> We know not where consciousness begins in the animal world. We know where it surely resides—in ourselves. . . . Beyond this point, for all we know, it may exist in simpler and simpler forms until we reach the very lowest of living things.
>
> —Margaret Floy Washburn, 1908

The revival of interest in animal emotions in American behavioral science rested in part on the passing of a generation. In the immediate postwar years, a group of rising scientists had anchored an American branch of animal study that was strenuously experimental in method and developmental in argument. They championed the plasticity of natural endowments during the growth of individual organisms in environments. They registered specific emotions as part of the process. Their empiricism also made them wary of foreign theories. Then, by happenstance, a number of leaders were gone. Daniel Lehrman died of a heart attack in 1972 on his way to Hawaii, where he had planned to combine birding with a paper delivered to the American Psychological Association. He was only fifty-three. His students felt keenly that his "scientific career extended less than 20 years," beginning with "his famous critique of Lorenz" in 1953. As hard, the loss came close on the heels of the death in 1968 of Lehrman's teacher, T. C. Schneirla, at age sixty-six. Both lived exemplary "lives as citizens and as scientists," admirers recalled at a 1974 conference.[1] Only Frank Beach remained of the psychologists affiliated in the 1940s with the Department of Animal Behavior at the American Museum of Natural History.

Never shy of controversy, their followers now faced charges of narrowness by fellow American scientists. In 1976 the bat specialist Donald R. Griffin rejected the "reductionism" of "the behavioristic tradition" in his case for "animal awareness." Creatures with complex minds, Griffin added, presumably also possessed "animal feelings."[2] Of all things, Griffin had studied under the psychologist Karl Lashley at Harvard in the late 1930s, as had Beach. His graduate work involved step-by-step restriction of bats' ability to hear or vocalize in an effort to determine how they navigate in the dark. It was a classic reduction experiment testing perception. Nearly three decades later, however, Griffin's *Question of Animal Awareness* showed an eagerness for synthesis and a comfort with inference unlike his teacher's methodical mindset. Griffin was enthusiastic about European ethology, too, making him a thorough internationalist. Nor was his transformation unique. A cohort of American animal scientists reaching academic maturity in the 1960s and 1970s became keen participants in a transatlantic conversation about the emotional experiences of animals.

This new American biology, eclectic and bold in temper, was nonetheless the outcome of the gradual restoration of the emotions as a fit subject for animal research in the

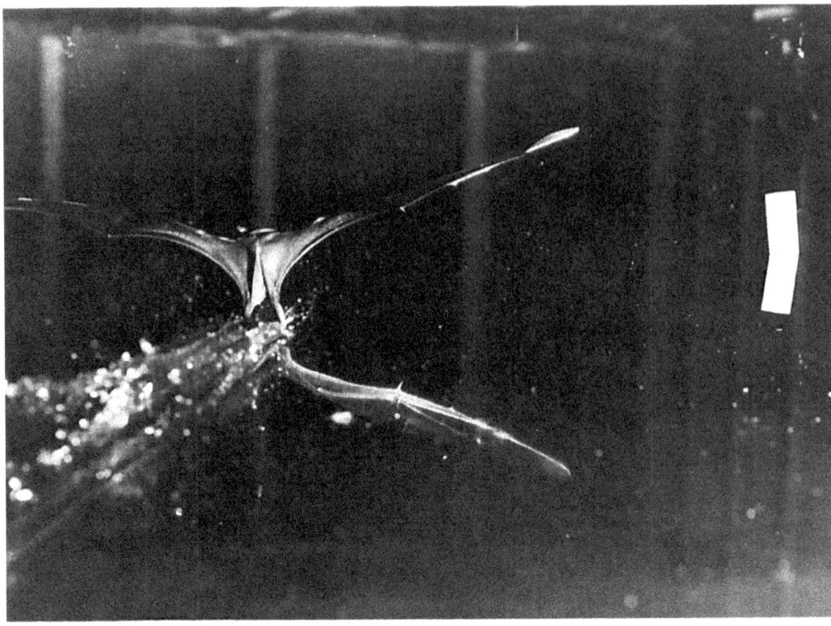

Figure 6.1 Donald Griffin's photograph of a bat fishing, Trinidad, ca. 1960. Courtesy of Donald R. Griffin Papers, Rockefeller Archive Center.

United States during the preceding half century. The fundamental intellectual change occurred within the older science, and not because it met able adversaries. To understand the animal learner, psychologists had jettisoned the idea of consciousness and turned after World War I to examine organs, hormones, and genes. They rediscovered affect in the body. Because they emphasized measurable sensations, and because their test subjects were unable to self-report, pleasure and pain became their practical analytical categories. Experimental disruptions of normal physiology by surgeries, injections, electric shock, restrained mobility, or social isolation produced in turn detectable changes in learned habits, understood to be visible indexes of emotions. Looking back in 1981, Beach saw this alliance of psychology and the life sciences as advantageous: "in other countries psychology had not become a biological science to the same extent as it had in the United States."[3]

Even so, the single-organism focus and nature-plus-nurture explanations of Beach and his colleagues were the tradition's weaknesses as well. Genetics, systematics, paleontology, and ecology—all wrapped into the modern evolutionary synthesis after World War II—made it increasingly tempting to explain individual behavior within emerging life systems. Over time, some American scholars trained in the older empiricism caught glimpses of this grander biology. Griffin was one of them, as was the primatologist George Schaller, trained in the 1950s at the University of Wisconsin by the ornithologist John T. Emlen, Jr., a specialist in the maturation of birds. There was a robust and unashamed scientific passion in 1963 to Schaller's confession of "my great liking for the gorillas" of central Africa that "became [his] acquaintances."[4] Schaller's fieldwork on these elusive creatures was patiently observational and in that sense empirical. But his exuberant breadth departed from the intellectual self-restraint of his American predecessors.

This gradual scientific development seemed revolutionary to those who made it, however. The willful neglect of earlier American animal behavior experimenters began with their descendants, who positioned themselves as rebels. It was Lashley who Griffin blamed publicly in 1976 for dismissing "the term consciousness" as "useless for scientific analysis." The year before, the ant expert Edward O. Wilson took aim at animal psychologists for using "nebulous independent variables" in their "nonevolutionary analysis." There was, to be sure, much justification in the men's impatience with their forebears' determination to refrain from theorizing until all the evidence was in. Rising scholars transferred their loyalty to European colleagues and, close to home, émigré teachers who gave them ideas and vocabulary that seemed otherwise lacking. Wholehearted internationalism made earlier native science seem provincial. Even Beach, who generously credited his American mentors and friends, devoted pages of his 1981 paper, "Historical Origins of Modern Research on Hormones and Behavior," to nineteenth-century Europe.[5] This was welcome breadth, but excitement about distant accomplishments obscured local traditions. American animal behavior researchers of the late twentieth century overlooked their own roots.

My discussion breaks down the emergence of theory-friendly and global American practice into three stages: migration of animal behavior questions from the field of psychology to evolutionary biology, conversion-like embrace of European ethology by the Americans, and preoccupation with animal emotions on the part of international scholars who absorbed the Americans' empirical rigor while rejecting their one-time dogmatism. For the sake of clarity, I treat these themes separately, but the historical changes were often concurrent as much as sequential. The pivotal era was the mid-1960s to the mid-1980s, and yet the terms of debate set at that time persist today. From this perspective, the work of Robert Yerkes, Walter Cannon, Karl Lashley, Frank Beach, and others was not mistaken but instead preliminary. Animal emotions belong to modern behavioral science because they took some initial steps.

The Animal Found

Speaking about "animals" as a single category has become commonplace, and the word's casual use suggests the value we place on understanding organisms besides humans. "The Status of Animals in the United States" was the opening chapter title of a 1990 book. "The Case for Animal Emotions" served similarly as a lead heading in 2007. Although these texts were aimed at nonspecialists and most likely argued broadly on purpose, scientific discussion of the last half century has favored principles of behavior applicable across species, such as altruism, communication, and awareness. As ambitiously, the blanket term "animals" seems a convenient fiction for an astounding variety of organisms and, if nothing else, registers the intensity of research into nonhuman life. The days when psychology seemed equivalent to rat experiments and genetics to fruit flies are over. In the United States, this intellectual expansion took off in the field of biology, a discipline no longer limited to bodily functions but encompassing relationships in the natural world.[6]

The search for analogous behavior across species accelerated along with the evolutionary "modern synthesis" announced by Julian Huxley in the 1940s. Even traditional psychology experiments became more comprehensive. In the hands of population ecologist John B. Calhoun in the 1950s, the modest rat-in-a-maze became a participant in catastrophic social collapse designed as a cautionary tale for humans. No longer simply tasked with learning a path, Calhoun's rats inhabited a labyrinth of "dwelling units," each equipped with "a drinking fountain, a food hopper, and an elevated artificial burrow," ideally housing eighty adults. When the experimenter altered space, food, and water to encourage crowded sectors, the result was aggression, reproductive failure, and "disorganization," epitomized by the soon-notorious "behavioral sink" as a state of barbarous chaos. Similarly, psychologist Donald Dewsbury expanded the method of species-by-species comparison pioneered by Robert Yerkes. Compiling data on male sexual performance in 118 types of mammals—including their thrusts, locks, and ejaculations of varying duration—Dewsbury aimed to devise a system assessing "adaptive" advantage in relation to "predation pressure." In this massive catalog, humans received a paragraph between gorillas and tree shrews.[7] Casting men in a minor role, Dewsbury called attention to the profusion of our warm-blooded relatives.

Almost predictably, breadth encouraged an eagerness for synthesis. By 1958, Donald Griffin had documented the sound-based navigation he named "echolocation" not only in bats but also water beetles, whales, nocturnal birds, and, in rudimentary form, blind humans. Affirming the "fruitful preoccupation with the universals of protoplasm" being demonstrated by his scientific peers, he reasoned "we can no longer expect to understand nature by merely describing and cataloguing her variety." Far more, "analytical biologists" have the skills to explain "the manifest ingenuity of natural machinery." Griffin's movement toward theorizing from a base of meticulous sensory experiments in the late 1930s suggests the logic behind his successive career stages. It should have surprised no one in 1976 that Griffin subtitled his book on animal consciousness "Evolutionary Continuity of Mental Experience." Critics did wonder whether this once "hard-nosed empiricist" was "slipping into premature senility," in the words of a former student. He offered "no new data, no figures, but a host of citations and arguments from philosophers as well as scientists that challenged the contemporary worldview of animals." The skeptics implied that Griffin's " 'philosopause,' " as he jokingly called it, betrayed responsible science.[8] More truly, its roots seemed to be his exuberance about what evolutionary biology was set to discover.

The push to identify commonalities had the circular effect of further nudging science out of its comfort zone in terms of what organisms to study. Rich data encouraged codification and then greater inclusiveness in turn. Griffin was proud that he rescued the "unpopular" bat from "superstition" that in the past made the species interesting only to "museum curators or pest exterminators." In the same spirit of stretching the meaning of "animal," Peter Singer asserted in *Animal Liberation* (1975) that he did not aim to create "sympathy toward 'cute' animals"—the pets of " 'animal lovers' "—but rather to outline a systematic morality toward living things. Other scientist-advocates applied this mandate by investigating the treatment of food-producing and laboratory animals normally kept out of sight in the industrial world.[9]

It was not just curiosity or ethics, however, that added research species. Wild animals living in proximity to humans could draw attention as perpetrators instead of victims. Control of feral pigs in the American South was the official reason for a 1959 study by veterinary bacteriologists. The report bore witness to the authors'

Figure 6.2 Experimental bat cage, Trinidad, ca. 1960. Courtesy of Donald R. Griffin Papers, Rockefeller Archive Center.

fascination, thoroughness, and also ambivalence. Introduced by sixteenth-century Spanish explorers, the pigs had since exacted "agricultural depredations" and harbored communicable diseases. "Man," though, "is the important predator of adult feral swine," customarily hunting with dogs at risk of being "disemboweled" by the animals' tusks.[10] Turning a keen eye on an elusive population, the paper demonstrated how closely science followed interspecies tensions.

The same socioeconomic development that was pushing southerners and wild pigs together also facilitated world travel and hence research access to rare animals. Lab or field no longer had to be a choice; exotic trips could be interspersed with campus experiments. The American-owned Creole Petroleum Corporation hosted Griffin in Venezuela in 1953, where he confirmed that cave-dwelling oilbirds, unusual avian nocturnal feeders, maneuver by echolocation. Earlier fieldwork was the precedent for such journeys, and the debt to forebears was sometimes direct. It was Harold J. Coolidge, a member of American zoological collecting expeditions to Africa and Asia between the wars and later an officer of the National Academy of Sciences, who approached John T. Emlen, Jr., in the 1950s about a comprehensive study of mountain gorillas. The project demonstrated the adaptability of the new biology. At Johns Hopkins in the 1940s, Emlen had worked on urban rat management just before the arrival of John Calhoun, architect of the dysfunctional rat city. Now Emlen was in Madison directing George Schaller's doctoral research on birds. Two years in Africa and several books later, Schaller emerged as an expert on "this giant ape in the forests," previously a "creature of mystery." Traversing physical and intellectual space was still not easy. "The path from an office in Wisconsin to the forest of Africa is not a straight one," Schaller wryly noted.[11] But the effort furthered a comparative view of animal behavior.

In the background of all this writing was a visible zest for animal research. Although nearly every scientist after 1900 seemed driven to know animals, the American academy's formalistic approach to experimentation and reporting had restrained personal expression. "Captivity cages minds as well as bodies, and rigid experimental procedure limits the range of performance," Konrad Lorenz had criticized in the 1940s.[12] Investigators trapped themselves, along with their animals. Now the increasing pace of international relocations during the careers of life scientists raised two questions: who exactly should be accounted an American, and did émigrés favor a new scientific style? Should Schaller, the author of affecting popular and scholarly accounts of gorillas, be classified as a European or an American, since he was born in prewar Germany and emigrated as a teenager to the United States, in time to attend an American college?[13] The simple answers were that the immigration of scientists did progressively blur national borders, and foreign-born scholars in the 1960s more often voiced their own reactions to animals as part of their science.

Compare, for example, the thick narratives of Jane Goodall with the abstract debates about the nature of life by American scientists. British-born, Cambridge-educated, and based in Tanzania, Goodall intertwined observations of her chimpanzee troop's maturation, reproduction, and social dynamics with her own life events and intuitions in her 1971 book, *In the Shadow of Man*. Reflecting on her work following her arrival in the Gombe Stream region and including her marriage and motherhood, she admitted that some of the intimacy she initially cultivated with the chimpanzees was a "grave mistake." Tickling and wrestling with young primates was "a delightful experience" that at first made them "trusting," but later "dangerous" as adults, when they "realized how much weaker humans really were." Goodall's second thoughts about physical contact, however, did not shake her conviction that common primate ancestry plus faithful attentiveness gave her reliable insight into the animals' minds. "The soft pressure of his [David's] fingers spoke to me not through my intellect but through a more primitive emotional channel." At another time, "I had become so attuned to William that I almost felt myself the chimps' instinctive distrust of strangers."[14] In effect, Goodall argued that empathy was scientifically valid.

A flurry of American papers in 1966–1967 on biological methods—really about life's essence—were just as impassioned, but read as high arguments over principle. The scientist kept out of the science. The paleontologist George Gaylord Simpson stated the basic issue in the Phi Beta Kappa Society's magazine, *The American Scholar*: the presumption of physical scientists, particularly chemists, that living beings are explainable by the laws of inanimate matter. Simpson responded that a life form is an interactive aggregation of molecules produced by a contingent "history both individual and racial" and, finally, "teleological" in the sense of seeking survival. Unvarying predictability, in contrast, characterized nonliving clusters of atoms. Richard Lewontin held similarly that memory, an inherently deficient guide for behavior, separated an organism from a mechanism. Its activity could never be expressed as "a large set of simultaneous differential equations." Months later, Theodosius Dobzhansky, Lewontin's one-time mentor at Columbia, tried to soothe controversy by entrusting the "venerable Cartesian tradition" of animal-as-machine to molecular biologists, while handing over " 'compositionist' " Darwinism to macro-oriented evolutionists.[15] The series of articles made heavy reading, but they left no doubt of the authors' devotion to explicating living things.

In the protean new biology, other advocates claimed that plants are nearly animals. Predictably, they met resistance. In 1973 *The Secret Life of Plants* summarized experimental evidence that plants have been underestimated and in fact resemble animals psychologically. Galvanometers and similar equipment measured plants' responsiveness to electricity, music, and damaging substances like hot coffee, indicating that they are "sentient," perhaps "capable of *intent*," and, if we dispel "our blindness," available "to cooperate with humanity" to restore nature.[16] Animal specialists jumped on the work, often performed by engineers and chemists, to defend animal uniqueness. Comparable "caloric value of plant or animal tissues" in a bomb calorimeter was irrelevant to whether animals experience awareness, in the judgment of Donald Griffin. Peter Singer declared that because plants lack a nervous system, "the belief that plants feel pain appears to be quite unjustified." Logically, these spokesmen for a comprehensive science of animals might be expected to side with the reformers of botany. Instead, they scrambled to protect their own innovative biology from potential skeptics.[17]

These skirmishes made animal researchers rethink their organisms' elemental properties time and again. Repeated scrutiny seemed further to raise estimates of animal capacities. Europeans, not unaware of the clashes in American publications and sometimes involved, more easily confessed naïve wonder at evolved behaviors. In 1949 the German zoologist Karl von Frisch recalled that he once considered bees' apparent use of varied dances to transmit information about food locations as "too fantastic for belief." Yet his experiments confirmed the bees' sophistication, and also revealed their sensitivity to polarized light, aiding orientation and communication. "Once again the bees appear to us miraculous," he affirmed in his Nobel lecture of 1973. At first glance, Richard Dawkins of Oxford was not likewise awed by the dynamics of animal societies or single-organism physiology. He saw physical bodies as "passive receptacles for the genes," serving as "walls to protect them from the chemical warfare of their rivals and the ravages of accidental molecular bombardment." But he marveled at the "ruthless selfishness" of the gene that made "blindly programmed" evolution resemble "science fiction," although "it is science." "This is a truth which still fills me with astonishment." Fantasy, miracle, fiction: unscientific vocabulary for nature's magnificence.[18]

In the United States, Edward O. Wilson's *Sociobiology* (1975) expressed similar wonder by the book's intellectual range and sheer bulk. In this review of animal relationships of nearly 700 oversized pages, the core argument was simple and total: the genetics of living matter determines all social behavior. "Ritualization," consisting in animals of inherited body signals that function to communicate with others of the same species, becomes language in people. In culture, "the equivalent of an important mutation is a new idea." Although a "population is bounded by a zone of sharply reduced gene flow" and a "society is bounded by a zone of sharply reduced communication," their parallelism implied that human groups were recent adaptations of animal organization. Methodologically, all proper scholarship was biology: "It may not be too much to say that sociology and the other social sciences, as well as the humanities, are the last branches of biology waiting to be included in the Modern Synthesis."[19] It is hard to imagine a more thoroughgoing endorsement of animal science.

Wilson's triumphalism predictably sparked yet another contest over the boundaries of biology: the sociobiology controversy. At stake was the age-old question of the difference between animals and humans. Looking back in the 1990s, Wilson emphasized how passions had distorted empirical science. The intellectual germ of sociobiology "began innocently" when he watched monkeys in Puerto Rico, making it hard to fathom why angry opponents later doused him with water at an academic conference, yelling, "you're all wet!'"[20] Not all of his critics were protestors, however. The anthropologist Marshall Sahlins countered Wilson's holistic analysis by citing principle and evidence: "Culture is the essential condition of the freedom of the human order from emotional or motivational necessity." For support, Sahlins presented field data on tribal customs, where the meanings attached to kinship categories were not identical to lines of biological descent. Even so, the plane of debate remained more theoretical than substantive, in part because neither man seemed well-read in the other's empirical literature. Sahlins made no mention of recent Nobel-winning ethologists, for example, visible symbols of a body of research that demonstrated animals' social complexity. Nor did Wilson discuss works of cultural anthropology even to refute them. Ernst Mayr, a bit older and more invested in his own causes than this one, dismissed the either-or mindset of the dispute. "Naturally, nearly all genetic programs are partly open programs," he wrote to a colleague in 1977, as if arguing was unproductive.[21] The exhilaration surrounding American animal biology nonetheless predisposed scholars to go at it.

In the midst of charged give-and-take, ideas migrated among disciplines, and the point most consequential for renewed discussion of animal emotions was that conscious experience is widespread. The impression that feelings are by definition subjective had long made them seem either hidden or absent in nonhuman creatures. The Vassar psychologist Margaret Washburn, who in 1908 devoted her *Animal Mind* to "the inner aspect of the behavior of animals," had all but vanished from scientific memory by the time Donald Griffin and others again proposed that animals have mental experiences.[22] The hunch that an unexpected number of species have awareness was nonetheless in the air.

In 1974, for example, the philosopher Thomas Nagel set out to show that "an objective concept of mind" is self-contradictory in his influential essay, "What Is It Like to Be a Bat?" Although Nagel's point that subjectivity is an irreducible quality of nature belonged to metaphysics, his example was the humble bat. If bats are aware, consciousness cannot be easily dismissed as a late evolutionary byproduct found only in primates. Rather, experience is a trait shared across a spectrum of organisms, even if we cannot enter those minds: "anyone who has spent some time in an enclosed space with an excited bat knows what it is to encounter a fundamentally *alien* form of life." Indeed, Nagel used his inability to answer his title's question to prove that subjectivity is inviolable: I can know "what it would be like for *me* to behave as a bat behaves," but I can never "know what it is like for a *bat* to be a bat." In the meantime, he had gone so far down the road toward endorsing a bat mind that he was willing to "believe that bats feel some versions of pain, fear, hunger, and lust," despite the impenetrability of bat emotions to him.[23] On the surface, his message was that bats will be bats, because their subjectivity defies external comprehension. But he also displayed empathy for bats as kindred beings, an attitude understandable in light of the vigorous discussions of animals going on around him.

Not surprisingly, American animal scientists still cast themselves modestly as faithful watchers, despite recent theoretical wrangling. When experimentalism was dominant in earlier decades, "naturalist" had been a suspect term. Now it stamped the protean range of evolutionary biologists. T. G. Dobzhansky reflected in the mid-1960s that "'naturalist' and 'natural history' do not any longer have precise meanings." *Naturalist* was the proud title of E. O. Wilson's memoir in 1994.[24] The charming image of the stalwart outdoor observer only partly explains, however, why American animal scientists of the later twentieth century fell in love all over again with their subject. Side by side with a passion for animal life was an excitement about their science—not only new locations and methods but also the academic jousting about how best to explain life forms. The morale boost that came along with bold thinking reached new levels when more and more American scientists stopped resisting European ideas about animal behavior and instead embraced them.

American Empiricism Forgotten

Ethology—the study of social behavior patterns in animals—was neither an unchanging nor a unified science, but from its inception under Konrad Lorenz in the 1930s it stressed natural endowment and biological transmission. In its logic, it connected individual habits with the survival of populations. This was a broader perspective than contemporaneous American concern with motivations conditioned by the body and developing in a single lifetime, an approach rooted in psychology. By the 1970s, the transatlantic theory-war once led by Schneirla and Lehrman ended less in concession by the Americans than in their conversion. "A ferment of constructive excitement is evident in ethology," declared Donald Griffin, being applied to problems ranging from "social organization" to "altruistic behavior" and "orientation and navigation." To an extent, this acclaim of ethology followed a shift in personnel specializing in animal behavior. Biologists were replacing psychologists. The Harvard biologists and physiologists who taught Griffin in the 1930s prepared him to appreciate ethology's emphasis on the descent of living matter. He owed his lifelong interest in perception and, later on, awareness, however, to the psychologist Karl Lashley. In 1976 Griffin nonetheless explicitly situated "complex cognitive functions" in the field of ethology.[25] His instinct to go over to the Europeans was not uncommon.

Over time, ethology convinced individual scholars. A comparison of 1972 and 1977 articles by Marc Bekoff, issued soon after he received his doctorate in psychology at Washington University in St. Louis, show the displacement of American authorities. In the first paper, Bekoff examined how play aids socialization, endorsing a "developmental, comparative approach" that included tests on "captive animals." He cited Frank Beach, reigning authority on animal play, along with Schneirla and Lehrman, defenders of nurture-oriented growth. He rejected "Lorenz's psychohydraulic model" of inner drives as "unsubstantiated" and called for evidence about "motor action patterns" that is "simple, observable, and measurable."[26]

Even Bekoff's 1972 report, however, deviated from the norm in American comparative psychology of empirically tracing behavioral changes. Although "development" was the first noun in Bekoff's title, "an ethological perspective" was his subtitle, and he included photographs of pairs of dogs in play postures. Ethologists explained such stances as typical of a species. Body positions expressed "the hereditary blueprint contained in the genome," Lorenz wrote, where the function was "ritualization" of communication.[27]

These patterns subsequently commanded more of Bekoff's attention. In his 1977 essay in *Science*, "stereotyped" body stance was no longer a component of individual maturation and instead the centerpiece of analysis. His subjects were still dogs and their evolutionary relatives at various ages, but his stated purpose was quantitative testing of the claim of "classical ethology" that a recurring "action pattern"—in this case, bowing—transmitted a message about the animals' intent. His timeframe was not the single life and its personal vicissitudes; instead, he considered species-level behavior as an asset for collective survival. A shift in his citations confirmed that his emphasis was not an organism's growth. He listed no publications by Beach, Schneirla, or Lehrman, while adding works by up-and-coming figures in Niko Tinbergen's circle at Oxford, including Richard and Marian Dawkins. Bekoff placed himself in the American empirical tradition by offering his data to evaluate ethological theory. But Bekoff's academic address, now at the University of Colorado, was the "Ethology Group."[28]

Similar instances of persuasion by ethology were ongoing, not clustered at a single chronological moment but generally earlier in time for American scholars who were more grounded in biology. Donald Griffin had already assigned the word "echolocation" to animals' sound-based navigation systems when he invited Karl von Frisch to lecture at Cornell in 1949. Helping von Frisch to plan his trip, which expanded to seventeen visits to US research programs, translating the lectures, and finding a publisher, Griffin gave practical evidence of the affirmation he repeated in his foreword to the resulting book, *Bees* (1950). To Griffin's satisfaction, von Frisch proved bees' "remarkable sensory capacities" and "elaborate social structure." Griffin had already argued that bats are active hunters, finding their way by emitting sounds and responding to echoes. Von Frisch supplied further evidence that creatures commonly judged unsophisticated were in fact adept communicators. His work now pushed Griffin to see bats as more than solitary sensing devices and, like bees, colony dwellers.[29]

Scientific logic and public endorsement did not immediately add up to Griffin's private conviction of ethology's merits, however. Suspicion of theorizing, insistence on evidence, and attentiveness to nurture—all from his American training—inflected his correspondence. Griffin confessed his "residue of skepticism" and "perhaps excessive conservatism" to Lashley soon after his European guest departed for home. Perhaps feeling a special need to inform his mentor about his rigorous repetition of von Frisch's experiments, Griffin remained hesitant about ethology a decade later. Writing in 1957 to his graduate-school friend, biologist John Tyler Bonner, Griffin noted the "tendency for ethologists to concentrate their attention and interest on behavior that is largely determined by genetic components, and not greatly influenced by learning." They also favored "jargon"—"ethologese," as he called it. Even so, "scientific 'father-figure'" was the phrase he chose for von Frisch in a letter to him in 1965, and for decades, Griffin reliably managed the European's American publications and visits. Surprisingly, the stalwart empiricist Robert Yerkes helped. Out of nowhere in 1949, Yerkes, now retired, sent Griffin "my very hearty thanks" for hosting von Frisch. "I never have met Professor von Frisch but for many years we exchanged reprints and I have a big bundle of his papers in my reprint collection." Yerkes suggested a contact for the German scholar at Yale University Press.[30] Slowly but seemingly surely, Griffin and his circle warmed up to ethology.

In another case, the unembarrassed application of "ethological interpretation" by George Schaller in *The Mountain Gorilla* (1963) was the outcome of a similarly uneven path toward acceptance of species-wide behavior patterns of his teacher, John T. Emlen,

Figure 6.3 Griffin in 1938, when he began graduate study at Harvard. Courtesy of Donald R. Griffin Papers, Rockefeller Archive Center.

Jr. In line with the idea of inherited signals, Schaller devoted fourteen pages to minute analysis of the gorillas' chest-beating display, which he believed was a "release for the tension which has accumulated in their system in an excitable situation." Schaller's focus on a typical sequence and his unapologetic use of the hydraulic metaphor allied him firmly with ethology. Emlen, remembered as a generous mentor of more than sixty graduate students and presumably influential for Schaller, was less consistent. Trained in ornithology at Cornell in the 1930s and later committed to studying the "ontogeny of behavior" in individuals, Emlen met and respected British and German ethologists in Africa and elsewhere. Moreover, he worked at Johns Hopkins during World War II with Curt Richter, the much-maligned American proponent of tension-release motivation. These contacts, along with open-mindedness perhaps traceable to Emlen's Quaker roots, help explain the easygoing ethology that flowered in his student's work. Schaller did not become an ideologue. The primatology of Robert Yerkes, including unquestioning respect for Yerkes's observational method, informed Schaller's gorilla research as well.[31]

For all these Americans, it was personal contact with European or European-trained animal specialists that encouraged them to view behavior more theoretically. Émigrés who relocated in the United States and subsequently advised doctoral students were especially influential. Bekoff, for one, acknowledged his mentoring at Washington University around 1970 by Michael W. Fox, the British-born veterinarian who soon left the academy for the Humane Society of the United States. Similarly, Robert Trivers at Harvard studied under Ernst Mayr. With interests spanning morphology, genetics, and ecology, Mayr was not focused on behavior, and yet he promoted the Europeans. In 1965 he wrote privately to his one-time colleague at the American Museum of Natural

History, Frank Beach, to ask Beach's help in recruiting psychologists on behalf of Konrad Lorenz, whose nomination as a foreign associate of the National Academy of Sciences was then pending.[32]

Trivers soon set off on a slightly different path to investigate the link between self-preservation and altruistic behavior preoccupying British evolutionists. His 1971 essay "The Evolution of Reciprocal Altruism" was indebted to the London-based theorist W. D. Hamilton and set the stage for Trivers's foreword to Richard Dawkins's *The Selfish Gene*, issued in New York five years later. Here he credited Dawkins with articulating "Darwinian social theory" to provide "a glimpse of an underlying symmetry and logic in social relationships." Integrated intellectually as Trivers was into British science, however, he did not blindly serve a new master. In 1978 Mayr advised him "to get there well-rested, and have a good detailed, logical outline" when Trivers addressed the Royal Society, "not only for your own career, but also for the sake of evolutionary biology." Trivers agreed, but he was sufficiently irritated by travel details to grumble about paying his own way to make a mark with these traditional gatekeepers.[33] There was a limit to how much Americans felt beholden to the Europeans.

Ethology nonetheless powerfully drew the Americans because it provided a synthesis of evolutionary biology and psychology, much as Julian Huxley had hoped years before. In the 1930s and 1940s, the tone of American evolutionary science was impersonal, as it tracked selection pressures in populations. Quantum physics was the acknowledged model for analyzing fossil sequences in George Gaylord Simpson's book, *Tempo and Mode in Evolution* in 1944. Yet here was Simpson in the mid-1960s arguing that an organism is a uniquely purposive kind of matter and that evolution on this planet was sufficiently complex to be virtually unrepeatable. A paleontologist who studied life through its material traces, he now deemed animals to be actors.[34] To the extent that Simpson's reorientation reflects a general shift among American evolutionists, it was fertile ground for the insertion of animal motivations into biological descent. This, in essence, was ethology: explanations of behavior as inherited survival strategies. Whereas opponents had long contended that reports of animals by the Europeans were too theory-laden to be truly empirical, ethology's coherence apparently overrode past objections. In a more cooperative spirit, Americans like Bekoff and Trivers set out to examine ethology's propositions experimentally.

Not all American animal behavior specialists accepted ethology, however, and yet the dissenters' diffuse reactions made an effective critique unlikely. One strategy was ideological counterattack. In 1974 a psychologist trained by Daniel Lehrman described "Lorenzian ethology" as foundational for "the Social Darwinists who today are the spokesmen for modern eugenics." The paper was weakened by repetitiousness, accusation, and, oddly, surrender. There was nothing new in the alternative proposal to study "the developmental history of the individual," a doctrine framed here in shrill generalities that mirrored the fervency of some advocates of genetic determinism. Falling back on the argument that biology could not explain human decisions ("Natural selection did not bring about the abolition of slavery"), the author all but conceded that inborn drives might be the causes of action in nonhuman animals.[35]

In a different mood, Frank Beach seemed inclined to compromise, in keeping with his career-long habit of building scientific bridges. He was so modest, though, that it was easy to miss his point about reconciling nature and nurture. Once committed to a view of evolution based on species comparisons, he now focused squarely on individual maturation. Although his 1974 lecture at Cambridge University bragged about the field of "behavioral endocrinology," he declined to analyze sex-related activity patterns in evolutionary terms. The claim that a hormone deficiency caused some girls to prefer "boys' wearing apparel to frilly dresses," along with assertions about what "every normal woman" feels, belonged to an ahistorical method. Seven years later, Beach's fifty-one-page review of experiments on "hormones and behavior" performed since 1849 was so fact-laden that a reader could barely see that Beach valued this historical work as evidence of the interplay between genetics and environment. A pigeon experimentally isolated might lay eggs "if she is exposed to a male or even if her head and neck are stroked repeatedly by her human caretaker." Perhaps Beach was fishing for a compliment in 1976 when he noted "my advanced age and declining intellectual powers" in a letter to Ernst Mayr.[36] More clearly, he did not want to enter controversy or much change his ideas.

Behavioral ecology, in contrast, was a new field with a different orientation from either developmental psychology or ethology. American ecologists did not oppose the turn toward ethology, but they did not join it. With a wide lens on the interplay of environments and populations, ecology in the 1960s had little directly to say about the maturation or awareness of single organisms. It applied the strict observational standard of early American researchers to animal societies. The mentors who trained Stephen T. Emlen at the University of Michigan for a 1966 PhD studied the aggregate effects on animal communities of measurable factors such as predators, parasites, and food supply. This view departed from the interest in "motivation" of Emlen's father, John, who tried to reconcile theories of motives in 1955 as "general or specific irritability" or "drives and appetites." Ecology was not static, however, and by the time Stephen Emlen published a memoir in 2010, his research had shifted from bird migration to altruism and mating. He still did not speculate about animal instincts or the emotions possibly found in consciousness. He did demonstrate that ecology could investigate emotion-like behavior.[37]

In the midst of these differing viewpoints, the international scholars who influenced the Americans were especially comfortable taking issues out of the academy and becoming advocates. Their style sharpened transatlantic debate. Marian Dawkins, who has spent her career at Oxford, impeccably reviewed the scholarship for her 1980 book, *Animal Suffering*, and yet aimed, too, at a "science of animal welfare," wording she chose for her subtitle. Pairing science and engagement, she followed one of her mentors, the Cambridge ethologist W. H. Thorpe, who contributed a paper on "pain and distress in animals" to the official *Brambell Report* (1965) that aimed to reform British husbandry. British-trained scholars who took posts in the United States chose even more provocative language. For Peter Singer, also Oxford-educated, the banality of evil was a theme running through his *Animal Liberation*, echoing Hannah Arendt. "How can a man who is not a sadist spend his working day heating an unanesthetized dog to death" in the laboratory and "then remove his white coat, wash his hands, and go home to dinner with his wife and children?" Michael Fox, resettled in the United States, bluntly denounced "the holocaust of the animal kingdom" in *Inhumane Society: The American Way of Exploiting Animals* (1990). Although "global technocracy" was finally at fault, in the near view "animal exploitation, suffering, and environmental desecration today are symptomatic of a dysfunctional American society."[38]

These books drew American readers. Rights agitation was in the air. Yet American animal behavior research did not become a provincial branch of a European school of thought. Although the meaning of "ethology" in American publications was not wholly plastic, the authors preserved some old ways, such as individual-oriented studies, under a new rubric. In this sense acceptance of evolutionary behavior theory was sincere but not blinding, and the initial flurry of declarations of loyalty to ethology seemed, if anything, to boost American eclecticism.

Robert Sapolsky, for one, has been a free force in primate neurobiology since the 1980s. His early biochemical measurement of stress in baboons began with a classic American question about a body-behavior connection. Kenya was the research site, however, dominance relations the source of psychic pressure, evolutionary kinship to humans the project's logic, and stress an amorphous experience. The emphasis on communal life in the wild clearly diverged from the one-time American experimental norm. Two decades later, stereotyped components of rat trials were as dramatically rearranged in Sapolsky's laboratory. Experimentally inducing memories of environments—that is, learning—to instill either fear or comfort led to this conclusion: rats make the best choices when they are apprehensive but not terrified. The rationale for the tests was the evolutionary premise that "the ability to discriminate between two spatial contexts is crucial for survival." In the year 2000, though, Sapolsky had warned against fixating on simple genetic explanations in his paper, "Genetic Hyping." "I'M NOT trying to bash genes," he reassured, but rather aiming to discourage single-mindedness: "The environment, even working subtly, can still more than hold its own in the biological interactions that shape who we are."[39] Elements of ethology as an evolutionary theory of social behavior mark Sapolsky's work, side by side with physiological psychology.

It was the step-by-step increase in American friendliness to European ideas about animal minds that seemed to jump-start innovation. Not only did the American ethos of plodding experimentalism lose steam but memories of it also dimmed. The process was more complicated than the replacement of one set of references by another. Scientific partisans actively shaped recollections of history that made the triumph of their group appear inevitable. Conference proceedings such as Ernst Mayr's *The Evolutionary Synthesis* (1980) and essays like Frank Beach's "Historical Origins of Modern Research on Hormones and Behavior" (1981) constructed the twentieth-century scientific past very differently. Said simply, esteem of Darwin prevailed over indebtedness to Lashley, because the scholarly wind was blowing toward theories of behavioral descent.[40]

Ethology in particular also encouraged habits of viewing time and facts that deflected attention from behavioral tests of proxy animals. The vast historical sweep of evolution and the need to recruit theories such as natural selection to interpret it may have made past achievements in American laboratories seem small. Whereas Yerkes expected to line up experiments on worms, frogs, crows, and pigs to find the truths of animal behavior, the new view was global and tolerant of imagination. Names of past giants of American animal research still showed up American syntheses, but in a selective way. In 1975 E. O. Wilson cited the once-formidable biologists William Morton Wheeler and W. C. Allee, along with Lorenz and the American mathematical psychologist Herbert A. Simon. The next year Donald Griffin favorably referenced arguments by Edward Tolman, along with more praise of Lorenz and allusion to the linguist Noam Chomsky. This breadth grafted itself onto American thinking already prodded out of localism by other evolutionary applications. Little wonder that in 1964 G. G. Simpson, the fossil expert, lauded Friedrich Wöhler's synthesis of an organic compound from inorganic elements in 1828 as if it was yesterday.[41] In these fragmentary perspectives on the past, the collective accomplishments of earlier American animal scholars did not stand out.

The Americans' welcome of European animal behavior science was undeniably a transformative event. It forced another look at inherited activity patterns shared by changing populations. As a body of research focused on the dynamics of inherited traits in societies, it helped make the emotions, already recognized as components of individual growth by American peers, more central to definitions of what an animal is. The emotional animal was not a simple import, however, but equally the legacy of a century of national science that took serious account of affective states because researchers aimed at thorough reporting. The practice of applying descriptors such as "phlegmatic," "hyperactive," and "good mothers" to observed animal habits was alive and well when John Calhoun explained the chaos in rat city in 1962. [42] More systematic discussion of animal feelings followed, and scientific debate continued as well.

The Emotional Animal

The pioneering ethologists were no more interested in ineffable private sensations than were Schneirla or Lehrman. Although they used the words "instinct" and "drive," their terms meant germs of innate behavior, with no implications for feelings. In 1951 Niko Tinbergen chose the title, *The Study of Instinct*, for his exposition of "the objective study of behavior." He deliberately omitted experiences "that can be known only by introspection." In 1966 Lorenz spoke in an equally unfriendly way about emotions in a chapter in *On Aggression*, "The Great Parliament of Instincts." His " 'big drives,' " all said to be inborn and expressed in activities, were "feeding, reproduction, flight, and aggression." The title of another chapter, "Social Organization without Love," suggests that it was easy to debate nature-nurture without introducing emotion.[43]

As an assertive movement of scientific revision, however, ethology spurred fresh thinking and new vocabulary. "The Evolution of Reciprocal Altruism" by Robert Trivers demonstrated how a combination of evolutionary and behavioral biology encouraged reconsideration of emotion. Although Trivers made individual feelings a subordinate part of his model, he did not omit them. Indeed, he stressed that Darwin and others who traced altruism to "*the emotions of liking and disliking*" merely put the cart before the horse. Natural selection produced altruistic behavior and, only subsequently, its psychic components: "emotions of friendship (and hatred) are not prerequisites for reciprocal altruism but may evolve *after* a system of mutual altruism has appeared, as important ways of regulating the system." This represented a rethinking of an old evolutionary puzzle: the survival of populations is often aided by individual self-sacrifice, such as a bird issuing a warning call when a predator approaches the flock. Selection genetically favors a mutual tit-for-tat expectation in the altruist, he argued. Moreover, in humans "strong positive emotions have evolved to motivate altruistic behavior." Trivers did not quite say that evolutionary continuity suggests that at least some other animals may also feel emotion. His essay left no doubt, though, that emotions are foundational in nature.[44]

Trivers thereby turned from observational analysis of actions in the direction of subjectivity. Identifying selfishness and altruism as contrasting behaviors was an inference nearly required by natural selection, which tended to imagine survival as a core drive. Laboratory experiments conducted on living organisms by developmental psychologists were too narrow either to confirm or deny why, across the ever-emerging animal kingdom, self-seeking sometimes gives way to selflessness. Moving beyond conventional genetics and psychology, Trivers took an intellectual leap, proposing that evolution produces internal emotional triggers. In this, he nearly replicated the theory of emotions as signals—to individual awareness and fellow creatures—stated in the 1920s by William McDougall. American psychologists had objected that McDougall was speculating about unknowable processes, whereas at the same time they were toying with terms such as appetites and drives. Like McDougall, Trivers did not shy away from holistic claims, titling one section of his article, "The Psychological System Underlying Human Reciprocal Altruism."[45] He began predictably enough by asserting that genes determine behavior. More unexpectedly, he posited affective experience.

As the word "biology" came to mean something and everything, self-identified biologists discussed emotions unhampered by the connotations of past language. In *Sociobiology*, a book that epitomized the field's ambition, E. O. Wilson swung easily from altruism to the proposition that emotions regulate internal conflicts. His background in entomology and population ecology, reaching back to the 1940s, appeared to free him from the psychologist's traditional worries about where emotions may be located in organisms, what catalyzes them physiologically, and how they develop. He made a frankly functional argument. The combination of DNA and natural selection produces an "efficient mixture of personal survival, reproduction, and altruism" in populations, motives that together understandably "tax the conscious mind with ambivalences" in individuals. In "a highly social species," brain chemistry mediates competing needs and permits plasticity: "Love joins hate; aggression, fear; expansiveness, withdrawal; and so on; in blends designed not to promote the happiness and survival of the individual, but to favor the maximum transmission of the controlling genes." Wilson's esteem of the social insects led him to imply that some kind of affect is prevalent across species. In termites and rhesus macaques alike, "group members communicate hunger, alarm, hostility, caste status or rank, and reproductive status," as well as organize according to kinship and division of labor.[46] This analysis of the service performed by emotions was educated guesswork, but his tone was assured. Taking a global view, Wilson seemed to reason: animals manifest responses that we recognize as emotions, making it wise to include feelings in evolutionary theory, even if the psychological mechanisms are unclear.

It was exponents of animal consciousness, in contrast, who laid a mental base for animal emotions. These were not mainly psychologists who might be imagined to be reclaiming a concept of mind last seen in the early work of Yerkes and Washburn. Rather, it was an eclectic group including the philosophers Thomas Nagel and Peter Singer and biologists Donald Griffin and Marian Dawkins. They seemed attentive to each other's writing, if not formally collaborators. Within two years, for example, Nagel chose the bat mind to exemplify irreducible selfhood, followed by a case for "animal awareness" by the bat expert Griffin.[47] The biologists left a deeper mark on animal science.

Sharing the sociobiologists' high estimate of animals—as actively sociable creatures genetically primed for survival—Griffin added cognition: species that communicate can reasonably be said to have "*mental experiences*." He had spent nearly four decades empirically tracking echolocation when he reached this conclusion, but it was American philosophical pragmatism that shaped it. Like William James, Griffin valued thoughts as akin to actions. Consciousness consists of "interrelated mental images of the flow of events," Griffin explained, and it serves "*intention*" when "the intender pictures himself as a participant and makes a choice as to which image he will bring to reality." Far from an idealist's passive concept of mind, Griffin's description was in essence James's "stream of thought" of 1890, where "attention" made "the voluntary life" possible.[48]

Although American psychologists had once excoriated McDougall for entertaining a similar notion of goal-driven action, "teleology" had made a steady comeback in the American academy. In 1943 Norbert Wiener and colleagues in cybernetics attributed teleology—"as synonymous with 'purpose controlled by feedback' "—to both organisms and machines. Five years later Edward Tolman stated that "a field map of the environment gets established in the rat's brain," because the animal is "bombarded by stimuli" and "surprisingly selective as to which of these stimuli it will let in at any given time." Repeatedly citing Tolman in *The Question of Animal Awareness*, Griffin extended the earlier thesis about a rat's private reconstruction of space to say that mental capacity to solve problems is inseparable from evolutionary adaptation.[49]

Marian Dawkins added feeling to the psychological proposition of adaptive minds in her 1980 book, *Animal Suffering*. Her analysis was historically informed. Having studied in the 1960s with the leaders of British ethology, she now cited Americans' empirical work. "Tolman, Yerkes, and Köhler, amongst others, many years ago pioneered the experimental study of the mental life of animals," she wrote to explain existing scholarship on "subjective experience." Here she better sensed the insistence on evidence of the two Americans than their hesitancy about consciousness as a concept. Dawkins herself sided openly with Griffin, affirming the likelihood that "many animals besides ourselves are conscious." More in tune with Griffin's American predecessors, however, she paused to consider how emotions work in the body. Judging emotion "to be the property of a highly developed nervous system," she referenced Walter Cannon's *Bodily Changes in Pain, Hunger, Fear and Rage* to suggest that hormones activate fight or flight. Throughout, she argued for animal suffering with a scrupulousness that would have pleased Yerkes. Subjective well-being should not be confused with either physical health or life in the wild, Dawkins warned, and she presented her book as just a starting point: "a guide to only a small part of the journey."[50]

Three decades later, Dawkins was more reserved about scientific analysis of animal suffering, although she was ever more committed to the goal of curtailing animal pain. Sounding perhaps even more American than these forebears, she insisted that science is by definition a process of empirical verification. Ironically, the success of her own thesis—that animals feel affliction—was one source of her growing caution. Her book had lent animal rights arguments a compelling evolutionary psychology and thereby helped catalyze a stream of publications that shifted thematically over time from suffering, the point of Michael Fox's *Inhumane Society* in 1990, to a celebration of sensibilities, Marc Bekoff's emphasis in *The Emotional Lives of Animals* in 2007. Writing the next year, Dawkins was skeptical that scientific evidence as yet warranted so much assurance. She spoke strategically and personally. She counseled that "animal welfare scientists are more likely to be taken seriously if they show that they understand the conceptual difficulties with attributing conscious experiences to animals." She continued modestly: "Personally, I do believe that many animals subjectively experience suffering but I also believe that my own belief is not scientific and I would not attempt to justify it on scientific grounds." Yet despite philosophical uncertainty, "an operational definition of suffering" is possible. GPS technology demonstrates that elephants in the wild, for example, choose longer routes around hills to avoid climbing them. The science-to-date sustained a "practical approach."[51]

The stream of scholarship descending from the revived premise of animal consciousness was not all about emotion, however, but also cognition and proto-morality. To set writing about feelings in this wider review of animal life underlines the persistence of rationalism in Western culture. Recent applications of the pleasure-pain dichotomy have been one-sided: pleasure-seeking as a kind of animal hedonism elicits little comment, in contrast to sustained focus on pain as the reason for ethical reform. Tying mind to feeling, Marian Dawkins identified rudimentary self-knowledge as a requisite for pain. "To say that an animal suffers implies that it is aware of its suffering," in her words. For the philosopher Peter Singer in 1975, relief of animal suffering rested on a justice argument to be implemented by rational actors. "Animal liberation," his book's title phrase, depends on "the fundamental moral principle of equality of considerations of interests." He did not "appeal to the reader's emotions where they cannot be supported by reason." A related line of inquiry traced ethics itself—a capacity for social behavior presumably beyond simple empathy—back to animals. Wilson's word choice in *Sociobiology* for his lead chapter, "The Morality of the Gene," implicitly subordinated emotions to traits with clearer evolutionary value.[52] The science of animal emotions of the era displayed no hedonistic temper celebrating liberated passions. Instead, varied allusions to reasonableness suggest why it had been so hard for American researchers to acknowledge the emotions of animals all along.

The very success of science also complicated the by-now old question of whether animal study is really about humans. Big science was at once facilitating and distracting. As a return visitor to central Africa around the year 2000, two decades after his first eager work with baboons, Robert Sapolsky confessed, "I am almost as afraid of not balancing the budget on my grant" as he once was of wild buffalo. Never quite away from the office, "I come here to think clearly about my lab work and to catch up on my sleep and to escape the demands of the endless academic committees."[53] Just as much, the ambition of many researchers to take investigation out of the lab or field in order to influence policy could easily make human behavior the subject. If the "dis-eased human condition" is the reason for habitat loss or industrial food production, animals can become mere victims.[54] To be sure, nearly the same dilemmas posed by the business and applications of science were part of the world of Yerkes, Cannon, Watson, and Lashley before World War I. But they have not been resolved.

Far more positively, there has been an enduring curiosity about how and why animals behave as they do. E. O. Wilson tells the story in his memoir of standing as a seven-year-old in 1936 in the shallows of the Gulf of Mexico, watching a sea nettle. "It came into my world abruptly, from I knew not where, radiating what I cannot put into words except—*alien purpose and dark happenings in the kingdom of deep water*." Since then, a distinguished career had taught him its Latinate name, explained its unity beyond being a cluster of tentacles, and much more. At the time, though, it was elusive: "The next morning the sea nettle was gone."[55] One imagines that Wilson has been pursuing his sea nettle ever since. When Margaret Nice banded song sparrow couples and Frank Beach played matchmaker to beagles, their mood was much the same.

The list of emotions cataloged has remained remarkably stable, too, despite the transformation of animal feelings from a surreptitious into a serious subject. Time and again recent animal scholars have reported feelings resembling Walter Cannon's 1915 title, *Bodily Changes in Pain, Hunger, Fear and Rage*. Only altruism as a discussion point, inseparable from appreciation of sociability, seems genuinely new. Modern genetics in the 1930s highlighted the activities of populations, although previous studies of ants, bees, birds, and primates by then had added cooperative gestures to expressions of individual mood. Cannon was similarly predictive in his choice of the body as the site of inquiry. The gradual relocation of research about emotion to the field of biology, redefined after World War II to cover all kinds of natural relationships including consciousness, seemed intellectually liberating. Although it is now unusual to equate emotions with simple states of nerves or hormones, living matter made scientific sense as the home of sensibility.

Throughout, American animal science has not been particularly welcoming to emotion. Readers might love a wilderness adventure story by Teddy Roosevelt, but for decades no American produced an equivalent to Darwin's *Expression of the Emotions in Man and Animals*. Twentieth-century science in the United States displayed a national character, where the most visible trait was practicality in principle and method. Empiricism was expected to yield general truths permitting prediction and control of behavior. To the researchers' credit, when they realized that learning was more than skill acquisition, they examined processes of reproduction, maturation, and homoeostasis. They rediscovered intent—or in old-fashioned terms, will—and the emotions that came along with shunning and seeking.

The animals were active partners, to the surprise of their keepers. Animals initially acquired value precisely because scientists took them to be perfect mute subjects of observation and, a bit later, exemplars of primal feelings. Working with them, however, drove home the temperaments and personalities of species and individuals. Alternatively frustrating and delighting their humans, the animals demonstrated their capacity for expression, albeit without language, and for emotion, feelings detected if not always reciprocated by people bent on explaining them. American appreciation of emotions in animals is due most profoundly to the animals.

Appendix

Major American-Born Figures (in alphabetical order)

Warder Clyde Allee	(1885–1955)	ecology
Frank A. Beach	(1911–1988)	psychology
Marc Bekoff	(b. 1945)	ethology
John B. Calhoun	(1917–1975)	biology
Walter B. Cannon	(1871–1945)	physiology
Leonard B. Carmichael	(1898–1973)	psychology
Clarence Ray Carpenter	(1905–1975)	psychology (primatology)
Harold J. Coolidge	(1904–1985)	zoology
Wallace Craig	(1876–1954)	biology (ornithology)
John T. Emlen, Jr.	(1908–1997)	biology (ornithology)
Donald R. Griffin	(1915–2003)	zoology
Harry Harlow	(1905–1981)	psychology (primatology)
William James	(1842–1910)	philosophy
Karl Lashley	(1890–1958)	psychology
Daniel Lehrman	(1919–1972)	psychology
Norman R. F. Maier	(1900–1977)	psychology
Thomas Hunt Morgan	(1866–1945)	genetics
Margaret Morse Nice	(1883–1974)	biology (ornithology)
Robert Sapolsky	(b. 1957)	neurobiology
Theodore Christian Schneirla	(1902–1968)	psychology (entomology)
George Gaylord Simpson	(1902–1984)	paleontology
Calvin Perry Stone	(1892–1954)	psychology
Edward L. Thorndike	(1874–1949)	psychology
Edward Chace Tolman	(1886–1959)	psychology
Robert Trivers	(b. 1943)	biology
Margaret Floy Washburn	(1871–1939)	psychology
John B. Watson	(1878–1958)	psychology
William Morton Wheeler	(1865–1937)	biology (entomology)
Charles O. Whitman	(1842–1910)	zoology
Norbert Wiener	(1894–1964)	mathematics (cybernetics)
Edward O. Wilson	(b. 1929)	biology (entomology)
Robert Mearns Yerkes	(1876–1956)	psychology (primatology)

Wives/Collaborators of Americans

Marjorie Hill Allee	(1890–1945)
Cornelia James Cannon	(1876–1969)
Ada Watterson Yerkes	(1873–1963)

International Figures and Internationally Born Émigrés

Marian Stamp Dawkins	(b. 1945)	ethology
Richard Dawkins	(b. 1941)	ethology
Theodosius G. Dobzhansky	(1900–1975)	genetics
Michael W. Fox	(b. 1937?)	veterinary science
Karl von Frisch	(1886–1982)	biology (entomology)
Jane Goodall	(b. 1934)	ethology (primatology)
Julian Huxley	(1887–1975)	biology
Zing-Yang Kuo	(1898–1970)	psychology
Konrad Z. Lorenz	(1903–1989)	ethology
William McDougall	(1871–1938)	psychology
Ernst Mayr	(1904–2005)	biology (ornithology)
Thomas Nagel	(b. 1937)	philosophy
W. H. R. Rivers	(1864–1922)	anthropology/ psychology
George B. Schaller	(b. 1933)	biology (primatology)
Peter Singer	(b. 1946)	philosophy
Nikolas (Niko) Tinbergen	(1907–1988)	ethology
Joseph Yoshioka	(1893–unknown)	psychology
Solomon Zuckerman	(1904–1993)	zoology (primatology)

Notes

Introduction

1. Walter B. Cannon to [Cannon family], June 10, 1917 (first four quotations); Cannon to [Cannon family], Thurs. [Aug. 16, 1917] (fifth and sixth quotations), box 185, letter book 1, Walter B. Cannon Papers (H MS c40), Center for the History of Medicine, Francis A. Countway Library of Medicine, Harvard Medical School, Boston, Mass.; W. B. Cannon, "A Consideration of Wound Shock," *Journal of the American Medical Association* 70:9 (1918): 614 (seventh quotation). Cannon named typical test animals as rabbits, guinea pigs, dogs, and cats, in *Bodily Changes in Pain, Hunger, Fear and Rage: An Account of Recent Researches into the Function of Emotional Excitement* (1915; reprint ed., New York: D. Appleton, 1920), p. 15. On his service abroad with the Harvard Medical Unit, see Elin L. Wolfe, A. Clifford Barger, and Saul Benison, *Walter B. Cannon: Science and Society* (Boston: Boston Medical Library in the Francis A. Countway Library of Medicine, 2000), chaps. 1–3.
2. Charles Darwin, *The Expression of the Emotions in Man and Animals* (1872; reprint ed., Chicago: University of Chicago Press, 1965). Darwin's high estimate of emotion as a force in personality was consistent with his Anglo-American contemporaries. See the British philosophical psychologist Alexander Bain, *The Emotions and the Will* (1859), ed. Daniel N. Robinson (Washington, D.C.: University Publications of America, 1977), and the American Thomas C. Upham, *Abridgement of Mental Philosophy* (1861; reprint ed., Delmar, N.Y.: Scholars' Facsimiles and Reprints, 1979), pp. 261–480. On the prevailing emotionalism of nineteenth-century Anglo-American culture, see Janet M. Davis, *The Gospel of Kindness: Animal Welfare and the Making of Modern America* (New York: Oxford University Press, 2016), chap. 1; Thomas Dixon, *From Passions to Emotions: The Creation of a Secular Psychological Category* (Cambridge: Cambridge University Press, 2003), chap. 5; and James Wharton Fay, *American Psychology before William James* (New York: Octagon, 1966), esp. pp. 91–109.
3. John B. Watson, "Psychology as the Behaviorist Views It," *Psychological Review* 20:2 (1913): 160 (first quotation); Margaret Floy Washburn, *The Animal Mind: A Text-Book of Comparative Psychology*, 2d ed. (New York: Macmillan, 1917), p. 4 (second quotation); George H. Daniels, *American Science in the Age of Jackson* (New York: Columbia University Press, 1968), p. 106 (third quotation); Ernst Mayr to Niko Tinbergen, HUGFP 14.7, Mar. 7, 1946, box 3, folder 142, Ernst Mayr Papers, Harvard University Archives, Cambridge, Mass. (hereafter Harvard University Archives) (fourth quotation). I have silently corrected Mayr's misspelled "abserved."
4. The classic German psychology experiment devised by Wilhelm Wundt (1832–1920) relied on self-reports of reactions by human subjects; see, e.g., Kurt Danziger, "Wundt's Psychology Experiment in the Light of His Philosophy of Science," *Psychological Research* 42:1–2 (1980): 109–22. British-born Edward Titchener (1867–1927), Wundt's student and later Cornell psychologist, championed Wundt's method in the United States. For conflict over the value of human versus animal subjects, see the dispute at Harvard in 1911 between Hugo Münsterberg, another Wundt protégé and émigré, and Robert Yerkes, in Anne C. Rose, "Animal Tales: Observations of the Emotions in American Experimental Psychology,

1890–1940," *Journal of the History of the Behavioral Sciences* 48:4 (2012): 304. For a sense of the prevalence of research animals in early American laboratories, see John O'Donnell, *The Origins of Behaviorism: American Psychology, 1870–1920* (New York: New York University Press, 1985), esp. chaps. 8–10; Philip J. Pauly, "The Loeb-Jennings Debate and the Science of Animal Behavior," *Journal of the History of the Behavioral Sciences* 17:4 (1981): 504–15; Richard W. Burkhardt, Jr., "Charles Otis Whitman, Wallace Craig, and the Biological Study of Animal Behavior in the United States, 1898–1925," in Ronald Rainger, Keith R. Benson, and Jane Maienschein, eds., *The American Development of Biology* (Philadelphia: University of Pennsylvania Press, 1988), pp. 185–218; Donald A. Dewsbury, "Triumph and Tribulation in the History of American Comparative Psychology," *Journal of Comparative Psychology* 106:1 (1992): 3–19; and Jill G. Morawski, ed., *The Rise of Experimentalism in American Psychology* (New Haven: Yale University Press, 1988). Scholars who highlight the special ardor of Americans for fact-based observation include Jonathan Harwood, "National Styles in Science: Genetics in Germany and the United States between the World Wars," *Isis* 78:3 (1987): 390–414; Stephen J. Cross and William R. Albury, "Walter B. Cannon, L. J. Henderson, and the Organic Analogy," *Osiris*, 2d ser., 3 (1987): 165–92; Donna Jeanne Haraway, *Crystals, Fabrics, and Fields: Metaphors of Organicism in Twentieth-Century Developmental Biology* (New Haven: Yale University Press, 1976), chap. 3; and Christopher D. Green, "Of Immortal Mythological Beasts: Operationism in Psychology," *Theory and Psychology* 2 (1992): esp. 297–310.

5. John Dewey, *How We Think: A Restatement of the Relation of Reflective Thinking to the Educative Process*, rev. ed. (Boston: D. C. Heath, 1933), pp. 170 (first and second quotations), 200 (third quotation); Watson, "Psychology as the Behaviorist Views It," p. 163 (fourth, fifth, and sixth quotations); Robert M. Yerkes, *Introduction to Psychology* (New York: Henry Holt, 1911), p. 275 (seventh and eights quotations); Dewey, *How We Think*, p. 200 (ninth and tenth quotations). Italics in quotations appear in the original texts unless otherwise noted. See also Edward L. Thorndike, "Animal Intelligence; an Experimental Study of the Associative Processes in Animals" (1898), in Thorndike, *Animal Intelligence: Experimental Studies* (New York: Macmillan, 1911), pp. 20–155, and John C. Burnham, "Thorndike's Puzzle Boxes," *Journal of the History of the Behavioral Sciences* 8:2 (1972): 159–67. The scientific preference for useful intelligence followed earlier philosophical loyalties of the nation's thinkers to Scottish common sense realism and Enlightenment utilitarianism. See Bruce Kuklick, *The Rise of American Philosophy: Cambridge, Massachusetts, 1860–1930* (New Haven: Yale University Press, 1977), pp. 1–126, and Benjamin Franklin, *The Autobiography of Benjamin Franklin*, ed. Leonard W. Labaree et al. (New Haven: Yale University Press, 1964), esp. pp. 240–46. On the American tendency to privilege intelligence as a trait, see Graham Richards, "'It's an American Thing': The 'Race' and Intelligence Controversy from a British Perspective," in Andrew W. Winston, ed., *Defining Difference: Race and Racism in the History of Psychology* (Washington: American Psychological Association, 2004), pp. 137–69. Despite Watson's iconoclasm, William James previously voiced the same goal of making psychology a reliable science and, surprising from the man who coined the phrase "stream of thought," expressed his own uncertainty about locating subjectivity. See William James, *The Principles of Psychology* (1890), vol. 1, in *The Works of William James*, electronic edition, ed. Frederick H. Burkhardt, Fredson Bowers, and Ignas K. Skrupskelis (Cambridge: Harvard University Press, 1975–1988), 8: 6, 219 (hereafter *James Works*), and William James, "Does 'Consciousness' Exist?" (1904), in *James Works*, 3: 3–20. Like Yerkes, James rued delayed scientific attention to the emotions, in *The Principles of Psychology*, vol. 2, in *James Works*, 9: 1086.

6. Donald R. Griffin, *The Question of Animal Awareness: Evolutionary Continuity of Mental Experience* (New York: Rockefeller University, 1976), pp. 5 (first and second quotations), 6

(third quotation); Charles G. Gross, "Donald R. Griffin, 1915–2003," *National Academy of Sciences Biographical Memoirs* (2005), http://www.nasonline.org/publications/biographical-memoirs/memoir-pdfs/griffin-donald.pdf (accessed June 1, 2019), p. 14 (fourth and fifth quotations). Michael S. Gazzaniga offers a similar chronology to mine—dating the "cognitive revolution" as beginning in the early 1970s—with a different analysis of reasons for revived interest in consciousness, in *The Consciousness Instinct: Unraveling the Mystery of How the Brain Makes the Mind* (New York: Farrar, Straus and Giroux, 2018), p. 71.

7. Calvin P. Stone, "William McDougall's Methodological Concept of Instinct," in Stuart A. Rice, ed., *Methods in Social Science: A Case Book* (Chicago: University of Chicago Press, 1931), pp. 186 (first and second quotations), 187 (third quotation); Dewey, *How We Think*, p. v (fourth quotation). In practice, the Americans breached their pledge to empirical concreteness, and this book includes their sometimes disputatious philosophizing. A prime example of an abstract (and a satiric) diatribe against theoretical abstraction is N. R. F. Maier, "Maier's Law," *American Psychologist* 15:3 (1960): 208–12. I equally argue that principles stated in polemics did not necessarily match actual research and that what the experimenters did was more influential in the long run than what they said. Hamilton Cravens similarly notes that argumentative positions of nature and nurture were less distinguishable than supposed if a person read only public controversy; see his *The Triumph of Evolution: American Scientists and the Heredity-Environment Controversy, 1900–1941* (Philadelphia: University of Pennsylvania Press, 1978), esp. pt. 4. See also Anne C. Rose, "William McDougall, American Psychologist: A Reconsideration of Nature-Nurture Debates in the Interwar United States," *Journal of the History of the Behavioral Sciences* 52:4 (2016): 325–48.
8. The balance of nationalism and internationalism in the twentieth-century life sciences merits reexamination. For example, Frank Biess and Daniel M. Gross, eds., *Science and Emotions after 1945: A Transatlantic Perspective* (Chicago: University of Chicago Press, 2014), argues for a shift in the direction of scientific influence from European to American dominance after World War II (p. 14). In our own era of global transmission, this and other studies also blend analysis of texts of disparate national origin with little attention to dissimilar cultural assumptions affected by place. But suppose national pride and jealousies once erected intellectual boundaries. The historical facts for American life scientists are clear: they made international contacts for a variety of reasons. I emphasize, however, that American attitudes toward foreigners and foreign science intervened to shape these actions. See n. 4 for additional sources on American distinctiveness.
9. Robert M. Yerkes, "Objective Nomenclature, Comparative Psychology and Animal Behavior," *Journal of Comparative Neurology and Psychology* 16:5 (1906): 383 (first quotation); Walter Cannon, "Dietetics in Disease," n.p. (the page follows p. 75 in the second series of pages; pagination erratic), handwritten lecture notes, post-1903, box 176, Cannon Papers (second and third quotations).
10. Richard W. Burkhardt, Jr., "The *Journal of Animal Behavior* and the Early History of Animal Behavior Studies in America," *Journal of Comparative Psychology* 101:3 (1987): 228; James Ewing to Walter Cannon, Jan. 9, 1909, box 28, folder 344, Cannon Papers; "Indian Monkeys for United States. Cargo of 500. Experimental Colony to be Formed," *Bombay Times*, Oct. 3, 1938, clipping, in Markle-Columbia Primate Journal, p. 128, Clarence Ray Carpenter Papers, Eberly Family Special Collections Library, Penn State University, University Park, Pa.; E. M. Smith, *The Investigation of Mind in Animals* (Cambridge: Cambridge University Press, 1915). Although animal experiments did not become as narrowly reliant on white rats and *Drosophila* as sometimes said, these species' widespread use in psychology and genetics laboratories, respectively, has elicited much comment. See Frank A. Beach, "The Snark Was a Boojum," (1950), in Donald A. Dewsbury, ed., *Foundations of Comparative Psychology*

(New York: Van Nostrand Reinhold, 1984), pp. 61–70 (rats); Bonnie Tocher Clause, "The Wistar Rat as the Right Choice: Establishing a Mammalian Standard and the Idea of a Standardized Mammal," *Journal of the History of Biology* 26:2 (1993): 329–49; and Robert E. Kohler, *Lords of the Fly: Drosophila Genetics and the Experimental Life* (Chicago: University of Chicago Press, 1994).

11. Minnie Ruth Stone, "Biography of Calvin Perry Stone," typescript, p. 33, Calvin Perry Stone Papers microfilm, Special Collections and University Archives, Stanford University Libraries, Stanford, Calif. (hereafter Stone Papers microfilm) (first quotation); Ernest R. Hilgard, *Calvin Perry Stone, 1892–1954: A Biographical Memoir* (1994), http://www.nasonline.org/publications/biographical-memoirs/memoir-pdfs/stone-calvin-p.pdf (accessed June 1, 2019), pp. 407–08 (second quotation); Wallace Craig, "Male Doves Reared in Isolation," *Journal of Animal Behavior* 4:2 (1914): 121 (third and fourth quotations), 132 (fifth and sixth quotations); Robert M. Yerkes, "The Heredity of Savageness and Wildness in Rats," *Journal of Animal Behavior* 3:4 (1913): 286 (seventh quotation), 287 (eighth quotation). See generally Hank Davis and Dianne Balfour, eds., *The Inevitable Bond: Examining Scientist-Animal Interactions* (Cambridge: Cambridge University Press, 1992). Although I share the scholarly view that tacit assumptions behind experimental designs in human environments—and specifically laboratories—affected conclusions, I argue that labs were interactive sites and hence that the intellectual outcomes were open-ended. For a more deterministic perspective, see Otniel E. Dror, "What Is Excitement?" in Biess and Gross, eds., *Science and Emotions*, pp. 121–38.

12. Yerkes, *Introduction to Psychology*, p. 275 (first quotation); Stone, "William McDougall's Methodological Concept of Instinct," in Rice, ed., *Methods in Social Science*, pp. 186 (second quotation), 187 (third quotation); Edward Chace Tolman, "Can Instincts Be Given Up in Psychology?" *Journal of Abnormal and Social Psychology* 17:2 (1922): 145 (fourth, fifth, and sixth quotations); Frank A. Beach, "Locks and Beagles," *American Psychologist* 24:11 (1969): 989 (seventh quotation); T. C. Schneirla, "A Comparison of Species and Genera in the Ant Subfamily *Dorylinae* with Respect to Functional Patterns (1) (2)," *Insectes Sociaux* 4:3 (1957): 263 (eighth and ninth quotations). The Americans' rejection or adoption of terminology did not occur in a historical vacuum, but with varying degrees of conscious awareness of established usages. On this past, see Ute Frevert, ed., *Emotional Lexicons: Continuity and Change in the Vocabulary of Feeling, 1700–2000* (Oxford: Oxford University Press, 2014). Eileen Crist argues that English-language scientific words shaped estimates of animals and rendered them as objects. See her *Images of Animals: Anthropomorphism and the Animal Mind* (Philadelphia: Temple University Press, 1999).

13. Jeremy Bentham, *An Introduction to the Principles of Morals and Legislation* (1781), in *The Collected Works of Jeremy Bentham*, electronic edition, ed. J. H. Burns and H. L. A. Hart (Oxford: Oxford University Press, 1970), chaps. 3–5. There was American debate about the nature of the emotions, although speculation did not dislodge the rudimentary pleasure-pain model in practice. For a sample, see Walter B. Cannon, "The James-Lange Theory of Emotions: A Critical Examination and an Alternative Theory," *American Journal of Physiology* 39:1/4 (1927): 106–24; Martin L. Reymert, ed., *Feelings and Emotions: The Wittenberg Symposium* (Worcester, Mass.: Clark University Press, 1928); David L. Krantz and David Allen, "The Rise and Fall of McDougall's Instinct Doctrine," *Journal of the History of the Behavioral Sciences* 3:4 (1967): 326–38; and William McDougall, *An Introduction to Social Psychology* (1908; reprint ed., London: Methuen, 1960), chap. 11. The discussions continue today. Some scholars stress the experiential quality of emotion, whereas others emphasize physiology. Among the former are Susanne K. Langer, *Mind: An Essay on Human Feeling* (Baltimore: Johns Hopkins University Press, 1967), 1: esp. chaps. 1–3; Robert C. Solomon,

The Passions: The Myth and Nature of Human Emotion (Notre Dame, Ind.: University of Notre Dame Press, 1983); and Jesse J. Prinz, *Gut Reactions: A Perceptual Theory of Emotion* (Oxford: Oxford University Press, 2004). Body-centered estimates include Paul Ekman and Richard J. Davidson, eds., *The Nature of Emotion: Fundamental Questions* (New York: Oxford University Press, 1994), and Otniel E. Dror, "The Affect of Experiment: The Turn to the Emotions in Anglo-American Physiology, 1900–1940," *Isis* 90:2 (1999): 205–37.

14. Robert M. Yerkes, *Chimpanzees: A Laboratory Colony* (New Haven: Yale University Press, 1943), p. 1.
15. C. O. Whitman, "Animal Behavior," in Jane Maienschein, ed., *Defining Biology: Lectures from the 1890s* (Cambridge: Harvard University Press, 1986), pp. 218–71; Burkhardt, "*Journal of Animal Behavior*"; Edward O. Wilson, *Sociobiology: The New Synthesis* (Cambridge: Harvard University Press, 1975), title page (first quotation), p. 4 (second quotation). Although Burkhardt states (p. 228) that Yerkes edited the Behavior Monographs, documentary evidence indicates that Yerkes recruited Watson to be at least the titular editor; see "Announcement of the Journal of Animal Behavior and the Behavior Monograph Series," p. 2, typescript, ca. 1910, HUGFP 87.10, box 39, folder "Robert M. Yerkes, 1908–1910," William Morton Wheeler Papers, Harvard University Archives. Like many widely used terms, "behavior" developed many meanings. For the distinction between "behavior" and "behaviorism," see James T. Todd and Edward K. Morris, "The Early Research of John B. Watson: Before the Behavioral Revolution," *Behavior Analyst* 9:1 (1986): 71–88. Historically, American animal researchers often explicitly disavowed the equation of the behavioral method and a behavioral nurture-only argument. The devoted experimenter Margaret Washburn distinguished herself in 1916 from "the most extreme 'behaviorist,' Professor Watson," for example; Calvin P. Stone, a pioneer of physiological psychology, complained in Europe in 1932, "so many foreigners think that Behaviorism is the rage in America." See Margaret Floy Washburn, *Movement and Mental Imagery: Outlines of a Motor Theory of the Complexer Mental Processes* (Boston: Houghton Mifflin, 1916), p. xii, and Calvin P. Stone, "Visit to the Psychological Laboratory at Edinburgh, April 26, [1932]," typescript, Stone Papers microfilm. The later prevalence of the word behavior in English-language texts (often translations) authored by European ethologists may be further testimony to the utility of the so-named method. See, e.g., N[iko]. Tinbergen, *Social Behavior in Animals, with Special Reference to Vertebrates* (1953; reprint ed., New York: Wiley and Sons, 1966). Despite historical evidence and scholarly arguments casting doubt on the thought that all mid-twentieth-century American psychologists were behaviorists, the conclusion remains widely repeated in secondary literature; see, e.g., Tania Munz, *The Dancing Bees: Karl von Frisch and the Discovery of the Honeybee Language* (Chicago: University of Chicago Press, 2016), pp. 173–74, 207–08, and passim, and Marilyn Bailey Ogilvie, *For the Birds: American Ornithologist Margaret Morse Nice* (Norman: University of Oklahoma Press, 2018), p. 134 and passim.
16. Whereas some of these lineages are well known, consider Leonard Carmichael to Wallace Craig, Feb. 3, 1936, and Wallace Craig to Leonard Carmichael, Feb. 15, 1936, folder "Wallace Craig, 1936," Leonard Carmichael Papers (Mss.B.C212), American Philosophical Society, Philadelphia, Pa. (reflections on Craig's Harvard course of the previous decade); James H. Jones, *Alfred C. Kinsey: A Public/Private Life* (New York: Norton, 1997), chap. 7 (Wheeler and Kinsey); Margaret Morse Nice, *Research Is a Passion with Me: The Autobiography of Margaret Morse Nice* (Toronto: Margaret Morse Nice Ornithological Club and Consolidated Amethyst Communications, 1979), chap. 4 (Hall and Nice); "Biographical Note," Finding Aid, Clarence Ray Carpenter Papers, Eberly Family Special Collections Library, Penn State University, http://www.libraries.psu.edu/findingaids/149.htm (accessed June 1, 2019) (McDougall and Carpenter); Gross, "Griffin," esp. pp. 6–10; James A. Simmons,

"Donald R. Griffin: 1915–2003," *Bat Research News* 45:1 (2004), http://www.batresearchnews.org/Miller/Griffin.html (accessed June 1, 2019), and correspondence between Lashley and Griffin, 1944–1958, box 7, folder 67, Donald R. Griffin Papers, RG 450G875, FA 164, Rockefeller Archive Center, Sleepy Hollow, N.Y. (Lashley and Griffin).

17. George Gaylord Simpson, "The Crisis in Biology," *American Scholar* 36:3 (1967): 363–77.
18. Walter B. Cannon, "The Body Physiologic and the Body Politic," *Science* 93:2401 (1941): 4 (first and second quotations), 2 (third quotation), 4 (fourth quotation). The lecture was Cannon's presidential address to the American Association for the Advancement of Science, Dec. 27, 1940.
19. Maria I. Diedrich, *Cornelia James Cannon and the Future American Race* (Amherst: University of Massachusetts Press, 2010); Anne C. Rose, "The Translator's Preface: Language Boundaries of English-Speaking Life Scientists, 1914–1945," unpublished paper. Cornelia's conservative opinions, documented by Diedrich, were counterbalanced by her personal unconventionality, eloquently recalled by her daughter, Marian Cannon Schlesinger, *Snatched from Oblivion: A Cambridge Memoir* (Boston: Little Brown, 1979). Studies of gender and science especially relevant to my principal scientists and their wives include Margaret W. Rossiter, *Women Scientists in America: Struggles and Strategies to 1940* (Baltimore: Johns Hopkins University Press, 1984); Donna Haraway, *Primate Visions: Gender, Race, and Nature in the World of Modern Science* (New York: Routledge, 1989); Marcia Myers Bonta, *American Women Afield: Writings by Pioneering Women Naturalists* (College Station: Texas A & M Press, 1995); and Ogilvie, *For the Birds.*
20. Edward C. Tolman to Richard C. Tolman, Mar. 22, 1918, box 2, folder 2.1, Richard Chace Tolman Papers, Archives, California Institute of Technology, Pasadena, Calif. (first and second quotations); David Kretch, *In Memoriam: Edward Chace Tolman, 1886–1959: Proceedings of a Meeting Held in Memory of Edward Chace Tolman, on the Berkeley Campus of the University of California, Dwinelle Hall, December 17, 1959*, p. 13, box 1, folder 1.19, Tolman Papers (third quotation); Edward Chace Tolman, *Purposive Behavior in Animals and Men* (New York: Appleton-Century-Crofts, 1967), p. xviii (fourth quotation). See also Edward Chace Tolman, *Drives toward War* (New York: Appleton-Century, 1942); Richard H. Parke, "Yale Honors Educator Dismissed by California over Loyalty Oath," no source, and anon., "Ousted Educator Honored by Yale," *New York Times*, June 12, 1951, clippings, box 2, folder 2.11, Tolman Papers. Works that constructively examine the implications of scientific ideas for social and political philosophy include Nadine Weidman, "Psychobiology, Progressivism, and the Anti-Progressive Tradition," *Journal of the History of Biology* 29:2 (1996): 267–308; Nadine M. Weidman, *Constructing Scientific Psychology: Karl Lashley's Mind-Brain Debates* (Cambridge: Cambridge University Press, 1999); and Marga Vicedo, *The Nature and Nurture of Love: From Imprinting to Attachment in Cold War America* (Chicago: University of Chicago Press, 2013). Sources that highlight unexpected and varied political outcomes are Gregg Mitman, *The State of Nature: Ecology, Community, and American Social Thought, 1900–1950* (Chicago: University of Chicago Press, 1992), and Nikolai Krementsov, *International Science between the World Wars: The Case of Genetics* (London: Routledge, 2005).
21. Robert M. Yerkes, "Inhibition and Reinforcement of Reaction in the Frog *Rana clamitans*," *Journal of Comparative Neurology and Psychology* 14:2 (1904): 137 (first quotation); T. C. Schneirla, "Instinct and Aggression," *Natural History* 75 (1966): 18 (second quotation); K. S. Lashley, "In Search of the Engram: Physiological Mechanisms in Animal Behavior" (1950), in *The Neuropsychology of Lashley: Selected Papers of K. S. Lashley*, ed. Frank A. Beach et al. (New York: McGraw-Hill, 1960), esp. pp. 503–05.
22. Walter B. Cannon, *Some Implications of the Evidence for Chemical Transmission of Nerve Impulses: Address at the Plenary Sitting of the Congress* (Moscow-Leningrad: State Publishing

House for Biological and Medical Literature, 1935), p. 4 (first and second quotations); Wheeler, July 28, 1914, diary "Trip to New Zealand and Australia, July 12th, 1914–Jan. 16th, 1915," HUGFP 87.45, box 2, Wheeler Papers (third quotation).

23. E. O. Wilson, *Naturalist* (Washington, D.C.: Island Press, 1994), p. 86 (first quotation); Theodore Roosevelt, *Through the Brazilian Wilderness* (1914; reprint ed., New York: Scribners, 1922), p. 328 (second quotation). The American popular romance with nature has been thoroughly examined; see, e.g., Roderick Nash, *Wilderness and the American Mind* (New Haven: Yale University Press, 1967), and Gregg Mitman, *Reel Nature: America's Romance with Wildlife on Film* (Cambridge: Harvard University Press, 1999). For Britain, the comparable masterwork remains Keith Thomas, *Man and the Natural World: A History of Modern Sensibility* (New York: Pantheon, 1983).
24. A number of recent scholars identify the post–World War II era and particularly the 1960s as the turning point in appreciation of the emotions in Western culture; see, e.g., Biess and Gross, eds., *Science and Emotions*, p. 10, and Erika Lorraine Milam, *Creatures of Cain: The Hunt for Human Nature in Cold War America* (Princeton: Princeton University Press, 2019), pp. 1–15. I agree that interest in animal emotions sharply gained visibility around 1970, but I also argue that World War I was the initial and indeed dramatic moment of change in the United States for scientific understanding of motivation, laying a foundation for gradual recognition of animal emotions.
25. Langer, *Mind*, 3 vols. (1967–1982); Solomon, *The Passions*; Thomas Nagel, "What Is It Like to Be a Bat?" *Philosophical Review* 83:4 (1974): 435–50; Julian Huxley, *Evolution: The Modern Synthesis* (1942; reprint ed., New York: Harper and Brothers, 1943); David Fraser, "Animal Behavior, Animal Welfare and the Scientific Study of Affect," *Applied Animal Behaviour Science*, 118:2–3 (2009): 108–117.
26. Dixon, *From Passions to Emotions*; Janet Browne, *The Secular Ark: Studies in the History of Biogeography* (New Haven: Yale University Press, 1983), chap. 8; Peter J. Bowler, *The Eclipse of Darwinism: Anti-Darwinian Theories in the Decades around 1900* (Baltimore: Johns Hopkins University Press, 1983); Vassiliki Betty Smocovitis, *Unifying Biology: The Evolutionary Synthesis and Evolutionary Biology* (Princeton: Princeton University Press, 1996); James Turner, *Reckoning with the Beast: Animals, Pain, and Humanity in the Victorian Mind* (Baltimore: Johns Hopkins University Press, 1980); Davis, *Gospel of Kindness*. Among many studies by Dewsbury and Dror, see, e.g., Donald A. Dewsbury, *Monkey Farm: A History of the Yerkes Laboratories of Primate Biology, Orange Park, Florida, 1930–1965* (Lewisburg, Pa. Bucknell University Press, 2006), and Dror, "Affect of Experiment" on animal emotions, as well as comments on emotions in human culture, in Otniel Dror et al., "An Introduction to History of Science and the Emotions," *Osiris* 31:1 (2016): 1–18. Because the Americans I analyze were rarely conventional Darwinians, I have been most helped by histories of alternative European theories to establish transatlantic connections, esp. Daniel L. Schacter, *Stranger behind the Engram: Theories of Memory and the Psychology of Science* (Hillsdale, N.J.: Lawrence Erlbaum, 1982); Laura Otis, *Organic Memory: History and the Body in the Late Nineteenth and Early Twentieth Centuries* (Lincoln: University of Nebraska Press, 1994); Anne Harrington, *Reenchanted Science: Holism in German Culture from Wilhelm II to Hitler* (Princeton: Princeton University Press, 1996); Charlotte Sleigh, *Six Legs Better: A Cultural History of Myrmecology* (Baltimore: Johns Hopkins University Press, 2007); and Egil Asprem, *The Problem of Disenchantment: Scientific Naturalism and Esoteric Discourse, 1900–1939* (Leiden: Brill, 2014).
27. Griffin, *Animal Awareness*, p. 25 (first quotation); Thomas Nagel, "What Is It Like to Be a Bat?," pp. 435 (second quotation), 436 (third and fourth quotations).

28. Marian Stamp Dawkins, *Why Animals Matter: Animal Consciousness, Animal Welfare, and Human Well-Being* (Oxford: Oxford University Press, 2012), pp. 6 (first quotation), 150 (second quotation); Stephen T. Emlen, "Ethics and Experimentation: Hard Choices for the Field Ornithologist," *Auk* 110:2 (1993): 406 (remaining quotations), as well as Marc Bekoff, "Experimentally Induced Infanticide: The Removal of Birds and Its Ramifications," *Auk* 110:2 (1993): esp. 5. See also Ruth Leys, *The Ascent of Affect: Genealogy and Critique* (Chicago: University of Chicago Press, 2017), and Daniel Horowitz, *Happier? The History of a Cultural Movement That Aspired to Transform America* (New York: Oxford University Press, 2017). Gazzaniga more forthrightly asserts the existence of consciousness than Dawkins, in *Consciousness Instinct*. For the experimental middle ground on animal consciousness, see, e.g., Daniel M. Weary, Paula Droege, and Victoria A. Braithwaite, "Behavioral Evidence of Felt Emotions: Approaches, Inferences, and Refinements," *Advances in the Study of Behavior* 49 (2017): 27–48.

Chapter 1

1. Margaret Morse Nice, *Research Is a Passion with Me: The Autobiography of Margaret Morse Nice* (Toronto: Margaret Morse Nice Ornithological Club and Consolidated Amethyst Communications, 1979), p. 107. A similar reminder of the sounds—and their animals—surrounding humans is John M. Marzluff, *Welcome to Subirdia: Sharing Our Neighborhoods with Wrens, Robins, Woodpeckers, and Other Wildlife* (New Haven: Yale University Press, 2015).
2. John B. Watson, "Recent Experiments with Homing Birds" (1915), in Donald A. Dewsbury, ed., *Foundations of Comparative Psychology* (New York: Van Nostrand Reinhold, 1984), pp. 75 (first quotation), 81 (second and third quotations). The quotations appeared in a popular article in *Harper's Monthly Magazine*; for the longer, scientific report, see Watson and K. S. Lashley, *Homing and Related Activities of Birds*, Papers from the Department of Marine Biology of the Carnegie Institution of Washington (Washington, D.C.: Carnegie Institution, 1915), 7: 5–84. On Watson's involvement with the new field station, see Richard W. Burkhardt, Jr., "Charles Otis Whitman, Wallace Craig, and the Biological Study of Animal Behavior in the United States, 1898–1925," in Ronald Rainger, Keith R. Benson, and Jane Maienschein, eds., *The American Development of Biology* (Philadelphia: University of Pennsylvania Press, 1988), pp. 207–08, and Alfred Goldsborough Mayer, "Marine Biological Laboratory at Tortugas, Florida," *Carnegie Institution of Washington Year Book* (Washington, D.C.: Carnegie Institution of Washington, 1904), 3: 50–54.
3. Watson, "Recent Experiments," in Dewsbury, ed., *Foundations of Comparative Psychology*, p. 75.
4. John Broadus Watson, "John Broadus Watson," in Carl Murchison, ed., *A History of Psychology in Autobiography* (Worcester: Clark University Press, 1936), 3: 276.
5. Watson, "Recent Experiments," in Dewsbury, ed., *Foundations of Comparative Psychology*, p. 75.
6. Margaret Floy Washburn, *The Animal Mind: A Text-Book of Comparative Psychology*, 2d ed. (New York: Macmillan, 1917), pp. 4–5 (first quotation), 5 (remaining quotations). Charles O. Whitman similarly ridiculed amateurs, in "Myths in Animal Psychology," *Monist* 9:4 (1899): 524–37.
7. Washburn, *Animal Mind*, 2d ed., p. 4. Eileen Crist offers an excellent critique of precisely the kind of professional insularity and technical language of analysis that Washburn fervently defended, in *Images of Animals: Anthropomorphism and the Animal Mind* (Philadelphia: Temple University Press, 1999). Crist argues that scientific terminology tends

to reduce animal subjects to objects, whereas commonplace usage recognizes them as fellow living beings. In this sense, specialized training transmits an invisible philosophy about animals.

8. C. Lloyd Morgan, *An Introduction to Comparative Psychology* (1894), ed. Daniel N. Robinson (Washington, D.C.: University Publications of America, 1977), pp. 211 (first and second quotations), 204 (third and fourth quotations).
9. Konrad Lorenz, "Preface," in Charles Darwin, *The Expression of the Emotions in Man and Animals* (1872; reprint ed., Chicago: University of Chicago Press, 1965), p. xi.
10. Darwin, *Expression of the Emotions*, p. 140. On Darwin's methods, see, e.g., pp. 14–26 (questionnaires), 118–19 (his pets), and 209–10 (his children). Despite Darwin's eclecticism as a data gatherer, he was painstakingly reflective and repeatedly assessed the comparative value of existing theories of biological evolution and distribution for his own conclusions. For example, on his new application of mathematics to theorizing in the 1850s, see Janet Browne, *The Secular Ark: Studies in the History of Biogeography* (New Haven: Yale University Press, 1983), chap. 8. On his ongoing attempts to explain genetic variation in the 1860s and 1870s, see Harriet Ritvo, "Foreword to Charles Darwin, *The Variation of Animals and Plants under Domestication*," in Ritvo, *Noble Cows and Hybrid Zebras: Essays on Animals and History* (Charlottesville: University of Virginia Press, 2010), chap. 8.
11. Darwin, *Expression of the Emotions*, p. 118.
12. William Morton Wheeler, "Small Artificial Ant-Nests of Novel Patterns," *Psyche* 17:2 (1910): 73.
13. William Morton Wheeler, "A Study of Some Texan Ponerinae," *Biological Bulletin* 2:1 (1900): 2–3 (first quotation); Feb. 27, 1923, diary "Trip to Panama, Feb. 8th–April 12th, 1923," HUGFP 87.45, box 3, William Morton Wheeler Papers, Harvard University Archives, Cambridge, Mass. (second quotation). On the feeding experiments, see Wheeler, "Some Texan Ponerinae," pp. 11–12. Wheeler first elaborated the concept of "trophallaxis"—the mutual nourishment of ants as social behavior—in William Morton Wheeler, "A Study of Some Ant Larvae, with a Consideration of the Origin and Meaning of the Social Habit among Insects," *Proceedings of the American Philosophical Society* 57:4 (1918): 293–343.
14. Alex. [*sic*] Nielson to Robert M. Yerkes, Feb. 22, 1903, series II, box 70, folder 1340, Robert Mearns Yerkes Papers (MS 569), Manuscripts and Archives, Yale University Library, New Haven, Conn.; James P. Porter, "A Preliminary Study of the English Sparrow," *American Journal of Psychology* 15:3 (1904): 315.
15. Walter B. Cannon to Charles B. Davenport, Mar. 16, 1901, folder "Walter B. Cannon 1," Charles B. Davenport Papers (Mss.B.C27), American Philosophical Society, Philadelphia, Pa. See also Cannon to Davenport, Mar. 27, 1901, in the same folder.
16. Mary A. Barker, College of Medicine, Syracuse University, Dec. 18, 1908, response to undated questionnaire sent to medical schools and hospitals by Walter B. Cannon on behalf of the Council in the Defense of Medical Research of the American Medical Association, box 28, folder 344, Walter B. Cannon Papers (H MSc 40), Center for the History of Medicine, Francis A. Countway Library of Medicine, Harvard Medical School, Boston, Mass.
17. Willard S. Small, "An Experimental Study of the Mental Processes of the Rat," *American Journal of Psychology* 11:2 (1900): 156. On Small's original experiment, see p. 135, n. 2. On Whitman's death, see Burkhardt, "Charles Otis Whitman," in Rainger et al., eds., *American Development of Biology*, pp. 194–95.
18. Hugo Münsterberg, "The Psychological Laboratory in Emerson Hall," *Harvard Psychological Studies* 2 (1906): 35 (first, second, and third quotations); C. O. Whitman, "Biological Instruction in Universities," *American Naturalist* 21:6 (1887): 517 (fourth quotation). For Münsterberg's opinion on the thematic boundaries of psychology, see Anne C. Rose, "Animal

Tales: Observations of the Emotions in American Psychology, 1890–1940," *Journal of the History of the Behavioral Sciences* 48:4 (2012): 304.

19. Nielson to Yerkes, Feb. 22, 1903. For an example of Yerkes's notes, see handwritten and untitled notes, beginning with *Cabomba caroliniana*, series II, box 70, folder 1340, Yerkes Papers.
20. Edwin Holt to Robert M. Yerkes, May 2, 1904, series I, box 26, folder 476, Yerkes Papers. Although Holt kept emotion in the center of his philosophy, he perhaps wisely discontinued empirical work with animals; see Edwin Holt, *The Freudian Wish and Its Place in Ethics* (New York: Henry Holt, 1915), and Bruce Kuklick, *The Rise of American Philosophy: Cambridge, Massachusetts, 1860–1930* (New Haven: Yale University Press, 1977), chap. 22.
21. C. O. Whitman, "A Biological Farm: For the Experimental Investigation of Heredity, Variation and Evolution and for the Study of Life-Histories, Habits, Instinct and Intelligence," *Biological Bulletin* 3:5 (1902): 214 (first and second quotations), 221 (third quotation), 222 (fourth quotation), 221 (fifth quotation), 218 (sixth quotation). Italics in quotations appear in the original texts unless otherwise noted.
22. John B. Watson, "The Need of an Experimental Station for the Study of Certain Problems in Animal Behavior," *Psychological Bulletin* 3:5 (1906): 154 (first and second quotations), 155 (fourth quotation); Robert M. Yerkes, "The Harvard Laboratory of Animal Psychology and the Franklin Field Station," *Journal of Animal Behavior* 4:3 (1914): 181 (third quotation). The relationship of field and laboratory study remained an open question. For a contemporaneous view, see Marjorie Hill Allee's 1937 novel about young women biologists at the University of Chicago, *The Great Tradition* (Boston: Houghton Mifflin), where Merritt wants " 'an outdoor problem' " for her dissertation, whereas Jenny prefers "those standardized creatures known a 'laboratory animals' " (pp. 47, 78). Allee was the wife of the ecologist Warder Clyde Allee, and the plot predictably favored Merritt's research instincts, although the book fairly vetted the issue of field versus laboratory. Robert E. Kohler documents the historical difficulty experienced by American biologists who attempted to apply laboratory methods to field studies during the first half of the twentieth century, in *Landscapes and Labscapes: Exploring the Lab-Field Border in Biology* (Chicago: University of Chicago Press, 2002).
23. Hugo Münsterberg to Robert M. Yerkes, Aug. 1, 1914, series I, box 35, folder 676, Yerkes Papers.
24. See, e.g., Edward L. Thorndike, "Animal Intelligence; an Experimental Study of the Associative Processes in Animals" (1898), in Thorndike, *Animal Intelligence: Experimental Studies* (New York: Macmillan, 1911), pp. 115–16, and Robert M. Yerkes and Charles A. Coburn, "A Study of the Behavior of the Pig *Sus scrofa* by the Multiple Choice Method," *Journal of Animal Behavior* 5:3 (1915): 222. On these methods, see also John C. Burnham, "Thorndike's Puzzle Boxes," *Journal of the History of the Behavioral Sciences* 8:2 (1972): 159–67, and Robert M. Yerkes, James W. Bridges, and Rose S. Hardwick, *A Point Scale for Measuring Mental Ability* (Baltimore: Warwick and York, 1915).
25. L. W. Cole, "Concerning the Intelligence of Raccoons," *Journal of Comparative Neurology and Psychology* 17:3 (1907): 218.
26. Margaret F. Washburn and I. Madison Bentley, "The Establishment of an Association Involving Color-Discrimination in the Creek Chub, *Semotilus atromaculatus*," *Journal of Comparative Neurology and Psychology* 16:2 (1906): 122 (first, second, and third quotations); Margaret Floy Washburn, *The Animal Mind: A Text-Book of Comparative Psychology*, 3d ed. (New York: Macmillan, 1926), p. 307 (fourth quotation).
27. Watson, "Need of an Experimental Station," p. 150. For Whitman's similar expectation, see "Biological Farm," p. 216.

28. John B. Watson and Mary I. Watson, "A Study of the Responses of Rodents to Monochromatic Light," *Journal of Animal Behavior* 3:1 (1913): 11. On Watson's dissections of rat brains performed at birth, see John B. Watson, *Animal Education: An Experimental Study of the Psychological Development of the White Rat, Correlated with the Growth of Its Nervous System* (Chicago: University of Chicago Press, 1903), p. 91. Medical researchers of the period were less interested in heredity than psychologists and hence less likely to sustain colonies of experimental animals. At Fordham University School of Medicine, for example, animals were "killed during the demonstrations or immediately afterwards" (E. E. [Seetle?] to Walter B. Cannon, questionnaire response, May 28, 1909, box 28, folder 344, Cannon Papers).
29. Wallace Craig, "The Expression of Emotion in the Pigeons. I. The Blond Ring-Dove (*Turtur risorius*)," *Journal of Comparative Neurology and Psychology* 19:1 (1909): 30; Robert M. Yerkes, "The Heredity of Savageness and Wildness in Rats," *Journal of Animal Behavior* 3:4 (1913): 288. On the relationship of Whitman and Craig, see also Burkhardt, "Charles Otis Whitman," p. 198.
30. K. S. Lashley and John B. Watson, "Notes on the Development of a Young Monkey," *Journal of Animal Behavior* 3:2 (1913): 114–39.
31. Laboratory notes, Apr. 23, [18]97, Walter B. Cannon, "Notes on the Study of Alimentation by X-Ray, 1896–1897," manuscript, box 180, Cannon Papers (first and second quotations); S. J. Holmes, "Sex Recognition among Amphipods," *Biological Bulletin* 5:5 (1903): 289 (third and fourth quotations), 290 (fifth quotation).
32. Quoted in Donald A. Dewsbury, "Early Interactions between Animal Psychologists and Animal Activists and the Founding of the APA Committee on Precautions in Animal Experimentation," *American Psychologist* 45:3 (1990): 321.
33. "Professor William James on Vivisection," *Boston Transcript*, May 24, 1909, clipping, box 1, folder 5, Cannon Papers. The definitive study of popular humane efforts in the United States is Janet M. Davis, *The Gospel of Kindness: Animal Welfare and the Making of Modern America* (New York: Oxford University Press, 2016), esp. chaps. 1–3.
34. The quoted words appear on each questionnaire form that was completed and returned to Cannon, of which there are nine total in a single folder in the Cannon Papers; see, e.g., S. B. Wolbach, Bender Hygienic Laboratory, Albany Medical College, Dec. 14, 1908, box 28, folder 344.
35. The word appears, e.g., in "Consciousness in Animals," Lecture 30, Notebook 2, Philosophy 13[indecipherable character], 1910–1911, series II, box 89, folder 1725, Yerkes Papers. See also "Extension Course," 1910–1911, notebook, and "Educational Psychology," Philosophy 6, notebook, 1910–1911, in the same folder.
36. William Morton Wheeler, "The Dry Rot of Our Academic Biology" (1923), in William Morton Wheeler, *Foibles of Insects and Men* (New York: Knopf, 1928), pp. 201 (first quotation), 203 (second and third quotations). The essay was Wheeler's presidential address to the American Association of University Professors in December 1922.
37. Quoted in Robert S. Woodworth, "Biographical Memoir of Margaret Floy Washburn, 1871–1939," *National Academy of Sciences Biographical Memoirs* (1948), p. 278, http://www.nasonline.org/publications/biographical-memoirs/alphabetical-listing/memoirs-w.html (accessed June 1, 2019).
38. [Cailey?] Palmer, MD, questionnaire response, Mar. 2, 1909, box 28, folder 344, Cannon Papers.
39. Craig, "Expression of Emotion. I," p. 74 (first quotation); Cole, "Intelligence of Raccoons," p. 236 (second quotation).
40. Watson, *Animal Education*, p. 9 (first and second quotations); Yerkes, "Heredity of Savageness and Wildness," p. 296 (third and fourth quotations).

41. Yerkes, "Educational Psychology," Philosophy 6, Lecture 9, notebook, 1910–1911, series II, box 89, folder 1725, Yerkes Papers (first and second quotations); C. V. Piper to William Morton Wheeler, Mar. 28, 1923, HUGFP 87.10, box 4, folder "B, Regular, 1923," Wheeler Papers (third quotation). Biologists seemed more willing than psychologists to speculate about the relation of domestication and temperament. C. O. Whitman argued in 1898 that domestication made organisms more flexible and hence intelligent, in "Animal Behavior," in Jane Maienschein, ed., *Defining Biology: Lectures from the 1890s* (Cambridge: Harvard University Press, 1986), esp. pp. 267–70. At a later time, both the American H. S. Jennings and Austrian Konrad Lorenz proposed that domestication adversely affects the gene pool by permitting survival of the less fit, leading to a weaker population. See Jennings, *The Biological Basis of Human Nature* (New York: Norton, 1930), p. 329, and Theodora J. Kalikow, "Konrad Lorenz's Ethological Theory: Explanation and Ideology, 1938–1943," *Journal of the History of Biology* 16:1 (1983): 39–73. On the slow and uneven conversion of scientists to Mendel's ideas, see Charlotte Sleigh, "'The Ninth Mortal Sin': The Lamarckism of W. M. Wheeler," in Abigail Lustig et al., eds., *Darwinian Heresies* (Cambridge: Cambridge University Press, 2004), pp. 151–72; Gregg Mitman, *The State of Nature: Ecology, Community, and American Social Thoughts, 1900–1950* (Chicago: University of Chicago Press, 1992), esp. pp. 28–47; and Ernst Mayr, *The Growth of Biological Thought: Diversity, Evolution, and Inheritance* (Cambridge: Harvard University Press, 1982), chaps. 16–18.
42. Robert M. Yerkes, *Chimpanzees: A Laboratory Colony* (New Haven: Yale University Press, 1943), p. 1.
43. Josiah Royce to Charles William Eliot, May 7, 1906, quoted in John M. O'Donnell, *The Origins of Behaviorism: American Psychology, 1870–1920* (New York: New York University Press, 1985), p. 193 (first quotation); Josiah Royce to Robert M. Yerkes, Dec. 13, 1912, series I, box 42, folder 806, Yerkes Papers (remaining quotations). For Royce's 1916 effort on Yerkes's behalf, see Royce to Dr. Walcott, June 10, 1916, series I, box 42, folder 806, Yerkes Papers. On contention in the behavioral sciences at this time, see Anne C. Rose, "The Invention of Uncertainty in American Psychology: Intellectual Conflict and Rhetorical Resolution, 1890–1930," *History of Psychology* 14:4 (2011): 356–82.
44. James McKeen Cattell to William Cattell (his father), Mar. 9, 1883, quoted in *An Education in Psychology: James McKeen Cattell's Journal and Letters from Germany and England, 1880–1888*, ed. Michael M. Sokal (Cambridge: MIT Press, 1981), p. 67. For his reflections on suicide, see his journal, e.g., April 8, 1883, *Education in Psychology*, p. 72.
45. Saul Benison, A. Clifford Barger, and Elin L. Wolfe, *Walter B. Cannon: The Life and Times of a Young Scientist* (Cambridge: Harvard University Press, 1987), pp. 17–26; Robert M. Yerkes, "Robert Mearns Yerkes," in Carl Murchison, ed., *A History of Psychology in Autobiography* (Worcester: Clark University Press, 1932), 2: 386–88. On Watson, who grew up in rural South Carolina as a widow's son and yet graduated from Furman University before entering the University of Chicago, see Watson, "Watson," in Murchison, ed., *Psychology in Autobiography* 3: 271–73.
46. Benjamin Franklin, *The Autobiography of Benjamin Franklin* (1791), ed. Leonard W. Labaree et al. (New Haven: Yale University Press, 1964), pp. 146 (first quotation), 145 (second quotation); William Morton Wheeler, "Carl Akeley's Early Work and Environment" (1927), quoted in G. H. Parker, "Biographical Memoir of William Morton Wheeler, 1865–1937," *National Academy of Sciences Biographical Memoirs* (1938), p. 207, http://www.nasonline.org/publications/biographical-memoirs/memoir-pdfs/wheeler-william.pdf (accessed June 1, 2019) (third and fourth quotations); Dec. 25, 1914, diary "W. M. Wheeler, Trip to New Zealand and Australia, July 12th, 1914-Jan. 16th, 1915," HUGFP 87.45, box 2, Wheeler Papers (fifth and sixth quotations); Ada Yerkes to Robert M. Yerkes, Sept. 26, 1920, series VI, box 159, folder 2494, Yerkes Papers (seventh quotation).

47. Edward Lee Thorndike, "Edward Lee Thorndike," in Murchison, ed., *Psychology in Autobiography* 3: 267 (first and second quotations); Watson, "Watson," in Murchison, ed., *Psychology in Autobiography*, 3: 275 (third quotation). Karl Lashley (1890–1958), who identified himself as a psychologist, exemplified the heterogeneous training that was commonplace before disciplinary boundaries became fairly definite: he earned a bachelors degree in comparative histology, masters in bacteriology, and doctorate in genetics. See Nadine M. Weidman, *Constructing Scientific Psychology: Karl Lashley's Mind-Brain Debates* (Cambridge: Cambridge University Press, 1999), p. 18.
48. Nice, *Research Is a Passion*, pp. 27 (first quotation), 35 (second quotation).
49. Thorndike, "Thorndike," in Murchison, ed., *Psychology in Autobiography* 3: 267. The book he dedicated to James was *The Original Nature of Man* (New York: Teachers College, Columbia University, 1913).
50. Frank M. Chapman, "Ornithology at the World's Fair," *Auk* 10:4 (1893): 319 (first quotation), 316 (second quotation).
51. Robert Cushman Murphy, "Frank Michler Chapman, 1864–1945," *Auk* 67 (1950): 307. Chapman's later writings gave full play to his literary flair. See esp. *My Tropical Air Castle: Nature Studies in Panama* (New York: D. Appleton, 1929), and *Life in an Air Castle: Nature Studies in the Tropics* (New York: Appleton-Century, 1938).
52. Frank M. Chapman, "On the Winter Distribution of the Bobolink (*Dolichonyx oryzivorus*) with Remarks on Its Routes of Migration," *Auk* 7:1 (1890): 39–45. On the relationship of popular and scientific ornithology, see Mark V. Barrow, Jr., *A Passion for Birds: American Ornithology after Audubon* (Princeton: Princeton University Press, 1998), as well as a memoir extract by later University of Wisconsin ornithologist John T. Emlem, Jr., about his summer work as a college student with Chapman and others at the American Museum of Natural History in 1929, in Sumner W. Matteson, "John T. Emlen, Jr.: A Naturalist for All Seasons, Part I: The Making of a Naturalist-Ornithologist, 1908–1934," *The Passenger Pigeon* 60:2 (1998): 153. For a sense of Chapman's cultural authority, see Theodore Roosevelt, *Through the Brazilian Wilderness* (New York: Scribners, 1922), p. 8, and Nice, *Research Is a Passion*, pp. 31–32, 47, 61, 70, 112–35.
53. The story of Washburn's struggle for graduate training appears in Woodworth, "Washburn," *National Academy of Sciences Memoirs*, p. 276. Woodworth's bibliography of her publications (pp. 287–95) indicates her choice of animal subjects. *The Animal Mind* was published in 1908, 1917, 1926, and 1936, in all cases by Macmillan; it was reprinted at intervening dates as well.
54. Examples of these techniques include Holmes, "Amphipods," pp. 288–89 (observation of natural behavior as a baseline); Small, "Mental Processes of the Rat," p. 3 (laboratory log); Porter, "English Sparrow," p. 336 (drawings of shapes); Yerkes and Coburn, "Behavior of the Pig," figs. 1–2 (photos and drawing of apparatus) and pp. 191–92, 196–97, 200–05, 208–12, 214–21 (quantified results). Even Yerkes's exhaustive reporting did not always satisfy a later generation. In the 1930s Calvin P. Stone justified his reexamination of wild and laboratory rats by claiming that Yerkes's 1913 paper, "Heredity of Savageness and Wildness," did not include "tabular data" for some of the genetic variants Yerkes used. See Calvin P. Stone, "Wildness and Savageness in Rats of Different Strains," in Karl S. Lashley, ed., *Studies in the Dynamics of Behavior* (Chicago: University of Chicago Press, 1932), p. 5.
55. Yerkes and Coburn, "Behavior of the Pig," contained only three citations in notes on p. 185 (including two early publications by the paper's authors) and no bibliography.
56. The quotation appears in Burkhardt, "Charles Otis Whitman," p. 193, as does the story about Watson, p. 208.
57. Wallace Craig, "The Expression of Emotion in the Pigeons. III. The Passenger Pigeon (*Ectopistes migratorius* Linn.)," *Auk* 28:4 (1911): 408 (quotations). Craig's dissertation was

"Expression of Emotion. I." See also his "The Voices of Pigeons Regarded as a Means of Social Control," *American Journal of Sociology* 14:1 (1908): 86–100. Craig's connection with the sociology periodical was probably the University of Chicago, where the journal was established in 1896 and where Craig earned his PhD in biology in 1908.

58. Craig, "Expression of Emotion. III," p. 408.
59. Craig to C. C. Adams, May 24, 1913, quoted in Burkhardt, "Charles Otis Whitman," p. 200.
60. [Robert M. Yerkes], "Announcement of the Journal of Animal Behavior and the Animal Behavior Monograph Series," p. 1, typescript, HUGFP 87.10, box 39, folder "Robert M. Yerkes, 1908–1910," Wheeler Papers. On the establishment of the journal as a professional milestone, see Richard W. Burkhardt, Jr., "The *Journal of Animal Behavior* and the Early History of Animal Behavior Studies in America," *Journal of Comparative Psychology* 101:3 (1987): 223–30, and Donald A. Dewsbury, "Triumph and Tribulation in the History of American Comparative Psychology," *Journal of Comparative Psychology* 106:1 (1992): 3–19. See also a critique of the subsequent thematic narrowing of behavioral studies of animals authored by Frank A. Beach, "The Snark was a Boojum" (1950), in Dewsbury, ed., *Foundations of Comparative Psychology*, pp. 61–70, and Dewsbury's response to Beach, Donald A. Dewsbury, "Animal Psychology in Journals, 1911–1927: Another Look at the Snark," *Journal of Comparative Psychology* 112:4 (1998): 400–05.
61. Yerkes, "Harvard Laboratory and Franklin Field Station"; Robert M. Yerkes, "Recent Progress and Present Tendencies in Comparative Psychology," *Journal of Abnormal Psychology* 2:6 (1908): 271–79.
62. [Indecipherable name], Harper and Brothers, to Robert M. Yerkes, Apr. 17, 1916, series VI, box 161, folder 2520, Yerkes Papers (quotations). See also an earlier letter from the same correspondent, Apr. 8, 1916, as well as a letter from "the editor" of the *Atlantic* to Robert M. Yerkes, Aug. 14, 1916, same folder. Accompanying manuscripts are Ada Watterson Yerkes, "Animal Individuality and Genius," 14-page typescript, and "How Ideas Look: A Glimpse of the Monkey Mind, the Ideas of Monkeys and Apes," 18-page typescript, same folder. A slightly altered version of the first paper appeared as Robert M. Yerkes and Ada W. Yerkes, "Individuality, Temperament, and Genius in Animals," *American Museum Journal* 17:4 (1917): 235–43.
63. John B. Watson, "How Animals Find Their Way Home," *Harper's Monthly Magazine* 119:713 (1909): 685; Watson, "Recent Experiments," in Dewsbury, ed., *Foundations of Comparative Psychology*, p. 78. For helpful background on the relationship of popular and professional nature study, see Charlotte Sleigh, *Six Legs Better: A Cultural History of Myrmecology* (Baltimore: Johns Hopkins University Press, 2007), chaps. 5–6, and Kevin C. Armitage, *The Nature Study Movement: The Forgotten Popularizer of America's Conservation Ethic* (Lawrence: University Press of Kansas, 2009).
64. Hugo Münsterberg to Robert M. Yerkes, series I, box 35, folder 677, Yerkes Papers.
65. Edwin Holt to Robert M. Yerkes, Dec. 16, [19]02, series I, box 26, folder 476 (first quotation); Hugo Münsterberg to Robert M. Yerkes, July 23, [19]09, series I, box 35, folder 676 (second and third quotations); Robert M. Yerkes to Hugo Münsterberg, Dec. 18, 1911, series I, box 35, folder 677 (fourth and fifth quotations); Hugo Münsterberg to Robert M. Yerkes, Dec. 19, 1911, series I, box 35, folder 677 (sixth and seventh quotations), Yerkes Papers.
66. Edwin Holt to Robert M. Yerkes, Sept. 13, 1909, series I, box 26, folder 476, Yerkes Papers.
67. Selma Münsterberg to Roberta Yerkes, Jan. 11, 1910, series I, box 35, folder 676 (first quotation); Selma Münsterberg to Robert M. Yerkes, Feb. 28, [1917], series I, box 35, folder 678 (second and third quotations), Yerkes Papers.
68. [James McKeen Cattell], "An Apology for My Life at Columbia," typescript, box 185, folder "Article File," James McKeen Cattell Papers, Library of Congress, pp. 4 (first quotation), 1

(second quotation), 3 (third quotation). See also Michael M. Sokal, "James McKeen Cattell, Nicholas Murray Butler, and Academic Freedom at Columbia University, 1902–1923," *History of Psychology* 12:2 (2009): 87–122.

69. The comparison of Harvard faculty salaries and the $10,000 annual salary of the football coach appears in Ralph Barton Perry, "Memorandum on Compulsory Physical Training at Harvard," undated typescript, HUG 4683.76, Ralph Barton Perry Papers, Harvard University Archives. On Thorndike, see Geraldine Jonçich, *The Sane Positivist: A Biography of Edward L. Thorndike* (Middletown, Conn.: Wesleyan University Press, 1968), chaps. 6–8. Note that Thorndike's interest in animals did not end abruptly. He served on the founding editorial board of the *Journal of Animal Behavior* in 1910, a position that entailed not only advisory support but also an initial financial commitment of one hundred dollars. See Robert M. Yerkes "To Members of the Editorial Board," Oct. 28, 191[0], HUGFP 87.10, box 39, folder "Robert M. Yerkes, 1908–1910," Wheeler Papers, and "Announcement," p. 2.
70. See Anne C. Rose, "An American Science of Feeling: Harvard's Psychology of Emotion during the World War I Era," *Journal of the History of Ideas* 73:3 (2012): 496–99.
71. On the Franklin community, see Yerkes, "Harvard Laboratory and Franklin Field Station," p. 181; Benison et al., *Cannon: Young Scientist*, p. 210; Marian Cannon Schlesinger, *Snatched from Oblivion: A Cambridge Memoir* (Boston: Little Brown, 1979), pp. 123–25; Yerkes and Coburn, "Behavior of the Pig"; Charles A. Coburn and Robert M. Yerkes, "A Study of the Behavior of the Crow *Corvus Americanus* Aud. by the Multiple Choice Method," *Journal of Animal Behavior* 5:2 (1915): 75–114; and Robert M. Yerkes and Blanche W. Learned, *Chimpanzee Intelligence and Its Vocal Expressions* (Baltimore: Williams and Wilkins, 1925). The Cannons seem to have been the first to discover Franklin in 1909 (Benison et al., p. 210). Learned was a musician and neighbor who collaborated with Yerkes on primate vocalization.
72. For a summary of "golden age" arguments, see Burkhardt, "*Journal of Animal Behavior* in America," p. 223.
73. "Lowell Institute, 1914–15, Lectures on Bodily Changes under Emotional Excitement by Walter B. Cannon, MD," loose notes, box 180, Cannon Papers; on Watson and Lashley, see Weidman, *Constructing Scientific Psychology*, pp. 32–40.
74. Robert M. Yerkes, "Objective Nomenclature, Comparative Psychology and Animal Behavior," *Journal of Comparative Neurology and Psychology* 16:5 (1906): 383.
75. Herbert Spencer, *The Principles of Psychology* (1855; reprint ed., New York: D. Appleton, 1890), 1: 182 (first quotation); Small, "Mental Processes of the Rat," p. 154 (second and third quotations).
76. George John Romanes, *Mental Evolution in Animals* (1884; reprint ed., New York: AMS Press, 1969), pp. 51 (first and second quotations), 52 (third quotation), 53 (fourth quotation); Edward Burnett Tylor, *The Origins of Culture* (1871; reprint ed., New York: Harper and Brothers, 1958), p. 187 (fifth quotation); William James, "The Feelings of Effort" (1884), in *The Works of William James*, electronic edition, ed. Frederick H. Burkhardt, Fredson Bowers, and Ignas K. Skrupskelis (Cambridge: Harvard University Press, 1975–1988), 13: 83–124 (hereafter *James Works*). Although the emotionalism, and oftentimes worry about excessive emotion, of nineteenth-century Anglo-American culture is a frequent scholarly premise, systematic analyses of the Victorian temper are less common. For suggestive background, see Thomas Dixon, *From Passions to Emotions: The Creation of a Secular Psychological Category* (Cambridge: Cambridge University Press, 2003), esp. chaps. 4–7; George W. Stocking, Jr., "Anthropology and the Science of the Irrational: Malinowski's Encounter with Freudian Psychoanalysis," in Stocking, ed., *Malinowski, Rivers, Benedict and Others: Essays on Culture and Personality* (Madison: University of Wisconsin Press, 1986), pp. 13–49; Walter E. Houghton, *The Victorian Frame of Mind, 1830–1870* (New Haven: Yale University Press,

1957); and Anne C. Rose, *Victorian America and the Civil War* (Cambridge: Cambridge University Press, 1992), esp. pp. 7–9.

77. John B. Watson, "Psychology as the Behaviorist Views It," *Psychological Review* 20:2 (1913): 159 (first quotation), 163 (second quotation); 161 (third quotation); 163 (fourth quotation).
78. Watson, "Need of an Experiment Station," p. 152.
79. Small, "Mental Processes of the Rat," p. 133.
80. Thorndike, *Animal Intelligence*, pp. 63 (first quotation), 64 (second quotation); Porter, "English Sparrow," p. 317 (third and fourth quotations); Cole, "Intelligence of Raccoons," p. 216 (fifth quotation); William Morton Wheeler, "A Solitary Wasp (*Aphilanthops frigidus* F. Smith) that Provisions Its Nest with Queen Ants," *Journal of Animal Behavior* 3:5 (1913): 378 (remaining quotations).
81. Small, "Mental Processes of the Rat," p. 137 (first and second quotations); Porter, "English Sparrow," p. 325 (third quotation).
82. Craig, "Expression of Emotion. III," p. 425 (on Whitman's pigeon feed); Morgan, *Comparative Psychology*, p. 207.
83. Washburn and Bentley, "Creek Chub," pp. 115 (first quotation), 125 (second quotation); Thorndike, *Animal Intelligence*, pp. 62 (third quotation), 64 (fourth quotation); Wesley Mills, "The Nature of Animal Intelligence and the Methods of Investigating it," *Psychological Review* 6:3 (1899), p. 265 (fifth and sixth quotations).
84. Milk and milk-soaked bread were the common food rewards for rats; see, e.g., Small, "Mental Processes of the Rat," pp. 138, 144, and Watson, *Animal Education*, p. 13. Porter fed corn to his sparrows ("English Sparrow," p. 318), and Cole added apples and sugar lumps for his raccoons ("Intelligence of Raccoons," p. 212).
85. Robert M. Yerkes and John D. Dodson, "The Relation of Strength of Stimulus to Rapidity of Habit Formation," *Journal of Comparative Neurology and Psychology* 18:5 (1908): 459–82; Norman R. F. Maier, *Studies of Abnormal Behavior in the Rat: The Neurotic Pattern and An Analysis of the Situation Which Produces It* (New York: Harper and Brothers, 1939), p. 21.
86. Porter, "English Sparrow," p. 335 (first quotation); Alexander Bain, *The Emotions and the Will* (1859), ed. Daniel N. Robinson (Washington, D.C.: University Publications of America, 1977), p. 32 (second quotation), 33 (third quotation); Small, "Mental Processes of the Rat," p. 40 (fourth and fifth quotations).
87. W. H. R. Rivers, *Instinct and the Unconscious: A Contribution to a Biological Theory of the Psycho-Neuroses* (Cambridge: Cambridge University Press, 1922), p. 5. For an excellent discussion of Rivers as an exponent of what Henrika Kuklick's identifies as the diffusionist school of British anthropology, see H. Kuklick, *The Savage Within: The Social History of British Anthropology, 1885–1945* (Cambridge: Cambridge University Press, 1991), chap. 4.
88. Holmes, "Amphipods," p. 289 (first and second quotations); William Morton Wheeler, "The Ant Colony as an Organism" (1910), in Wheeler, *Foibles*, p. 133 (third quotation); Wallace Craig, "Male Doves Reared in Isolation," *Journal of Animal Behavior* 4:2 (1914): 125 (fourth quotation). Wheeler did think about possible sexual feelings in insects, albeit inconclusively, in an undated lecture for his Harvard course, Zoology 7b. "Intensity of pairing instinct" in insects, he informed students, was connected experimentally to their "apparent insensibility to pain." "It is, however, possible that this insensibility to pain may characterize the whole life of an insect" (HUGFP 87.65, box 1, folder "Courtship in Insects," Wheeler Papers).
89. Robert M. Yerkes, "Inhibition and Reinforcement of Reaction in the Frog *Rana clamitans*," *Journal of Comparative Neurology and Psychology* 14:2 (1904): 137.
90. Washburn, *Animal Mind*, 2d ed., pp. 1 (first and second quotations), 4 (third quotation).
91. Adele M. Fielde, "A Cause of Feud between Ants of the Same Species Living in Different Communities," *Biological Bulletin* 5:6 (1903): 329.

92. Craig, "Male Doves," p. 121; Cole, "Intelligence of Raccoons," p. 251.
93. Small, "Mental Processes of the Rat," p. 157; Cannon, "Dietetics in Disease," p. 72, lecture, post-1903, manuscript and typescript pages, box 176, Cannon Papers.
94. David L. Krantz and David Allen, "The Rise and Fall of McDougall's Instinct Doctrine," *Journal of the History of the Behavioral Sciences* 3:4 (1967): 326–38. The return of the idea of purpose is a subject of Chapter 2.
95. William James, *The Principles of Psychology*, vol. 1, in *James Works*, 8: 219 (first quotation); "Does Consciousness Exist?" in *James Works*, 3: 20 (second and third quotations).
96. Robert M. Yerkes, *Introduction to Psychology* (New York: Henry Holt, 1911), pp. 184 (quotation), 409–17 (eugenics). On Yerkes's interest in eugenics and connection with Davenport, see also Rose, "American Science of Feeling," esp. 498–99.
97. William McDougall, *An Introduction to Social Psychology* (1908; reprint ed., London: Methuen, 1960), p. 40.
98. Yerkes, *Introduction to Psychology*, p. 382 (first quotation); McDougall, *Social Psychology*, p. 14 (second quotation); Walter B. Cannon, *Bodily Changes in Pain, Hunger, Fear and Rage: An Account of Recent Researches into the Function of Emotional Excitement* (1915; reprint ed., New York: D. Appleton, 1920), p. 188 (third quotation).
99. Walter B. Cannon to Cannon family, May 28, 1917, box 185, letter book 1, Cannon Papers. Yerkes's final letter to McDougall was dated Jan. 30, 1932, series I, box 32, folder 616, Yerkes Papers.
100. Watson, "Watson," in Murchison, ed., *Psychology in Autobiography*, 3: 273.
101. Watson, *Animal Education*, pp. 49 (first quotation), 62 (second and third quotations), 26 (fourth and fifth quotations), 29 (sixth quotation), 107 (seventh quotation).

Chapter 2

1. William Morton Wheeler, "The Termitodoxa, or Biology and Society," *Scientific Monthly* 10:2 (1920): 124 (first quotation), 121 (second and third quotations), 122 (fourth quotation), 115 (fifth quotation), 118 (sixth quotation), 117 (seventh quotation).
2. Wheeler, "Termitodoxa," pp. 116 (first quotation), 115 (second, third, and fourth quotations), 113 (fifth quotation).
3. W. H. R. Rivers, "Psychology and the War," Address to the British Association [for the Advancement of Science], Sept. 1919, in W. H. R. Rivers, *Instinct and the Unconscious: A Contribution to a Biological Theory of the Psycho-Neuroses*, 2d ed. (Cambridge: Cambridge University Press, 1922), pp. 252 (first quotation), 248 (second quotation), 252 (third quotation); Ralph Barton Perry, *The Present Conflict of Ideals: A Study of the Philosophical Background of the World War* (1918; reprint ed., New York: Longmans, Green, 1922), p. 64 (fourth and fifth quotations).
4. Diary, Aug. 17, 1917, "Journal 1917–1918, Especially Important Events in the Life History of Robert Mearns Yerkes," p. 92, series VI, box 171, folder 2663, Robert Mearns Yerkes Papers (MS 569), Manuscripts and Archives, Yale University Library, New Haven, Conn. (first quotation); W. McDougall, "Four Cases of 'Regression' in Soldiers," *Journal of Abnormal Psychology* 15:2–3 (1920): 136 (second quotation), 156 (third quotation); Wheeler, "Termitodoxa," p. 114 (fourth quotation). The social experience of war of the animal behavior scientists studied here conformed to the muted engagement of most Americans, as explained by David M. Kennedy, *Over Here: The First World War and American Society* (1980; reprint ed., Oxford: Oxford University Press, 2004). Jennifer D. Keene surveys enduring American ambivalence toward World War I, in "Remembering the 'Forgotten War': American Historiography on World War I," *Historian* 78:3 (2016): 439–68. Her

account of historical questions, which have mainly focused on economics and politics, helps explain the dearth of work on American science and the war. Daniel J. Kevles cites the war as a catalyst of scientific nationalism, in "'Into Hostile Political Camps': The Reorganization of International Science in World War I," *Isis* 62:1 (1971): 47–60. More generally, Carol Gruber analyzes the limits of academic freedom during wartime, in *Mars and Minerva: World War I and the Uses of Higher Learning in America* (Baton Rouge: Louisiana State University Press, 1975). British science has been more comprehensively examined; see, e.g., Mark Harrison, *The Medical War: British Military Medicine in the First World War* (Oxford: Oxford University Press, 2010), and Anne Hardy, "'Straight Back to Barbarism': Anti-Typhoid Inoculation and the Great War, 1914," *Bulletin of the History of Medicine* 74:2 (2000): 265–90.

5. See these patterns of publications in bibliographies, Ernest R. Hilgard, "Robert Mearns Yerkes, 1876–1956," *National Academy of Sciences Biographical Memoirs* (1965), pp. 418–20, http://www.nasonline.org/publications/biographical-memoirs/memoir-pdfs/yerkes-robert-m.pdf (accessed June 1, 2019); Robert S. Woodworth, "Biographical Memoir of Margaret Floy Washburn, 1871–1939," *National Academy of Sciences Biographical Memoirs* (1948), pp. 290–93, http://www.nasonline.org/publications/biographical-memoirs/memoir-pdfs/washburn-margaret.pdf (accessed June 1, 2019). The books that marked Yerkes's return to animal studies were Robert M. Yerkes, *Almost Human* (New York: Century, 1925), and Robert M. Yerkes and Blanche W. Learned, *Chimpanzee Intelligence and Its Vocal Expressions* (Baltimore: Williams and Wilkins, 1925).
6. K. S. Lashley, "A Simple Maze: With Data on the Relation of the Distribution of Practice to the Rate of Learning," *Psychobiology* 1:5 (1918): 362 (first quotation), 360 (second quotation).
7. Lashley, "A Simple Maze," pp. 353 (first and second quotations), 357 (third and fourth quotations), 353 (fifth quotation), 357 (sixth quotation), 365 (seventh quotation), 366 (eighth quotation), 364 (ninth quotation). Nadine M. Weidman provides the most thorough study of Lashley's work; for this period, see esp. *Constructing Scientific Psychology: Karl Lashley's Mind-Brain Debates* (Cambridge: Cambridge University Press, 1999), chap. 2.
8. Lashley, "A Simple Maze," p. 355.
9. Charles A. Coburn and Robert M. Yerkes, "A Study of the Behavior of the Crow *Corvus Americanus Aud.* by the Multiple Choice Method," *Journal of Animal Behavior* 5:2 (1915): caption to figure 2 between 78 and 80 (first quotation), 107 (second quotation), 112 (third and fourth quotations). Yerkes used nearly identical apparatus, physically adapted to each species, for pigs and primates; see Robert M. Yerkes and Charles A. Coburn, "A Study of the Behavior of the Pig *Sus scrofa* by the Multiple Choice Method," *Journal of Animal Behavior* 5:3 (1915): 185–225, and Robert M. Yerkes, *The Mental Life of Monkeys and Apes*, ed. George M. Haslerud (1916; reprint ed., Delmar, N.Y.: Scholars' Facsimiles and Reprints, 1979).
10. Robert M. Yerkes, "The Harvard Laboratory of Animal Psychology and the Franklin Field Station," *Journal of Animal Behavior* 4:3 (1914): 176 (first quotation); Marian Cannon Schlesinger, *Snatched from Oblivion: A Cambridge Memoir* (Boston: Little Brown, 1979), p. 123 (second and third quotations).
11. For Yerkes's activities at Boston Psychopathic, see Robert M. Yerkes to Elmer Ernest Southard, Jan. 29, 1914, series I, box 44, folder 859, Yerkes Papers; additional correspondence between Yerkes and Southard in box 44, folders 859–60, Yerkes Papers; and [E. E. Southard] to A. Lawrence Lowell, Sept. 17, 1914, series II, box 57, folder 1088, Yerkes Papers. On the background of Yerkes's California research, see Haslerud, "Introduction," in Yerkes, *Mental Life*, pp. v–xiv.
12. [Robert M. Yerkes,] "To the President and Fellows of Harvard College," typescript, July 1914, series II, box 57, folder 1088, Yerkes Papers (first quotation); Coburn and Yerkes,

"A Study of the Behavior of the Crow," p. 77 (second quotation); Yerkes, "Harvard Laboratory and Franklin Field Station," p. 181 (third and fourth quotations).

13. A full record of Yerkes's nearly single-handed efforts to establish and maintain the *Journal of Animal Behavior* may be found in folders of his correspondence with William Morton Wheeler: "Robert M. Yerkes, 1907–1914," "Robert M. Yerkes, 1908–1910," and "Robert M. Yerkes, 1913–1914," HUGFP 87.10, box 39, William Morton Wheeler Papers, Harvard University Archives, Cambridge, Mass. In future references to these folders, note that they are all located in box 39. Wheeler was a member of the editorial board, which Yerkes routinely consulted. See also Richard W. Burkhardt, Jr., "The *Journal of Animal Behavior* and the Early History of Animal Behavior Studies in America," *Journal of Comparative Psychology* 101:3 (1987): 223–30. On the varied research interests of psychologists, see Donald A. Dewsbury, "Triumph and Tribulation in the History of American Comparative Psychology," *Journal of Comparative Psychology* 106:1 (1992): 5–6.
14. See biographical information in Robert H. Wozniak, "Gilbert Van Tassel Hamilton and *An Introduction to Objective Psychopathology*," 1997, http://www.brynmawr.edu/psychology/rwozniak/gilbert.html (accessed June 1, 2019). For Hamilton's work at McLean Hospital, see G. Van T. Hamilton, "An Experimental Study of an Unusual Type of Reaction in a Dog," *Journal of Comparative Neurology and Psychology* 17:4 (1907): 329–41, as well as Hamilton and Yerkes's recollections of their early relationship, in G. V. Hamilton, *An Introduction to Objective Psychopathology* (St. Louis: C. V. Mosby, 1925), pp. 8 (Hamilton), 13 (Yerkes). For Hamilton's work with primates, see G. V. Hamilton, "A Study of Sexual Tendencies in Monkeys and Baboons," *Journal of Animal Behavior* 4:5 (1914): 295–318.
15. Robert M. Yerkes to William Morton Wheeler, Apr. 1, 1908, folder "Robert M. Yerkes, 1908–1910" (quotation); Robert M. Yerkes to editorial board, Jan. 27, 1915, folder "Robert M. Yerkes, 1907–1914" (announcement of acting editor), Wheeler Papers. For a sample of their exchanges of scientific views, see William Morton Wheeler to Robert M. Yerkes, Feb. 4, 1911, and Robert M. Yerkes to William Morton Wheeler, May 4, 1910, folder "Robert M. Yerkes, 1908–1910," Wheeler Papers.
16. Mary Alice Evans and Howard Ensign Evans, *William Morton Wheeler, Biologist* (Cambridge: Harvard University Press, 1970), chap. 13; Richard W. Burkhardt, Jr., "Charles Otis Whitman, Wallace Craig, and the Biological Study of Animal Behavior in the United States, 1898–1925," in Ronald Rainger, Keith R. Benson, and Jane Maienschein, eds., *The American Development of Biology* (Philadelphia: University of Pennsylvania Press, 1988), p. 194.
17. Emerson Hall in Cambridge, where Yerkes's office was located, was sufficiently far from Wheeler's office at the Bussey Institution in Jamaica Plain for the men to note in writing when they missed each other's visits; see Robert M. Yerkes to William Morton Wheeler, Mar. 26, 1910, folder "Robert M. Yerkes, 1908–1910," and William Morton Wheeler to Robert M. Yerkes, Mar. 2, 1914, folder "Robert M. Yerkes, 1907–1914." On the unusual larvae, see Diary, Aug. 14, Aug. 25, Sept. 7, Sept. 24, 1917, and passim, HUGFP 87.45, box 3, folder "Western Trip June 28 to Sept. 6, 1917 (and Other Notes, 1918–1919)," Wheeler Papers. Typical notes on travel arrangements for the 1917 trip may be found in the same diary, July 9 (arrived at railroad station), July 15 (purchased groceries and auto tires for camping), July 19 (ate in a Chinese restaurant), Aug. 7 (traveled by Buick), and Aug. 14 (visited family in San Diego). See similar information in other travel diaries, HUGFP 87.45, boxes 1–5, Wheeler Papers. For an excellent general discussion of Wheeler's scientific ideas, see Charlotte Sleigh, *Six Legs Better: A Cultural History of Myrmecology* (Baltimore: Johns Hopkins University Press, 2007), pt. 2.

18. On these academic conflicts, see Anne C. Rose, "The Invention of Uncertainty in American Psychology: Intellectual Conflict and Rhetorical Resolution, 1790–1930," *History of Psychology* 14:4 (2011): 356–82, and Michael M. Sokal, "James McKeen Cattell, Nicholas Murray Butler, and Academic Freedom at Columbia University, 1902–1923," *History of Psychology* 12:2 (2009): 87–122.
19. Edward Chace Tolman, *Purposive Behavior in Animals and Men* (1932; reprint ed., no place: Appleton, Century, Crofts, 1967); Edward Chace Tolman autobiography, quoted in Benbow F. Ritchie, "Edward Chace Tolman, 1886–1959," *National Academy of Sciences Biographical Memoirs* (1964), pp. 294, 297, http://www.nasonline.org/publications/biographical-memoirs/memoir-pdfs/tolman-edward.pdf (accessed June 1, 2019).
20. Apsley Cherry-Garrard, *The Worst Journey in the World* (1922; reprint ed., New York: Carroll and Graf, 1989), p. 564 (first, second, and third quotations); Theodore Roosevelt, *Through the Brazilian Wilderness* (1914; reprint ed., New York: Scribners, 1922), preface [n.p.] (fourth quotation), p. 347 (fifth and sixth quotations).
21. Edgar Rice Burroughs, *Tarzan of the Apes* (New York: Buccaneer Books, 1914), pp. 146 (first quotation), 18 (second quotation), 16 (third quotation), 107 (fourth quotation). Behind this and similar works of fiction stand persistent popular theories of a natural hierarchy of human races, an application of evolutionary ideas akin to the primate-human comparison here. See William Stanton, *The Leopard's Spots: Scientific Attitudes toward Race in America, 1815–1859* (Chicago: University of Chicago Press, 1960); Carl N. Degler, *In Search of Human Nature: The Decline and Revival of Darwinism in American Social Thought* (New York: Oxford University Press, 1991); and Elazar Barkan, *The Retreat of Scientific Racism: Changing Concepts of Race in Britain and the United States between the World Wars* (Cambridge: Cambridge University Press, 1992).
22. Hamilton, "Sexual Tendencies," pp. 296 (quotations), 295 (reference to Freud).
23. Robert M. Yerkes, "Maternal Instinct in a Monkey," *Journal of Animal Behavior* 5:5 (1915): 403–05; Yerkes, *Mental Life*; [Robert M. Yerkes] to E. E. Southard, May 29, 1916, series I, box 44, folder 860, Yerkes Papers (first, second, and third quotations); Diary, Jan. 17, 1918, "Journal 1917–1918," p. 100, Yerkes Papers (fourth quotation).
24. William McDougall, *An Introduction to Social Psychology* (1908; reprint ed., London: Methuen, 1960), chap. 11. Although Darwin explained natural selection as an impersonal interaction between species variations and environment—thereby anticipating modern evolutionary understanding—he remained enough of a Victorian to consider animal willfulness in survival as well. As background to human war-making, see his discussion of the evolutionary uses of "Hatred and Anger," in *The Expression of the Emotions in Man and Animals* (1872; reprint ed., Chicago: University of Chicago Press, 1965), chap. 10.
25. Ralph Barton Perry, *The Plattsburg Movement: A Chapter in America's Participation in the World War* (New York: Dutton, 1921), p. 25; Robert M. Yerkes, "Psychology in Relation to the War," *Psychological Review* 25:2 (1918): esp. 85–87. Perry clearly articulated his goals for the military training of Harvard students, in "Memorandum on Compulsory Physical Training at Harvard," typescript, [ca. 1915–1916], HUG 4683.76, Ralph Barton Perry Papers, Harvard University Archives. He was asked to write the history of the Plattsburg military preparedness movement; see Grenville Clark to Ralph Barton Perry, Jan. 16, 1920, HUG 4683.75, Perry Papers. For background on the formation of the National Research Council in 1916, which drew scientists closer to the war effort, see Gregg Mitman, *The State of Nature: Ecology, Community, and American Social Thought, 1900–1950* (Chicago: University of Chicago Press, 1992), esp. pp. 58–59.
26. Robert M. Yerkes to E. E. Southard, May 23, 1917, series I, box 44, folder 860, Yerkes Papers (first quotation); Diary, Dec. 30, 1918, "Journal 1917–1918," p. 172, Yerkes Papers (second quotation). Perry's lectures were published as *The Present Conflict of Ideals*.

27. Paul Fussell, *The Great War and Modern Memory* (New York: Oxford, 1975), p. 180 (first quotation); Diary, Aug. 17, 1917, "Journal 1917–1918," p. 90, Yerkes Papers (second quotation).
28. Walter B. Cannon to Cornelia Cannon, Oct. 8, 1918, box 185, Walter B. Cannon Papers (H MS c40), Center for the History of Medicine, Francis A. Countway Library of Medicine, Harvard Medical School, Boston, Mass.
29. For excellent discussions of semipublic readings of Cannon's letters and also his wife's publications plans, see Elin L. Wolfe, A. Clifford Barger, and Saul Benison, *Walter B. Cannon: Science and Society* (Boston: Boston Medical Library in the Francis A. Countway Library of Medicine, 2000), p. 50, and Maria I. Diedrich, *Cornelia James Cannon and the Future American Race* (Amherst: University of Massachusetts Press, 2010), esp. pp. 97, 103–04.
30. Cannon Schlesinger, *Snatched from Oblivion*, p. 63.
31. Cannon Schlesinger, *Snatched from Oblivion*, pp. 63 (first quotation), 88 (second quotation).
32. Walter B. Cannon to Cornelia Cannon, June 10, 1917 (first quotation); Sept. 28, 1918 (second quotation); Sept. 1, 1918 (third quotation); [Aug. 16, 1917] (fourth quotation); July 4, 1918 (fifth quotation); [Aug. 16, 1917] (sixth quotation); Sept. 28, 1918 (seventh quotation), box 185; untitled and undated notes, box 136, folder 1936 (eighth quotation), Cannon Papers. Cannon wrote to Cornelia that the officers in the Harvard medical unit, then on shipboard, had agreed to abstain from alcohol for the duration; see May 15, 1917, box 185, Cannon Papers.
33. Diary, Aug. 15, 1918, "Journal 1917–1918," p. 131 (first quotation); Oct. 3, 1917, p. 93 (second quotation); Mar. 12, 1918, p. 105 (third quotation); Jan. 4, 1918, p. 99 (fourth quotation); Sept. 28, 1918, p. 145 (fifth quotation); Dec. 22, 1917, p. 96 (sixth quotation); May 9, 1918, p. 113 (seventh quotation), Yerkes Papers.
34. The birth date and year of completion of PhD for each man are as follows: Edward C. Tolman (1886; Harvard University, 1915); Calvin P. Stone (1892; University of Minnesota, 1921); Leonard Carmichael (1898; Harvard University, 1924); and Warder C. Allee (1885; University of Chicago, 1912).
35. Calvin Perry Stone, "Two Years of Service: Being a Reminiscent Account of the Military Service of Captain Calvin P. Stone, Sanitary Corps, U.S.A., Presented to James Herbert Stone on the Event of His First Birthday, Sept. 10, 1919," typescript, Calvin Perry Stone Papers microfilm (SC0440), University Archives, Special Collections, Stanford University Library, Stanford, Calif. (hereafter Stone Papers microfilm). A headnote explains that the typescript was prepared from three-by-four cards, presumably after Stone's death in 1954.
36. Raleigh Webster Stone, "Calvin Perry Stone: Biographical Notes by Raleigh Webster Stone," p. 47, typescript, ca. 1954, Stone Papers microfilm.
37. Calvin Stone, "Two Years of Service," pp. 2 (quotations), 10 (total number of recruits examined), Stone Papers microfilm. Note that Stone was initially released from general military service as a private, but he later achieved the rank of captain in the Sanitary Corps (p. 5).
38. James McKeen Cattell, "An Apology for My Life at Columbia," p. 8, typescript, box 185, folder "An Apology for My Life at Columbia," James McKeen Cattell Papers, Library of Congress, Washington, D.C.
39. W. C. [Warder Clyde] and M. H. [Marjorie Hill] Allee, "Adventures of a College Professor," *American Friend*, Twelfth Month [Dec.] 5, 1919, p. 1133 (first and second quotations); Edward C. Tolman to Richard C. Tolman, Mar. 22, 1918, box 2, folder 2.1, Richard Chace Tolman Papers, Archives, California Institute of Technology, Pasadena, Calif. (remaining quotations). On Richard Tolman's bolder pacifism, see [Richard C. Tolman?], "Academic Freedom at the University of Illinois," typescript, no date, box 2, folder 2.1, Tolman Papers. Allee had already held two one-year positions by the time the war began, and during the conflict he taught at the University of Oklahoma and Lake Forest College, in addition to annual

summer-instructor work at the Marine Biological Laboratory at Woods Hole, Massachusetts. See Karl Patterson Schmidt, "Warder Allee, 1885–1955," *National Academy of Sciences Memoirs* (1957), pp. 6–7, http://www.nasonline.org/publications/biographical-memoirs/memoir-pdfs/allee-warder.pdf (accessed June 1, 2019). Threats at Lake Forest, where he spent most of the war years, included losing salary and being excluded from the chapel speaker series if he voiced his pacifism; see Allee and Allee, "Adventures," p. 1133.

40. W. C. Allee, "The Biology of Peace," *Quaker* 1 (1920): 40. See also W. Clyde Allee, "The Biology of Disarmament," *Quaker* 2 (1921): 173–74.
41. "Minutes of Second Meeting of Psychology Committee of the National Research Council, Held at Hotel Walton, Philadelphia, Thursday, June 28, [1917], 7:30 p.m." typescript, box 153, folder "Edward L. Thorndike," Cattell Papers. Cattell's participation was apparently limited, because he was not a member of the subcommittee formed to plan military psychological testing, which was chaired by Yerkes. See "Report of Subcommittee on Methods of Examining Recruits," July 14, 1917, same folder.
42. Diary, June 28, 1917, HUGFP 87.45, box 3, folder "Western Trip 1917," Wheeler Papers (first quotation); William Morton Wheeler, "The Organization of Research," *Science*, new series, 53:1360 (1921): 62 (second quotation). The latter article was Wheeler's address as chair of the Zoological Science section of the American Association for the Advancement of Science, delivered in Chicago in 1920. Ralph Emerson Wheeler was born in 1901, graduated from Harvard in 1922, and completed Harvard Medical School in 1926.
43. Cattell, "An Apology," p. 7, box 185, folder "An Apology," Cattell Papers. On the contribution of Cattell's son McKeen, see also Walter B. Cannon, *Traumatic Shock* (New York: D. Appleton, 1923), p. x. McKeen (born 1891) received an MA under Cannon at Harvard in 1917 and a PhD there in 1920.
44. Walter B. Cannon to Cornelia Cannon, May 28, 1917, box 185, Cannon Papers; Rivers, *Instinct and the Unconscious*, p. v. The year of Rivers's lectures at Johns Hopkins was 1920. On the shift in American scholarly emphasis during the war from the German tradition to the British, see Perry, *Present Conflict of Ideals*, chap. 35, and Mitman, *State of Nature*, pp. 59–62. For excellent background on these British psychologists, see Henrika Kuklick, *The Savage Within: The Social History of British Anthropology, 1885–1945* (Cambridge: Cambridge University Press, 1991), chap. 4.
45. W. H. R. Rivers, "Psychiatry and the War," *Science*, new series, 49:1268 (1919): 367–69, and W. H. R. Rivers, "Psychology and the War," *Scribners* 68:2 (1920): 161–66. See also W. H. R. Rivers, "Affect in the Dream," *British Journal of Psychology* 12:2 (1921): 113–24; McDougall, "Four Cases of 'Regression'"; William McDougall, *Outline of Abnormal Psychology* (New York: Scribners, 1926), chaps. 7 (dreams), 25 (Freud).
46. A correspondent, C. A. Strong, identified McDougall's new position as "William James's chair" in a letter to McDougall, Dec. 22, 1920, box 1, folder 1, William McDougall Papers, David M. Rubenstein Rare Book and Manuscript Library, Duke University, Durham, N.C. Bruce Kuklick stresses McDougall's connection with Münsterberg, in *The Rise of American Philosophy: Cambridge, Massachusetts, 1860–1930* (New Haven: Yale University Press, 1977), p. 459. Ralph Barton Perry implied that McDougall took over the psychology laboratory from Yerkes, in "Psychology, 1876–1929," in Samuel Eliot Morison, ed., *The Development of Harvard University since the Inauguration of President Eliot, 1869–1929* (Cambridge: Harvard University Press, 1930), pp. 220–21. McDougall most consistently linked himself with James, dedicating his *Outline of Psychology* (New York: Scribners, 1923), for example, to "the honored memory of William James, great philosopher, great psychologist, and great man" (dedication page).
47. Alexander F. Shand, *The Foundations of Character: Being a Study of the Tendencies of the Emotions and Sentiments*, 2d ed. (London: Macmillan, 1920), p. 2 (quotations); Shand's

acknowledgment of McDougall appears in the preface to the first edition, p. ix. To estimate American awareness of Shand and McDougall, I surveyed texts of the war period and immediate postwar years by Walter Cannon, Wallace Craig, G. V. Hamilton, Edwin Holt, Karl Lashley, Ralph Barton Perry, Edward Tolman, Margaret Washburn, William Morton Wheeler, and Robert Yerkes. Although the Americans almost always cited McDougall, only Craig mentioned Shand—and enthusiastically so—in "A Note on Darwin's Work on the Expression of the Emotions in Man and Animals," *Journal of Abnormal and Social Psychology* 16:5–6 (1922): 359. Yerkes's praise of McDougall had appeared in *Introduction to Psychology* (New York: Henry Holt, 1911), p. 382.

48. William McDougall, "William McDougall," in Carl Murchison, ed., *A History of Psychology in Autobiography* (1930; reprint ed., New York: Russell and Russell, 1961), 1: 210–12; see also Major Greenwood and May Smith, "William McDougall, 1871–1938," *Obituary Notices of Fellows of the Royal Society* 3:8 (1940): esp. 49–51. For McDougall's observational records of his children (Angus, Duncan, Kenneth, Lesley, and Janet), see box 11, folders 114–16, McDougall Papers.
49. Shand, *Foundations of Character*, p. xiii (first quotation); Charles S. Myers, *Shell Shock in France, 1914–1918, Based on a War Diary Kept by Charles S. Myers* (Cambridge: Cambridge University Press, 1940), pp. 141 (second quotation), 26 (third quotation).
50. Kuklick, *Rise of American Philosophy*, p. 459 (quotations). I take issue with the image of McDougall as an intractable alien, in "William McDougall, American Psychologist: A Reconsideration of Nature-Nurture Debates in the Interwar United States," *Journal of the History of the Behavioral Sciences* 52:4 (2016): 325–48. The best account of American criticism of McDougall in the 1920s is David L. Krantz and David Allen, "The Rise and Fall of McDougall's Instinct Doctrine," *Journal of the History of the Behavioral Sciences* 3:4 (1967): 326–38. At the same time, some of the same authors offer an insightful and appreciative analysis of McDougall's emphasis on purpose-seeking behavior: David L. Krantz, Richard Hall, and David Allen, "William McDougall and the Problem of Purpose," *Journal of the History of the Behavioral Sciences* 5:1 (1969): 25–38.
51. Clarence S. Yoakum and Robert M. Yerkes, *Army Mental Tests* (New York: Henry Holt, 1920).
52. Wallace Craig, "Appetites and Aversions as Constituents of Instincts," *Biological Bulletin* 34:2 (1918): 92 (first quotation), 99 (second quotation), 91 (third quotation). Italics in quotations appear in the original texts unless otherwise noted. Craig has been the subject of fine scholarship, including Burkhardt, "Whitman, Craig, and Animal Behavior," and Theodora J. Kalikow and John A. Mills, "Wallace Craig (1876–1954), Ethologist and Animal Psychologist," *Journal of Comparative Psychology* 103:3 (1989): 281–88.
53. William Morton Wheeler, "A Study of Some Ant Larvae, with Consideration of the Origin and Meaning of the Social Habit among Insects," *Proceedings of the American Philosophical Society* 57:4 (1918): 293 (first quotation), 296 (second quotation), 319 (third quotation). The word "altruistic" appears in quotation marks as the term of a German scholar, with whom Wheeler agreed (p. 319). Wheeler discussed Craig's ideas on pp. 334–35 and listed Craig's 1918 paper in his bibliography, p. 340.
54. Wheeler, "Some Ant Larvae," p. 322.
55. William Morton Wheeler, "On Instincts," *Journal of Abnormal Psychology* 15:5–6 (1920–1921): 295–318. The paper was delivered to the Royce Club in May 1917.
56. Wheeler, "On Instincts," p. 295 (first quotation); William Ernest Hocking, *Human Nature and Its Remaking* (New Haven: Yale University Press, 1918), p. x (second quotation). See also Hocking, *Human Nature*, chaps. 7–10.
57. Ralph Barton Perry, "Docility and Purposiveness," *Psychological Review* 25:1 (1918): 4 (first, second, and third quotations), 12 (fourth and fifth quotations), 11 (sixth quotation). In 1846 the Harvard philosopher Francis Bowen expressed the prevailing view of the time that

instincts dictate rigid behaviors: "Another peculiarity of instinct, and one of the broadest grounds of distinction between it and reason, is that it is not susceptible of improvement or education. It is complete from the beginning; it makes no progress either in the individual or the race" ([Francis Bowen], "Instinct and Intellect," *North American Review* 63:132 [1846]: 97). Some twentieth-century Europeans, on the whole favorable to the idea of biological inheritance, repeated this judgment. In 1941 the British evolutionary theorist Julian Huxley wrote about the cuckoo: "He acts blindly, because he is a machine constructed to act thus and not otherwise" (*The Uniqueness of Man* [London: Chatto and Windus, 1941], p. 214). In 1950, though, the Cambridge ethologist W. H. Thorpe envisioned a continuum between instinct and learning, encouraging study of "the relation between fixity and plasticity in behaviour" ("The Concepts of Learning and Their Relation to Those of Instinct," *Physiological Mechanisms in Animal Behaviour*, Symposia of the Society for Experimental Biology, no. IV [Cambridge: Cambridge University Press 1950], p. 406). It is safe to say that the Americans more consistently saw instincts as modifiable, however; see, e.g., William Morton Wheeler, *The Social Insects: Their Origins and Evolution* (New York: Harcourt Brace, 1928), p. 23. Changing ideas about instinct among European and American myrmecologists is a theme in Sleigh, *Six Legs Better*, esp. pp. 16, 39–40, 60–61.

58. Perry, "Docility and Purposiveness," p. 7 (first, second, and third quotations); Perry, *Present Conflict of Ideals*, p. 48 (fourth, fifth, and sixth quotations).

59. Robert Sessions Woodworth, *Dynamic Psychology* (New York: Columbia University Press, 1918), pp. 157 (first, second, and third quotations), 179 (fourth quotation); Perry, "Docility and Purposiveness," p. 7 (fifth quotation); Woodworth, *Dynamic Psychology*, pp. 61 (sixth quotation). 58 (seventh quotation). Woodworth's idea that human beings stimulate expression of instincts in one another is much like Konrad Lorenz's later, and better known, concept of conspecific responses. See Konrad Lorenz, "Companions as Factors in the Bird's Environment: The Conspecific as the Eliciting Factor for Social Behavior," in Lorenz, *Studies in Animal and Human Behavior*, trans. Robert Martin (Cambridge: Harvard University Press, 1970), 1: 101–258. Woodworth further developed his 1918 discussion of instincts and emotions, in Robert S. Woodworth, *Psychology: A Study of Mental Life* (New York: Henry Holt, 1921), esp. chaps. 5–9.

60. Edward Chace Tolman, "Can Instincts Be Given Up in Psychology?" *Journal of Abnormal Psychology and Social Psychology* 17:2 (1922): 144. For a list of citations for Tolman's numerous articles on the relevance of instinct, emotion, and purpose to behaviorism, 1920–1928, see Ritchie, "Tolman," pp. 319–20.

61. Zing Yang Kuo, "How Are Our Instincts Acquired?" *Psychological Review* 29:5 (1922): esp. 352. Tolman announced that he would explain Kuo's provocative views (first expressed in Zing Yang Kuo, "Giving Up Instincts in Psychology," *Journal of Philosophy* 18 [1921]: 645–64), in "Can Instincts Be Given Up," p. 139. Kuo never completed his PhD under Tolman, although he claimed later in China that he had. The two men maintained a positive relationship. See Geoffrey H. Blowers, "'To Be a Big Shot or To Be Shot': Zing-Yang Kuo's Other Career," *History of Psychology* 4:4 (2001): 367–87.

62. Walter B. Cannon, "New Evidences for Sympathetic Control of Some Internal Secretions," *American Journal of Psychiatry* 79:1 (1922): 19.

63. Walter B. Cannon, "The James-Lange Theory of Emotions: A Critical Examination and an Alternative Theory," *American Journal of Psychology* 39:1/4 (1927): 110. As evidence of Cannon's broad interdisciplinary interests and contacts, consider also arrangements made for his 1923 lecture on emotional physiology at St. Elizabeths Hospital, Washington, D.C., the principal federally supported hospital for the mentally ill, at the invitation of Shepherd Ivory Franz, box 103, folder 1416, Cannon Papers, and his 1927 paper delivered at the Wittenberg

Symposium held at Wittenberg College in Ohio, mainly composed of psychologists and psychiatrists, "Neural Organization for Emotional Expression," in Martin L. Reymert, ed., *Feelings and Emotions: The Wittenberg Symposium* (Worcester: Clark University Press, 1928), pp. 257–69.

64. Warder C. Allee and Marjorie Hill Allee, *Jungle Island* (Chicago: Rand McNally, 1925), pp. 17 (first quotation), 159 (second quotation). For another popular account of the site's history, see Frank M. Chapman, *My Tropical Air Castle: Nature Studies in Panama* (New York: D. Appleton, 1929), chap. 1.
65. Wheeler, "Organization of Research," p. 62 (first quotation); William Morton Wheeler, "The Dry Rot of Our Academic Biology," in Wheeler, *Foibles of Insects and Men* (New York: Knopf, 1928), p. 195 (second quotation), 199 (third quotation); Diary, Mar. 2 and Mar. 12, 1923, HUGFP 87.45, box 3, folder "Trip to Panama, Feb. 8th to April 12th, 1923," Wheeler Papers (fourth and fifth quotations). Extensive manuscript correspondence on the development of Barro Colorado Island as a scientific site may be found in Wheeler's correspondence with his Harvard colleague Thomas Barbour and others, HUGFP 87.10, box 4, folder B, Regular 1923.
66. Allee and Allee, *Jungle Island*, p. 166.
67. Edward Chace Tolman, "A New Formula for Behaviorism," *Psychological Review* 29:1 (1922): 51 (first and second quotations); Tolman, "Can Instincts Be Given Up," p. 148 (third, fourth, and fifth quotations); Calvin P. Stone, "William McDougall's Methodological Concept of Instinct," in Stuart A. Rice, ed., *Methods in Social Science: A Case Book* (Chicago: University of Chicago Press, 1931), p. 187 (sixth quotation).
68. C. K. Ogden and I. A. Richards, *The Meaning of Meaning: A Study of the Influence of Language upon Thought and of the Science of Symbolism*, 8th ed. (New York: Harcourt, Brace and World, [1946]), title page (first quotation); McDougall, *Outline of Psychology*, p. 326. (second, third, and fourth quotations). On the influence of Ogden and Richards's ideas about communication on the behavioral sciences during the interwar period, see Sleigh, *Six Legs Better*, chap. 7.
69. Craig, "A Note on Darwin's Work," pp. 365 (first quotation), 359 (second and third quotations).
70. Kalikow and Mills, "Wallace Craig," p. 282.
71. Lashley, "A Simple Maze," p. 355.
72. Donald Keith Adams, "Experimental Studies of Adaptive Behavior in Cats," *Comparative Psychology Monographs*, no. 27, 6:1 (Baltimore: Johns Hopkins University Press, 1929): 2 (first quotation), 3 (second quotation), 2 (third quotation); William McDougall, "An Experiment for the Testing of the Hypothesis of Lamarck," *British Journal of Psychology* 17:4 (1926–1927): 269 (fourth quotation), 271 (fifth quotation); Adams, "Adaptive Behavior in Cats," p. 2 (sixth quotation). The narrowing range of experimental animals has been widely noted; see, e.g., Frank Beach, "The Snark Was a Boojum" (1950), in Donald A. Dewsbury, *Foundations of Comparative Psychology* (New York: Van Nostrand Reinhold, 1984), esp. pp. 62–63; Burkhardt, "*Journal of Animal Behavior*," p. 223; Dewsbury, "Triumph and Tribulation," pp. 3–4. Donald A. Dewsbury argues, however, that the conclusion about rat dominance depends on how one surveys the journals and classifies articles; see "Animal Psychology in Journals, 1911–1927: Another Look at the Snark," *Journal of Comparative Psychology* 112:4 (1998): 400–05.
73. John B. Watson and Rosalie Raynor, "Conditioned Emotional Reactions," *Journal of Experimental Psychology* 3:1 (1920): 1 (first quotation), 2 (second quotation).
74. McDougall, *Outline of Abnormal Psychology*, p. 39 (quotations). Here McDougall quoted from his own address to the American Psychiatric Association in 1921 (see p. 30). In " 'Four Cases of 'Regression' " (1920), McDougall wrote that he read Aesop's fables to shell-shocked soldiers, p. 142. W. H. R. Rivers believed that humans known to a dreamer took the form of

animals in the person's dreams; see "Affect in the Dream," p. 114. It is my view that anthropological theories and fieldwork influenced these British psychologists to stress a connection between animals and the human imagination. Both Rivers and McDougall had been members of the Torres Straits Expedition to Borneo, sponsored by Cambridge University, in 1898. McDougall explained animals as symbols connected with rituals, in Charles Hose and William McDougall, "The Relations between Men and Animals in Sarawak," *Journal of the Anthropological Institute of Great Britain and Ireland* 31 (1901): 173–213. On the anthropological ideas behind this expedition, see Kuklick, *Savage Within*, pp. 133–49. Although Watson was by no means as rigid a behaviorist as he is often remembered, I do not think he was similarly affected by anthropology, although he was an evolutionist. See a well-argued response to the commonplace image of Watson as narrowly focused on learned behavior: James T. Todd and Edward K. Morris, "The Early Research of John B. Watson before the Behavioral Revolution," *Behavior Analyst* 9:1 (1986): 71–88.

75. Bowen, "Instinct and Intellect"; Leonard Carmichael, "Heredity and Environment: Are They Antithetical?" *Journal of Abnormal and Social Psychology* 20:3 (1925): 245–60.
76. Adams, "Adaptive Behavior in Cats," p. 40 (first quotation); William McDougall, *The Group Mind: A Sketch of the Principles of Collective Psychology with Some Attempt to Apply Them to the Interpretation of National Life and Character* (Cambridge: Cambridge University Press, 1921), p. 299 (second, third, and fourth quotations), Walter B. Cannon, *The Wisdom of the Body* (New York: Norton, 1932), pp. 268 (fifth quotation), 282 (sixth quotation).
77. Diary, July 22, 1924, pp. 41 (first and second quotations), 42 (third quotation), 41 (fourth quotation); Aug. 2, 1924, p. 63 (fifth quotation); Aug. 15, 1924, p. 103 (sixth quotation), "Diaries. Record of Anthropoid Work, Havana, Cuba, 1924 July 1–Aug. 22," series VI, box 172, folder 2666, Yerkes Papers.

Chapter 3

1. C. R. Carpenter, "Psychobiological Studies of Social Behavior in Aves. I. The Effect of Complete and Incomplete Gonadectomy on the Primary Sexual Activity of the Male Pigeon," *Journal of Comparative Psychology* 16:1 (1933): 38 (first quotation); C. Ray Carpenter to Calvin Perry Stone, Apr. 5, 1934, box 8, folder "S Correspondence," group 19, Clarence Ray Carpenter Papers, Eberly Family Special Collections Library, Penn State University, University Park, Pa. (second quotation). Stone offered advice in a letter to Carpenter, Jan. 29, 1935, located in the same folder. All Carpenter manuscripts are found in group 19.
2. [C. Ray Carpenter], "Notes on Field Studies at Camp La Vaca [indecipherable word], June 1– [*sic*], 1932, The Clark Expedition" (hereafter Clark Expedition Journal), June 2, p. 10 (first quotation), June 17, p. 52 (second quotation), June 13, p. 42 (third quotation), box 5, folder "Primatology, Research Observation Journals, The Clark Expedition, Panama, June to July, 1932," Carpenter Papers. Carpenter recorded the total number of females killed from June through August under June 13, p. 43. For a full study of Carpenter's role in developing the field of primate sociology, see Donna Haraway, "Signs of Dominance: From a Physiology to a Cybernetics of Primate Society, 1930–1970," *Studies in the History of Biology* 6 (1982): 129–219.
3. Carpenter, "Psychobiological Studies of Social Behavior in Aves, I," p. 38. Carpenter also used the wording "primary sexual activity" (Jan. 16, [19]32, p. 135) and "secondary sexual activity" (Dec. 28, [19]31, p. 140) as section headings in an untitled field notebook, box 5, folder "Primatology, Research Observation Journals, Dec. 1931–Jan. 1932, Howlers Book 1" (hereafter Howlers Book 1), Carpenter Papers. Italics in quotations appear in the original texts unless otherwise noted.

4. Calvin P. Stone, "The Retention of Copulatory Ability in Male Rats following Castration," *Journal of Comparative Psychology* 7:5 (1927): 370 (first quotation); Carpenter, June 11, 1932, Clark Expedition Journal, p. 33, Carpenter Papers (second and third quotations); Edward Chace Tolman, "A New Formula for Behaviorism," *Psychological Review* 29:1 (1922): 51 (fourth quotation). Recognition of the difficulty for humans of understanding animal minds has been a constant in animal behavior science over a century. See, e.g., Margaret Floy Washburn, *The Animal Mind: A Text-Book of Comparative Psychology*, 2d ed. (New York: Macmillan, 1917), pp. 1–4, and Marian Stamp Dawkins, *Why Animals Matter: Animal Consciousness, Animal Welfare, and Human Well-Being* (Oxford: Oxford University Press, 2012), esp. chap. 3.
5. James Zetek to C. Ray Carpenter, June 19, 1935, box 8, folder "Z Correspondence," Carpenter Papers.
6. The intersection of gender identity, sexuality, and animal science has elicited an unusually diverse literature. Donna Haraway underscores the influence of social values on twentieth-century biology, in *Primate Visions: Gender, Race, and Nature in the World of Modern Science* (New York: Routledge, 1989). Erika Lorraine Milam connects the sociology of science with biologists' conclusions about female mate selection across the evolutionary spectrum, in *Looking for a Few Good Males: Female Choice in Evolutionary Biology* (Baltimore: Johns Hopkins University Press, 2010). On women in scientific research, see Margaret W. Rossiter's classic, *Women Scientists in America: Struggles and Strategies to 1940* (Baltimore: Johns Hopkins University Press, 1982). More generally, Elaine Tyler May affirms that a domestic role for American women remained alive and well beyond World War II; see *Homeward Bound: American Families in the Cold War Era*, rev. ed. (New York: Basic, 2008). I have called for an analytical focus on families—not simply gender—and the impact of private relationships on public life, in *Beloved Strangers: Interfaith Families in Nineteenth-Century America* (Cambridge: Harvard University Press, 2001).
7. [William Morton Wheeler], "Courtship in Insects," manuscript notes, Zoology 7b, p. 1, HUGFP 87.65, box 1, folder "Courtship in Insects," William Morton Wheeler Papers, Harvard University Archives, Cambridge, Mass. (first and second quotations); Stone, "Retention of Copulatory Ability," p. 371 (third quotation); Leonard Carmichael, "The Experimental Embryology of Mind," *Psychological Bulletin* 38:1 (1941): 17 (third quotation). Carmichael's paper was his presidential address at the annual meeting of the American Psychological Association, Sept. 6, 1940.
8. Walter B. Cannon to Robert M. Yerkes, Oct. 8, 1931, box 78, folder 1046, Walter B. Cannon Papers (H MS c40), Center for the History of Medicine, Francis A. Countway Library of Medicine, Harvard Medical School, Boston, Mass. Although I infer that Cannon meant the CRPS because he used the word "committee," Yerkes replied obliquely that Rockefeller consented to public identification with the Bureau of Social Hygiene, another allied agency (see below), when Rockefeller and the BSH were in agreement; see Oct. 22, 1931, same folder. Either way, the bureaucratic flow of money was mazelike. In a 1920 lecture, William Morton Wheeler called the NRC, for example, the "superorganization of superorganizers"; see "The Organization of Research," *Science*, new series, 53:1360 (1921): 62. For Carpenter's acknowledgment of NRC support, see his "Psychobiological Studies of Social Behavior in Aves, I," p. 25. Yerkes informed CRPS committee members about the amount of annual funding, Nov. 11, 1930, box 78, folder 1046, Cannon Papers. See also Norma S. Thompson (Rockefeller Foundation) to Yerkes, May 9, 1931, where she indicated that the annual amount would be 75,000 dollars from July 1, 1931–June 30, 1933, same folder. See also Wade Pickren, "Robert Yerkes, Calvin Stone, and the Beginning of Programmatic Sex Research by Psychologists, 1921–1930," *American Journal of Psychology* 110:4 (1997): 603–19, and, on the NRC, Gregg

Mitman, *The State of Nature: Ecology, Community, and American Social Thought, 1900–1950* (Chicago: University of Chicago Press, 1992), esp. pp. 58–59.

9. Robert M. Yerkes to James R. Angell, Dec. 5, 1923, series I, box 2, folder 32, Robert Mearns Yerkes Papers (MS 569), Manuscripts and Archives, Yale University Library, New Haven, Conn. Ironically, Angell, a psychologist and also Yale's new president, was otherwise unsuccessful in his own efforts to secure funding from the Rockefeller Foundation for Yerkes's proposed primate research. See Angell to Yerkes, May 25, 1923, June 7, 1923, same folder.
10. Robert M. Yerkes to Walter B. Cannon, Dec. 24, 1931, box 78, folder 1046, Cannon Papers (quotation). For a sample of the animal subjects, see R. K. Burns, Jr., to [Robert M. Yerkes], Apr. 27 1932, box 78, folder 1049 (amphibians); S Bayne-Jones to [Robert M. Yerkes], May 12, 1933, box 78, folder 1049 (horses); and Frank Beach to Robert M. Yerkes, Dec. 12, 1940, box 78, folder 1061 (birds), Cannon Papers.
11. Walter B. Cannon to Robert M. Yerkes, Oct. 23, 1930, box 78, folder 1046, Cannon Papers (first and second quotations); "Questions for Discussion at a Birth Control Luncheon Conference, January 26, 1931," box 78, folder 1046, Cannon Papers (third and fourth quotations). The secretary, Ruth Topping, writing on BSH letterhead, subsequently commended Cannon's comments at the luncheon as "magnificent"; see Ruth Topping to Walter B. Cannon, Jan. 30, 1931, same folder. On these events, see also Yerkes to Cannon, Oct. 24, 1930; Ruth Topping to Cannon, Jan. 17, 1931, Jan. 21, 1931, same folder. On the Bureau of Social Hygiene, see http://www.rockarch.org/collections/rockorgs/bsh.php and http://rockefeller100.org/exhibits/show/health/bureau-of-social-hygiene (accessed June 1, 2019). On Cornelia Cannon's eugenic views, see Maria I. Diedrich, *Cornelia James Cannon and the Future American Race* (Amherst: University of Massachusetts Press, 2010), esp. chap. 5, and more generally, Daniel J. Kevles, *In the Name of Eugenics: Genetics and the Uses of Human Heredity* (New York: Knopf, 1985).
12. William Morton Wheeler, "On Instincts," *Journal of Abnormal Psychology* 15:5–6 (1920–1921): 315 (first and second quotations), 317 (third and fourth quotations).
13. Minnie Ruth Stone, "Biography of Calvin Perry Stone," p. 35, Calvin Perry Stone Papers microfilm (SC0440), University Archives, Special Collections, Stanford University Library, Stanford, Calif. (hereafter Stone Papers microfilm) (first quotation); K. S. Lashley, "A Note on the Persistence of an Instinct," *Journal of Animal Behavior* 4:4 (1914): 293 (second quotation), 294 (third quotation). On Stone's European tour in connection with the International Congress of Psychology in Copenhagen, see "Letters to Family While on Sabbatical Leave in England and Europe, 1932," unpaginated typescripts, Stone Papers microfilm. On his meeting with Hirschfeld, see "Der Institut fur Sexual Wissenschaft [*sic*]"; on his visit in Vienna with Charlotte Buhler, see "Visit to the Child Research Laboratory of Frau Buhler." On Freudianism in American popular culture between the wars, see Nathan G. Hale, Jr., *The Rise and Crisis of Psychoanalysis in the United States: Freud and the Americans, 1917–1985* (New York: Oxford University Press, 1995), pt. 1.
14. Robert M. Yerkes, *Introduction to Psychology* (New York: Henry Holt, 1911), p. 275 (quotation); K. S. Lashley, "A Simple Maze: With Data on the Relation of the Distribution of Practice to the Rate of Learning," *Psychobiology* 1:5 (1918): 353–67.
15. Calvin Perry Stone, "Sexual Drive," in Edgar Allen, ed., *Sex and Internal Secretions* (Baltimore: Williams and Wilkins, 1932), pp. 835 (rat puberty), 860 (reusing rats in new experiments), and passim.
16. Leonard Carmichael, "The Development of Behavior in Vertebrates Experimentally Removed from the Influence of External Stimulation," *Psychological Review* 33:1 (1926): 51–58; Leonard Carmichael, "The Genetic Development of the Kitten's Capacity to Right Itself in the Air When Falling," *Pedagogical Seminary and Journal of Genetic Psychology* 44

(1934): 453–58. In an equally rudimentary strategy, Carmichael used a flashlight as a stimulus for salamander embryos, in "A Further Experimental Study of the Development of Behavior," *Psychological Review* 35:3 (1928): 257.

17. Walter B. Cannon to Clarence C. Little, Feb. 9, 1940, box 78, folder 1061 (first quotation); C. D. Turner to Robert M. Yerkes, July 20, 1937, box 78, folder 1058 (second quotation); Frederick L. Hisaw to Robert M. Yerkes, June 24, 1933, box 78, folder 1049 (third quotation); Franklin F. Snyder, "Application for Grant-in-Aid," Sept. 28, 1936, box 78, folder 1058 (fourth quotation), Cannon Papers.
18. [Carpenter], Howlers Book I, Jan. 19, [19]32, p. 190 (first quotation); Dec. 26, [19]31, p. 230 (second quotation); Dec. 28, [19]31, p. 260 (third quotation); Jan. 16, [19]32, p. 178 (fourth quotation), Carpenter Papers. On the collection of reproductive organs, see [Carpenter], Clark Expedition Journal, June 6 and 11, 1932, pp. 21–22, 36, Carpenter Papers. For a list of participants in the summer 1932 research, all of them faculty members at Johns Hopkins Medical School and Harvard Medical School, see [Carpenter], Clark Expedition Journal, p. 295. The organizer was the pathologist Herbert C. Clark; see https://www.facs.org/about%20acs/archives/pasthighlights/clarkhighlight (accessed June 1, 2019). Although the Carpenter Finding Aid names 1932 as the year he completed his PhD at Stanford, Stone, his dissertation adviser, cited 1931 as the date of the dissertation's submission, in "Sex Drive," in Allen, ed., *Sex and Internal Secretions*, p. 877. I infer that Carpenter completed his laboratory work on castrated pigeons in 1931, and then traveled in December to the Canal Zone, where he participated in two separate field studies until late summer 1932.
19. [Carpenter], Clark Expedition Journal, June 2, 1932, p. 9 (quotations), Carpenter Papers. For other details, see the same journal, pp. 4 (trip to camp), 15 (showering by animals), 31, 53, 62 (tracking subjects). The monograph was "A Field Study of the Behavior and Social Relations of Howling Monkeys," *Comparative Psychology Monographs*, no. 48, 10:2 (Baltimore: Johns Hopkins University Press, 1934). By the time the book was published, Carpenter was a research assistant at Yerkes's Laboratories of Comparative Psychobiology at Yale ("Field Study of Howling Monkeys," p. 1).
20. For discussions of Carpenter's job prospects and grant activity, see C. Ray Carpenter to Calvin Perry Stone, Apr. 5, 1934, and Stone to Carpenter, June 2, 1937, box 8, folder "S Correspondence," Carpenter Papers, as well as Robert M. Yerkes to C. Ray Carpenter, Feb. 6 and Apr. 1, 1935, Feb. 28, 1936, box 8, folder "Robert Yerkes," Carpenter Papers. My sense is that Stone did not favor Carpenter in the process of doctoral-student placement because Carpenter was not strictly committed to laboratory studies.
21. Margaret Morse Nice, *The Watcher at the Nest* (New York: Macmillan, 1939), pp. 4 (first, second, third, and fourth quotations), 74 (fifth quotation), 72 (sixth and eighth quotations), 11 (seventh quotation). Interpont, located near where Nice's husband was employed at Ohio State University Medical School, was an undeveloped tract of forty to sixty acres (Nice named different sizes in her account, pp. 2, 74). In 1933 the New Deal appropriated the land for garden acreage for the unemployed, p. 74. See also Margaret Morse Nice, *Research is a Passion with Me: The Autobiography of Margaret Morse Nice* (Toronto: Margaret Nice Ornithological Club and Consolidated Amethyst Communications, 1979), chaps. 12, 14. Her early essays on child development occasionally mention nature; see esp. "A Child and Nature," *Pedagogical Seminary* 28 (1921): 22–39. Books by women "amateurs" were by no means uninformed or uninformative; see, e.g., Mabel Osgood Wright, *Birdcraft: A Field Book of Two Hundred Song, Game, and Water Birds* (New York: Macmillan, 1897). On popular ornithology, see Mark V. Barrow, Jr., *A Passion for Birds: American Ornithology after Audubon* (Princeton: Princeton University Press, 1998). On the nature study movement, see Kevin C. Armitage, *The Nature Study Movement: The Forgotten Popularizer of America's Conservation*

Ethic (Lawrence: University Press of Kansas, 2009), and Charlotte Sleigh, *Six Legs Better: A Cultural History of Myrmecology* (Baltimore: Johns Hopkins University Press, 2007), chap. 5. On popular science more broadly, see David Elliston Allen, *The Naturalist in Britain: A Social History* (1976; reprint ed., Princeton: Princeton University Press, 1994).

22. W. C. Allee, *Animal Aggregations: A Study in General Sociology* (Chicago: University of Chicago Press, 1931), p. 342. See Nice's scholarly books, *Studies in the Life History of the Song Sparrow, I*, as *Transactions of the Linnean Society of New York*, vol. 4 (1937), and *Studies in the Life History of the Song Sparrow, II: The Behavior of the Song Sparrow and Other Passerines*, as *Transactions of the Linnean Society of New York*, vol. 6 (1943). She served as associate editor of the *Wilson Bulletin* (begun 1889), issued by the Wilson Ornithological Society, and *Bird-Banding* (begun 1930), published by the Northeastern Bird-Banding Association. Her correspondence with Konrad Lorenz may be found in box 1, folders 3–6, Margaret Morse Nice Papers (Collection 2993), Division of Rare and Manuscript Collections, Cornell University Library, Ithaca, N.Y. She presented Lorenz's theories to American readers in, e.g., Margaret Morse Nice, "The Social Kumpan and the Song Sparrow," *Auk* 56:3 (1939): 255–62. See also Richard W. Burkhardt, Jr., *Patterns of Behavior: Konrad Lorenz, Niko Tinbergen, and the Founding of Ethology* (Chicago: University of Chicago Press, 2005), esp. pp. 161–63, 214–18.
23. William McDougall, *Outline of Psychology* (New York: Scribners, 1923), esp. chaps. 11–13. Robert W. Woodworth incorporated affective concepts into his *Psychology: A Study of Mental Life* (New York: Henry Holt, 1921), a book "intended for the beginner," p. iii. See discussions of purpose, instinct, emotion, and feeling, chaps. 4–9.
24. Carpenter, "Psychobiological Studies of Social Behavior in Aves, I," p. 56 (first, second, third, and fourth quotations); Nice, *Watcher at the Nest*, p. 53 (fifth quotation). Although Nice's narrative style makes it difficult to follow life events of specific sparrows, 4M's successive mates, as named by Nice, were Quarta (also spelled Quarte), pp. 38–39; Chatvar, pp. 39, 53; and Rosemary, p. 53.
25. Carpenter, "Psychobiological Studies of Social Behavior in Aves, I," p. 56 (quotation). Note that the second half of Carpenter's dissertation was published in the same journal immediately following the first half and in the same issue: "Psychobiological Studies of Social Behavior in Aves, II. The Effect of Complete and Incomplete Gonadectomy on Secondary Sexual Activity with Histological Studies," *Journal of Comparative Psychology* 16:1 (1933): 59–97. The emphasis here on billing, preening, charging, and wing plucking did not substantially extend the conclusions stated in the first part. Carpenter credited McDougall, under whose supervision he completed bachelors and masters degrees at Duke in 1928 and 1929, respectively, with the idea for his doctoral dissertation, that is, measuring hormonal responses under various physiological conditions; see "Psychobiological Studies of Social Behavior in Aves, I," p. 25. Carpenter probably chose Duke because he was a native of North Carolina. Although Nice implied in her autobiography (p. 39), *Research Is a Passion with Me*, that Hall misdirected her to focus about child development, presumably the appropriate subject for a woman, she was also complimentary: "Dr. [Clifton F.] Hodge, Dr. Hall, and my fellow students [at Clark University] showed me that the world was full of problems crying to be solved; at every turn there was a challenge—nature waiting to be studied and understood" (p. 26).
26. [Ada Yerkes], Research Journal, May 14, 1933, "Experiments with Gua, 2 yrs. & 3 yrs. [*sic*] by Ada Yerkes," p. 66, series VI, box 161, folder 2523, Yerkes Papers. Ada also noted that Gua manipulated a stick, Mar. 30, 1933, p. 20; played with blocks, Apr. 20, 1933, p. 43; and played with a cup in the sand, May 10, 1933, p. 63. A number of experiments concerned tool use; see, e.g., notes under heading "Summary of Exp. [*sic*] by Gua on Instrumentation 1933," pp. 180–81. Robert Yerkes explained the purpose of his month-long Cuba trip as observing the "large collection of animals," housed in fifty cages, owned privately by Rosalía Abreu of

Havana; see *Almost Human* (New York: Century, 1925), pp. xiii (quotation), 204. In 1929 Yerkes remained in Dakar from Aug. 22 until Sept. 8, just long enough to establish contacts to help him acquire primates; during his stay, he traveled minimally in the countryside. See [Robert M. Yerkes], "African Portion of European-African Journey, June-October, 1929. Log and Notes of Robert M. Yerkes," typescript made by Roberta Yerkes Blanshard, series VI, box 172, folder 2668, Yerkes Papers (hereafter Yerkes Africa Journal).

27. [Robert M. Yerkes], Research Journal, July 8, 1924, "Daily Record, Anthropoid Work, Havana, Cuba, July–[Aug.], 1924. Robert M. Yerkes" (hereafter Yerkes Cuba Journal), p. 9, series VI, box 172, folder 2666, Yerkes Papers (first and second quotations); Robert M. Yerkes, *Chimpanzees: A Laboratory Colony* (New Haven: Yale University Press, 1943), pp. vii (third quotation), 15 (fourth quotation). Yerkes was deeply involved emotionally with Chim, who died in Cuba on July 22 presumably as a result of exposure to tropical parasites (see entries in Yerkes Cuba Journal, pp. 15–54). Yerkes brought Chim to Havana with him from New Hampshire, where the chimpanzee had been living with the Yerkes family, and he blamed himself for Chim's death. He arranged to have Chim's remains preserved and shipped to his laboratory in New Haven; see entries for July 28 and Aug. 4–5, Yerkes Cuba Journal, pp. 54, 69. His idealized image of Chim, focusing on Chim's intelligence, was the concluding chapter (chap. 13) of *Almost Human*, "The Light That Failed: A Tribute to Prince Chim."
28. Carmichael, "Experimental Embryology of Mind," p. 9.
29. Anon., "In Memoriam. M. H. A.," clipping, [1945], no source, p. 193, box 25, folder 3, Warder Clyde Allee Papers, Special Collections Research Center, University of Chicago Library, Chicago, Ill. (quotation). Marjorie Allee used the Woods Hole setting for *Jane's Island* (1931; reprint ed., Woods Hole, Mass.: Woods Hole Historical Collection, 1988). She made the animal biological laboratory at the University of Chicago the background for *The Great Tradition* (Boston: Houghton Mifflin, 1937). The Barro Colorado Island field site in the Canal Zone was the subject of Marjorie's coauthored book with her husband, Warder C. Allee, *Jungle Island* (Chicago: Rand McNally, 1925). On Cornelia Cannon's profits and the Italian tour, see Diedrich, *Cornelia Cannon*, pp. 161, 175, and Marian Cannon Schlesinger, *Snatched from Oblivion: A Cambridge Memoir* (Boston: Little Brown, 1979), chap. 8. By all reports, Cornelia, a busy mother and an activist in local reform, wrote on the fly: her desk in Cambridge was in a corner under the stairs (Cannon Schlesinger, *Snatched from Oblivion*, p. 9), and she also wrote in the bathroom as a warm place during winter months, as well as in the family car and a shack on the Cannons' New Hampshire farm (Diedrich, *Cornelia Cannon*, p. 161; Cannon Schlesinger, *Snatched from Oblivion*, pp. 123–26).
30. Margaret Morse Nice recalled the production of her coauthored book with Leonard Blaine Nice, *The Birds of Oklahoma* (Norman: University of Oklahoma, 1924), in *Research Is a Passion with Me*, chaps. 7–9. On Ada's scientific competency in Spanish, Italian, Dutch, German, and Latin, see Robert W. Yerkes and Ada W. Yerkes, *The Great Apes: A Study of Anthropoid Life* (New Haven: Yale University Press, 1929), p. x. As early as 1914 Ada was translating for the *Journal of Animal Behavior*, edited by her husband; see mention of her translation of an article from Italian, in Robert M. Yerkes to William Morton Wheeler, Mar. 27, 1914, HUGFP 87.10, box 39, folder "Robert M. Yerkes, 1907–1914," Wheeler Papers.
31. Besides some excellent published biographical sources, information about the women is spotty. Online finding aids are helpful; see esp. Warder C. Allee Papers, University of Chicago; Cannon Family Papers, Arthur and Elizabeth Schlesinger Library, Harvard University; Margaret Morse Nice Papers, Cornell University; Robert Mearns Yerkes Papers, Yale University; and Robert M. Yerkes and Ada W. Yerkes Papers, Whitney Library, New Haven Museum. Biographical information, in chronological order, is as follows: Ada Yerkes

(1873–1963), married 1905; Cornelia Cannon (1876–1969), married 1901; Margaret Nice (1883–1974), married 1908; Marjorie Allee (1890–1945), married 1912.

32. Reports of the twenty-fifth anniversary dinner honoring Margaret Washburn (1871–1939), held at the American Psychological Association meeting in Columbus, Ohio, Dec. 28, 1927, are found in "Release as Soon as Possible," typescript, Jan. 6, 1928, as well as clippings from newspapers that picked up the story, in Margaret Floy Washburn Folders, Biographical Files, folder 1, Archives and Special Collections Library, Vassar College, Poughkeepsie, N.Y. Wheeler warmly cited the work of Adele Fielde (1839–1916) in a number of his publications, including "Ant Larvae, with Consideration of the Origin and Meaning of the Social Habit among Insects," *Proceedings of the American Philosophical Society* 57:4 (1918): 300–01. Over time, male scientists seemed increasingly to view Fielde's work as foundational. T. C. Schneirla, for example, included her in a list of pioneering ant researchers who discovered the socializing function of odor, in "'Cruel Ants'—and Occam's Razor," *Journal of Comparative Psychology* 34:1 (1942): 81.
33. Althea Sherman to Margaret Nice, Dec. 30, 1921 (first and second quotations); Mar. 26, 1921 (third quotation); Nov. 13, 1924 (fourth quotation), box 1, folder 1, Nice Papers; Margaret Morse Nice, "Some Letters of Althea R. Sherman," *Iowa Bird Life* 22 (1952): 51 (fifth quotation). Despite her lack of advanced degree or academic position, Althea R. Sherman (1853–1943) was a major force in American ornithology; see http://althearsherman.org/AltheaRSherman.Palimpsest.html (accessed June 1, 2019). She and Nice enjoyed a mentor-protégé relationship; see Sherman's letters to Nice, 1921–1932, box 1, folders 1–3, Nice Papers.
34. On Lucile Quarry Mann (1897–1986), see Pamela M. Henson, "A Life on the Wild Side: Lucile Quarry Mann," *Smithsonian Institution Archives*, Mar. 15, 2011, http://siarchives.si.edu/blog/life-wild-side-lucile-quarry-mann (accessed June 1, 2019). William M. Mann's readable memoir of his adventures in search of insects prior to accepting a position at the Smithsonian, and prior to his marriage in 1926, is *Ant Hill Odyssey* (Boston: Little, Brown, 1948). Information on the other wives is fragmentary; these data come from various published and unpublished sources: Kathleen Drew Tolman (1886–1963), married 1915; Minnie Ruth Kemper Stone (1892–1980), married 1917; Pearl Kidston Carmichael (birth and death dates unknown), married 1932; and Marianna Evans Carpenter (d. 1963), married 1932.
35. Betty Friedan, *The Feminine Mystique* (1963; reprint ed., New York: Dell, 1983), p. 76. Friedan was a freshman at Smith College in 1938.
36. [Ada Yerkes], 'My Prayer," in Journal, "House 4 St. R. Terr." [*sic*], p. 26, box 1, folder J, Robert M. Yerkes and Ada W. Yerkes Papers (MSS 139) (hereafter Yerkes Family Papers), Whitney Library, New Haven Museum, New Haven, Conn. (first and second quotations); "House 4 St. R. Terr.," p. 20 (third, fourth, and fifth quotations). Her account of the family's successive residences appears as "History," pp. 7–17, in the same journal. See also Robert Yerkes's letter to a Boston firm, Home Builders' Service Bureau, Aug. 24, 1925, about the proposed design and cost, as well as a separate handwritten list of fifteen essential components of the projected house, plus three "luxuries": a first-floor bathroom, billiard room, and second chimney (box 1, folder A, Yerkes Family Papers). The well-known New Haven architect, J. Frederick Kelly, in the end designed the house. On the Cannons' rented houses in Cambridge, located amid more substantial homes staffed by uniformed servants, see Diedrich, *Cornelia Cannon*, p. 88, and Cannon Schlesinger, *Snatched from Oblivion*, p. 16. The Cannons purchased their farmhouse in Franklin, New Hampshire, in 1911.
37. [Ada Yerkes], "House 4 St. R. Terr.," p. 20, Yerkes Family Papers. On the death of Warder Clyde Allee, Jr., see Karl Patterson Schmidt, "Warder Allee, 1885–1955," *National Academy*

of Sciences Biographical Memoirs (1957), p. 9, http://www.nasonline.org/publications/biographical-memoirs/memoir-pdfs/allee-warder.pdf (accessed June 1, 2019). The Allees dedicated their coauthored book, *Jungle Island* (1925), to their son's memory; the father also dedicated his *Animal Aggregations* (1931) to him. On Eleanor Nice, see Nice, *Research Is a Passion with Me*, p. 91.

38. Marjorie Hill Allee, "Alcohol and Racial Degeneration," unpaginated typescript, [p.1], box 5, folder 1, Allee Papers (first and second quotations); Cornelia James Cannon, "American Misgivings," *Atlantic Monthly* 129:2 (1922): 147 (third quotation), 150 (fourth and fifth quotations). Cornelia wrote frequently on biology and social morality; see, e.g., Cornelia James Cannon, "Can Our Civilization Maintain Itself?" *Atlantic Monthly* 126:5 (1920): 633–36, and "Prohibition and the Younger Generation," *North American Review* 222:828 (1925): 65–69.
39. [Robert M. Yerkes], Diary, Dec. 9, 1918, "Journal 1917–1918, Especially Important Events in the Life History of Robert Mearns Yerkes," p. 164, series VI, box 171, folder 2663, Yerkes Papers (first quotation); Cornelia James Cannon, [letter to Walter Cannon, Aug. 16, 1917], "Summer at Hudson, Wisconsin, 1917 (Selections from the Letters of Cornelia James Cannon to Walter Bradford Cannon), May-September, 1917, from Cambridge and Hudson to England and France," p. 16, typescript dated Sept. 19, 1934, series II, subseries B, 12.2, Cannon Family Papers (MC 553), Schlesinger Library, Harvard University, Cambridge, Mass., http://oasis.lib.harvard.edu/oasis/deliver/~sch01170 (accessed June 1, 2019) (second, third, and fourth quotations); Ada Yerkes, "Bungalow Land: An Appreciation," typescript, ca. 1915–1916, pp. 4 (fifth quotation), 3 (sixth quotation), series VI, box 161, folder 2520, Yerkes Papers; Ada Yerkes to Robert Yerkes, Aug. 6, [1924], series VI, box 159, folder 2494, Yerkes Papers (driver's license).
40. Ada Yerkes, "Bungalow Land," p. 6 (quotations). On Allee's training, see Schmidt, "Allee," p. 6. Marjorie Allee explored women's socialization to choose a humanities education, rather than science, in her character Ellen in *Jane's Island*. On Nice, see *Research is a Passion with Me*, chaps. 3–4. Ada Yerkes earned a bachelors degree at Barnard and a masters at Columbia, both in botany, and also taught at Columbia before her marriage in 1905. For an excellent discussion of Cornelia Cannon's experience at Radcliffe as a case of the academic options of highly educated women, see Diedrich, *Cornelia Cannon*, esp. pp. 45–48.
41. Allee, *Great Tradition*, pp. 172 (first and second quotations), 204 (third quotation), 205 (fourth quotation). Similar praise of science appears in the character Matts [*sic*] Swenson of Cornelia Cannon's bestselling novel, *Red Rust* (New York: Grosset and Dunlap, 1928), esp. pp. 26–27, 186–87, 298, 315–20.
42. Allee and Allee, *Jungle Island*, pp. ix (first quotation), 194 (second quotation); Cannon, "American Misgivings," pp. 145 (third and fourth quotations), 156 (fifth quotation). Although the Allees' book demonstrated Marjorie's literary facility, the couple chose a first-person narrative in her husband's voice. After they located together near Panama City, W. C. Allee camped in primitive conditions in the interior jungle; presumably Marjorie stayed behind with their six-year-old daughter and later wrote up his notes. Use of a first-person voice identified as that of the better known of multiple authors was not uncommon; see discussion that follows of William and Kenneth McDougall. On Robert Yerkes's wartime testing, see Clarence S. Yoakum and Robert M. Yerkes, *Army Mental Tests* (New York: Henry Holt, 1920). On the acquaintance of the Cannon and Yerkes families, see Cannon Schlesinger, *Snatched from Oblivion*, pp. 123–24.
43. Yerkes and Yerkes, *Great Apes*, pp. 435 (first quotation), 440 (second quotation). Further examples of Ada's continuing role in her husband's fieldwork may be seen in "How Ideas Look: A Glimpse of the Monkey Mind, the Ideas of Monkeys and Apes," typescript, ca.

1915–1916, series VI, box 161, folder 2520, Yerkes Papers, and her research journal, "Experiments with Gua," 1933.

44. Cornelia James Cannon, "A Middle-Aged Adventure," typescript, Aug. 1917, pp. 10 (first and second quotations), 3 (third quotation), 5 (fourth quotation), series II, subseries B, 12.3, Cannon Family Papers, http://oasis.lib.harvard.edu/oasis/deliver/~sch01170 (accessed June 1, 2019).
45. Nice, *Watcher at the Nest*, pp. 89 (first and second quotations), 90 (third quotation), 154 (fourth and fifth quotations).
46. [Indecipherable name], Harper and Brothers, to Robert M. Yerkes, Apr. 8, 1916, series VI, box 161, folder 2520, Yerkes Papers (first quotation); Ada Yerkes, "Animal Individuality and Genius," typescript, ca. 1915–1916, pp. 11 (second quotation), 10 (third and fourth quotations), series VI, box 161, folder 2520, Yerkes Papers; Robert M. Yerkes, *The Mental Life of Monkeys and Apes*, ed George M. Haslerud (1916; reprint ed., Delmar, N.Y.: Scholars' Facsimile Reprints, 1979), pp. 21 (fifth quotation), 64 (sixth quotation). A slightly revised version of Ada's paper later appeared as Robert M. Yerkes and Ada W. Yerkes, "Individuality, Temperament, and Genius in Animals," *American Museum Journal* 17:4 (1917): 235–43. During the 1930s, Ada increasingly published both as sole author and coauthor on the subject of the couples' work with chimpanzees; see, e.g., her "Experiments with an Infant Chimpanzee," *Pedagogical Seminary and Journal of Genetic Psychology* 46 (1935): 171–80.
47. Kenneth D. McDougall and William McDougall, "Insight and Foresight in Various Animals—Monkey, Raccoon, Rat, and Wasp," *Journal of Comparative Psychology* 11:3 (1931): 237 (first and second quotations), 273 (third quotation). Kenneth (1908–1944) died in combat the year after his Duke dissertation was published as "Sessile Marine Invertebrates of Beaufort, North Carolina," *Ecological Monographs* 13:3 (1943): 321–74.
48. Nice, *Watcher at the Nest*, pp. 119 (first quotation), 145 (Constance); [Robert Yerkes], Yerkes Africa Journal, Aug. 31, p. 10 (second quotation), Aug. 22, p. 4 (third quotation). Yerkes repeated that he would be silent without Roberta's help with language, in a letter to Ada and their son David, Aug. 17, 1929, series VI, box 163, folder 2566, Yerkes Papers. For the coauthored paper, see Robert M. Yerkes and David N. Yerkes, "Concerning Memory in the Chimpanzee," *Journal of Comparative Psychology* 8:3 (1928): 237–71. McDougall's private journals on the development of Duncan, Lesley, and Janet are found in box 11, folders 114–15, William McDougall Papers, David M. Rubenstein Rare Book and Manuscript Library, Duke University, Durham, N.C.
49. Daughters far outnumbered sons among the children who lived to adulthood of my four principal women: Allees, two daughters; Cannons, four daughters and one son; Nices, four daughters; and Yerkeses, one daughter and one son. Among them, only two children clearly chose the sciences: Bradford Cannon (1907–2005), who became a doctor specializing in public health, and Helen Cannon Bond (1915–unknown), who married a psychiatrist, Douglas Danford Bond. Both Allee daughters earned masters degrees in the social sciences and married social scientists: Barbara became the wife of social worker Steve Angell, and Molly wed the Norwegian anthropologist Fredrik Barth. Two Cannon daughters married men who became famous historians: Wilma (1909–2002) became the wife of John K. Fairbank, and Marian (1912–2017), Arthur M. Schlesinger, Jr. Roberta Yerkes (1907–2001) worked for many years for Yale University Press before marrying the Yale philosopher Brand Blanshard. David Norton Yerkes (1911–2011) was an architect and artist in Washington, D.C.; see http://articles.washingtonpost.com/2011-11-07/local/35280900_1_architect-world-war-ii-first-wife (accessed June 1, 2019). There is less public information about the Nice daughters: Constance, born in 1911; Marjorie, 1912; Barbara, 1916; and Janet, 1922.
50. Robert Yerkes to Ada Yerkes, Nov. 6, 1949 (first quotation); Robert Yerkes to [Ada Yerkes], Dec. 25, 1946 (second quotation); Robert Yerkes to Ada Yerkes, Sept. 16, 1947 (third and

fourth quotations); Robert Yerkes to Ada Yerkes, Sept. 12, 1934 (fifth quotation); series VI, box 163, folder 2566, Yerkes Papers.

51. Schmidt, "Allee," p. 11 (first quotation); Diedrich, *Cornelia Cannon,* p. 174 (second quotation). Allee had been publishing in Quaker periodicals since the war's end; e.g., W. C. And M. H. [*sic*] Allee, "Adventures of a College Professor," *American Friend,* Twelfth Month [Dec.] 5, 1919, pp. 1132–133, and Marjorie Hill Allee, "Educating Joe," *Messenger of Peace* 45:10–11 (1920): 145–49, 161–65. Diedrich implies that the friend was the philosopher Ralph Barton Perry, who dedicated his book, *The General Theory of Value* (New York: Longmans, Green, 1926), to the Cannons ("W. B. C. and C. J. C."), and to whom Walter Cannon in turn dedicated his book, *The Wisdom of the Body* (New York: Norton, 1932).
52. Allee, *Animal Aggregations,* pp. 339 (first quotation), 338 (second quotation), 337 (third quotation). In this passage Allee credited the British philosopher Herbert Spencer with the idea that grouping together offers an evolutionary advantage. Here, and throughout his career, Allee opposed the theory of Thomas Malthus that dense population necessarily triggers harmful competition; see *Animal Aggregations,* esp. chap. 20. See also Mitman, *State of Nature,* chap. 4. In unpublished retrospective notes, Allee identified *Animal Aggregations* as his most important book; see Warder Clyde Allee vitae, typescript, untitled (begins "Name: Warder Clyde Allee"), box 25, folder 8, Allee Papers.
53. Yerkes, *Almost Human,* p. 265.
54. W. C. Allee, "Where Angels Fear to Tread: A Contribution from General Sociology to Human Ethics," *Science,* new series, 97:2528 (1943): 523 (first and second quotations); [W. C. Allee], "Group Action among Animals and Its Social Implications," typescript, p. 2, box 1, folder 1, Allee Papers (third and fourth quotations); William Morton Wheeler, quoted in Allee, *Animal Aggregations,* p. 353 (fifth quotation).
55. Ralph H. Masure and W. C. Allee, "The Social Order in Flocks of the Common Chicken and the Pigeon," *Auk* 51:3 (1934): 306 (first quotation); [W. C. Allee], "Observations on Group Organization," typescript, p. 2, box 1, folder 1, Allee Papers (second, fifth, sixth, seventh and eighth quotations); Allee, *Animal Aggregations,* p. 355 (third and fourth quotations). In his coauthored article with Masure, Allee credited the Norwegian biologist Thorleif Schjelderup-Ebbe with the concept of peck order.
56. Cannon, *Wisdom of the Body,* pp. viii (first quotation), 268 (third quotation), 275 (fourth and fifth quotations), 282 (sixth quotation); [Walter B. Cannon], "Dietetics in Disease," p.22, handwritten lecture notes, post-1903, box 176, Cannon Papers (second quotation).
57. Cannon, *Wisdom of the Body,* pp. 296 (first and second quotations), 303 (third quotation), 305 (fourth and fifth quotations).
58. On Allee's illness, see Schmidt, "Allee," pp. 15–16, 18–20. See also Dr. Paul C. Bucy to Allee, Oct. 20, 1942; Allee to Bucy, Nov. 22, 1949; Allee to Veterans Administration (hereafter VA), Sept. 14, 1948; Thomas H. Hickey (VA) to Allee, Oct. 7, 1948, Oct. 12, 1948, box 25, folder 8, Allee Papers. Although Allee was a pacifist, he approached the Veterans Administration in hopes of borrowing plans being developed by the agency for house design for disabled veterans. On Cannon's illness, see Diedrich, *Cornelia Cannon,* pp. 176, 180–81, and Elin Wolfe, A. Clifford Barger, and Saul Benison, *Walter B. Cannon: Science and Society* (Boston: Boston Medical Library in the Francis A. Countway Library of Medicine, 2000), pp. 292–98.
59. W. C. Allee to "F.C." [Allee] (brother), Dec. 8, 1949, box 25, folder 8, Allee Papers (first quotation); Walter B. Cannon to Knight Dunlap, editor of the *Journal of Comparative Psychology,* Feb. 16, 1924, box 105, folder 1446, Cannon Papers (second and third quotations). The offending researcher was identified as Calvin Perry Stone, in Dunlap to Cannon, Feb. 14, 1924, same folder. On trends in CRPS funding for the biological sciences at the University of Chicago, including Allee's research, see Mitman, *State of Nature,* pp. 96, 105–06.

60. Clifford Moore to William Morton Wheeler, Feb. 20, 1926, HUGFP 87.10, box 24, folder "Regular 1926" (first quotation); William Morton Wheeler to W. M. MacNair, Dec. 9, 1926, box 24, folder "Miscellaneous 1926," Wheeler Papers (second quotation). Wheeler's first biographers wrote more guardedly about his hospitalization than he did in letters at the time; see brief mention in Mary Alice Evans and Howard Ensign Evans, *William Morton Wheeler, Biologist* (Cambridge: Harvard University Press, 1970), p. 286. On Dora Bay Emerson Wheeler, whose inherited wealth contributed to their family's comfort, see Evans and Evans, *Wheeler*, esp. pp. 108–13.
61. Jerome Bruner, preface to William McDougall, *Body and Mind: A History and Defense of Animism* (1911; reprint ed., Boston: Beacon Press, 1961), p. xiii. Although Bruner studied with McDougall in the 1930s, earlier commentators dated McDougall's deafness either to his arrival at Duke in 1927 (Major Greenwood and May Smith, "William McDougall, 1871–1938," *Obituary Notices of Fellows of the Royal Society* 3:8 [1940]: 51), or, in the view of McDougall himself, before his arrival in the United States in 1920 ("William McDougall," *History of Psychology in Autobiography*, ed. Carl Murchison [1930; reprint ed., New York: Russell and Russell, 1961], 1: 212). Although McDougall was an outspoken eugenicist from the turn of the century, his wife is all but invisible in both his published and unpublished writings. On his eugenic commitments, see Anne C. Rose, *Psychology and Selfhood in the Segregated South* (Chapel Hill: University of North Carolina Press, 2009), pp. 84–86.
62. William Morton Wheeler, "Emergent Evolution and the Social," *Science*, new series, 64:1662 (1926): 438. The most immediate forerunner on the subject was C. Lloyd Morgan, *Emergent Evolution: The Gifford Lectures* (New York: Henry Holt, 1923). Wheeler further developed his ideas, in *Emergent Evolution and the Development of Societies* (New York: Norton, 1928).
63. William Morton Wheeler, "A New Word for an Old Thing," *Quarterly Review of Biology* 1:3 (1926): 439 (first, second, and third quotations), 440 (fourth quotation); William Morton Wheeler, *The Social Insects: Their Origin and Evolution* (New York: Harcourt, Brace, 1928), p. 308 (fifth, sixth, and seventh quotations); William Morton Wheeler, "Biology and Society: Animal Societies," *Scientific Monthly* 39:4 (1934): 293 (eighth, ninth, and tenth quotations). In light of Wheeler's view of sexuality, it is noteworthy that Alfred C. Kinsey was Wheeler's doctoral student at Harvard, completing his degree in 1920 as an expert on gall wasps. See James H. Jones, *Alfred C. Kinsey: A Public/Private Life* (New York: Norton, 1997), chap. 7.
64. William McDougall, "Religion and the Sciences of Life," in William McDougall, *Religion and the Sciences of Life* (Durham: Duke University Press, 1934), p. 8 (first and second quotations); William McDougall, "Theory of Hypnosis and Suggestion" (1926), in Raymond Van Over and Laura Oteri, eds., *William McDougall: Explorer of the Mind, Studies in Psychical Research* (New York: Garrett Publications, 1967), p. 278 (third quotation); William McDougall, "World Chaos," in McDougall, *Religion and the Sciences of Life*, p. 193 (fourth quotation). Edward Chace Tolman argued for the utility of a concept of purposiveness in individual organisms, in "Can Instincts Be Given Up in Psychology?" *Journal of Abnormal Psychology and Social Psychology* 17:2 (1922): 139–52.
65. William McDougall, "Psychical Research as a University Study" (1926), in Van Over and Oteri, eds., *McDougall*, p. 76 (first quotation); William McDougall, foreword to J. B. Rhine, *Extra-Sensory Perception* (1934; reprint ed., Boston: Branden Press, 1964), p. xiii (second quotation); William McDougall to Dean R. B. House, Nov. 19, 1935, box 1, folder 1, McDougall Papers (third quotation); McDougall, "World Chaos," in McDougall, *Religion and the Sciences of Life*, p. 209 (fourth quotation).
66. Rhine, *Extra-Sensory Perception*, p. 173 (quotations). On later experiments, see Pierre Duval and Evelyn Montredon, "ESP Experiments with Mice"; Robert Brier, "PK Effect on a Plant

Polygraph System"; and Jean Barry, "Retarding Fungus Growth by PK," in J. B. Rhine, ed., *Progress in Parapsychology* (Durham: Parapsychology Press, 1971), pp. 17–27, 102–17, 118–21.

67. Among European animal behavior scholars who did not emphasize sexuality were the British ecologist, Charles Elton, and the Austrian biologist, Konrad Lorenz. For samples of their contemporaneous works, see Charles Elton, *Animal Ecology* (1927; reprint ed., London: Sidgwick and Jackson, 1935), and Konrad Lorenz, "Contributions to the Study of the Ethology of Social Corvidae" (1931), and "Companions as Factors in the Bird's Environment: The Conspecific as the Eliciting Factor in Social Behaviour" (1935), in Konrad Lorenz, *Studies in Animal and Human Behaviour*, trans. Robert Martin (Cambridge: Harvard University Press, 1970), 1: 1–56, 101–258.
68. S. [*sic*] Zuckerman, *The Social Life of Monkeys and Apes* (London: Kegan, Paul, Trench, Trubner, 1932), pp. 232 (first quotation), 238 (remaining quotations).
69. I base my conclusions on both textual citations and bibliographic entries in *Social Life of Monkeys and Apes*, pp. 317–40. Not surprisingly, C. Ray Carpenter struck up a correspondence with Zuckerman; see Zuckerman to Carpenter, Aug. 8, 1937, box 8, folder "Z Correspondence," Carpenter Papers. In contrast, Allee corresponded extensively with Charles Elton from 1928 to 1951, and although they had differences of opinion, neither emphasized sexual behavior; see box 16, folder 8, Allee Papers.
70. Bronislaw Malinowski, *Sex and Repression in Savage Society* (New York: Harcourt, Brace, 1927), p. 277 (quotation). Zuckerman described his interaction with Malinowski, in *Social Life of Monkeys and Apes*, p. xi. For an excellent discussion of Malinowski's psychological interests, see George W. Stocking, Jr., "Anthropology and the Science of the Irrational: Malinowski's Encounters with Freudian Psychoanalysis," in George W. Stocking, Jr., *Malinowski, Rivers, Benedict, and Others: Essays on Culture and Personality* (Madison: University of Wisconsin Press, 1986), pp. 13–49. Malinowski wrote pointedly against scholars who insisted on the biological basis of culture, as many of the American animal behavior scientists did. See his review of Auguste Forel's book, *Le Monde Social des Fourmis, Comparé à Celui de L'Homme*, "Instinct and Culture in Human and Animal Societies," *Nature* 114:2855 (1924): 79–82.
71. W. C. Allee to VA, Sept. 14, 1948 (first quotation); [W. C. Allee] to Bill (his nephew), Aug. 3, 1950 (second quotation), box 25, folder 8, Allee Papers. See also Evans and Evans, *Wheeler*, pp. 299–300; Greenwood and Smith, "McDougall," p. 51; [William McDougall], [Notes on his Terminal Illness], typescript, box 5, McDougall Papers; Wolfe, Barger, and Benison, *Cannon: Science and Society*, pp. 521–22; and Schmidt, "Allee," pp. 25–26.
72. W. C. Allee to Margaret Nice, May 24, 1949, box 20, folder 12, Allee Papers. Documents about Nice's campaign, such as "Christmas 1946," typescript, Nov. 14, 1946, are also found in box 5, folder 11, Nice Papers.
73. S. Bayne-Jones (of the NRC), [Report of Meeting with Dr. Herbert M. Evans], May 12, 1933 (first and second quotations); S. Bayne-Jones to Robert M. Yerkes, May 12, 1933 (third and fifth quotations); Frederick L. Hisaw to Robert M. Yerkes, June 24, 1933 (fourth quotation), box 78, folder 1049, Cannon Papers. Bayne-Jones informed Evans on June 6, 1933, that his $8,000 grant for 1933–1934 was being rescinded (same folder). Although CRPS committee members agreed that recipients and their institutions should not profit from commercial contracts, researchers felt differently. See Yerkes to Hisaw, May 30, 1933, as well as Hisaw's letter to Yerkes cited here. For committee opinions, see Adolf Meyer to Robert M. Yerkes, May 8, 1933, and Frank R. Lillie to Robert M. Yerkes, May 9, 1933, same folder.
74. William Walter Greulich, "Application for Grant for Research in Problems of Sex," Dec. 20, 1939, typescript, box 78, folder 1058, Cannon Papers (quotations). Although there is no record in the Cannon Papers of whether the committee funded Greulich's proposal, he seems

to have incorporated the measurements he outlined into a 1941 article on basal body temperature; see Alex F. Roche, "Obituary: William Walter Greulich, 1899–1986," *American Journal of Physical Anthropology* 72:3 (1987): 275.

75. Yerkes, *Chimpanzees*, p. 83 (quotations). Yerkes's judgment on female primates has not gone unnoticed by feminists. See Ruth Herschberger, "Josie Takes the Stand," in Ruth Herschberger, *Adam's Rib* (New York: Harper and Row, 1948), pp. 5–14, and Donna Haraway, *Primate Visions: Gender, Race, and Nature in the World of Modern Science* (New York: Routledge, 1989), pp. 79–83.
76. A photo of Ada gardening, as well as of a ribbon she won in 1929 for her tulips, are included in the Yerkes Family Papers, http://newhavenmuseum.org/library/catalog-of-manuscript-collections/?album=1&gallery=155 (accessed June 1, 2019). Her college major in botany most likely contributed to her gardening skill.
77. Robert M. Yerkes to Calvin P. Stone, Nov. 18, 1953, Stone Papers microfilm (quotation). Yerkes's paper was Robert M. Yerkes and John D. Dodson, "The Relation of Strength of Stimulus to the Rapidity of Habit-Formation," *Journal of Comparative Neurology and Psychology* 18 (1908): 459–82.

Chapter 4

1. Norman R. F. Maier, *Studies of Abnormal Behavior in the Rat: The Neurotic Pattern and an Analysis of the Situation Which Produces It* (New York: Harper and Brothers, 1939), title page (first quotation), p. 33 (second quotation).
2. On Maier's extended argument with Clifford T. Morgan, see Donald A. Dewsbury, "On Publishing Controversy: Norman R. F. Maier and the Genesis of Seizures," *American Psychologist* 48:8 (1993): 869–77. For a revealing sketch of Morgan, see Eliot Stellar and Gardner Lindzey, "Clifford T. Morgan: 1915–1976," *American Journal of Psychology* 91:2 (1978): 343–48. Although Dewsbury makes a compelling argument that unremitting professional conflict damages an academic field, and in this case drove Maier away from animal studies and into industrial psychology, I suggest that public distaste for these high-profile and graphic discussions of animals in distress also discouraged Maier from continuing the work. His disillusionment with psychology as he once practiced it was profound in any case; see N. R. F. Maier, "Maier's Law," *American Psychologist* 15:3 (1960): 208–12.
3. Robert M. Yerkes and Ada W. Yerkes, "The Comparative Psychopathology of Infrahuman Primates," in Madison Bentley and E. V. Cowdry, eds., *The Problem of Mental Disorder: A Study Undertaken by The Committee on Psychiatric Investigations, National Research Council* (New York: McGraw-Hill, 1934), title page (first quotation), p. 331 (second quotation); Robert M. Yerkes and Ada W. Yerkes, "Nature and Conditions of Avoidance (Fear) Response in Chimpanzee," *Journal of Comparative Psychology* 21:1 (1936): 53 (third quotation). Yerkes's earlier experiment using electric shock has been remembered as a classic study, yielding the "Yerkes-Dodson" law; but its use of a painful disincentive only became current practice in the field decades after the tests were performed. See Robert M. Yerkes and John D. Dodson, "The Relation of Strength of Stimulus to Rapidity of Habit-Formation," *Journal of Comparative Neurology and Psychology* 18:5 (1908): 459–82.
4. Shepherd Ivory Franz, "On the Functions of the Cerebrum: I. The Frontal Lobes in Relation to the Production and Retention of Simple Sensory-Motor Habits," *American Journal of Physiology* 8:1 (1902): 21 (first and second quotations); Carlyle F. Jacobsen, J. B. Wolfe, and T. A. Jackson, "An Experimental Analysis of the Functions of the Frontal Association Areas in Primates," *Journal of Nervous and Mental Disease* 82:1 (1935): 9 (third quotation), 10 (fourth quotation).

5. Walter B. Cannon, "The Body Physiologic and the Body Politic," *Science*, new series, 93:2401 (1941): 1 (first quotation), 8 (second quotation), 6 (third quotation), 7 (fourth quotation). Cannon delivered the speech in Philadelphia on Dec. 27, 1940.
6. Walter B. Cannon, "The Rôle of Emotion in Disease," *Annals of Internal Medicine* 9:11 (1936): 1465. Returning home to Cambridge in the fall of 1935, Cannon delivered this speech to the American College of Physicians in Detroit, Mar. 4, 1936.
7. Robert M. Yerkes, "Walter Bradford Cannon, 1871–1945," *Psychological Review* 53:3 (1946): 143.
8. Calvin S. Hall, "Emotional Behavior in the Rat: I. Defecation and Urination as Measures of Individual Differences in Emotionality," *Journal of Comparative Psychology* 18:3 (1934): 385 (first and second quotations), 387 (third quotation). For a description of the enclosure, see p. 390, and on the rat that starved, p. 394. This essay was the second of four articles by Hall that reported his dissertation; see also Calvin S. Hall, "Drive and Emotionality: Factors Associated with Adjustment in the Rat," *Journal of Comparative Psychology* 17:1 (1934): 89–108; "Emotional Behavior in the Rat. II. The Relationship between Need and Emotionality," *Journal of Comparative Psychology* 22:1 (1936): 61–68; and "Emotional Behavior in the Rat. III. The Relationship between Emotionality and Ambulatory Activity," *Journal of Comparative Psychology* 22:3 (1936): 345–52.
9. S. I. Franz and K. S. Lashley, "The Retention of Habits by the Rat after the Destruction of the Frontal Portion of the Cerebrum," *Psychobiology* 1:1 (1917): 3–18. For Franz's earlier surgeries, see "On the Functions of the Cerebrum," and Shepherd Ivory Franz, "Observations on the Functions of the Association Areas (Cerebrum) in Monkeys," *Journal of the American Medical Association* 47:18 (1906): 1464–67. For biographical information on Franz, see Roger K. Thomas, "Shepherd Ivory Franz (1874–1933)," in *Encyclopedia of Psychology*, ed. Alan E. Kazdin (Washington, D.C.: American Psychological Association, 2000), 3: 396–98. For biographical background on Lashley, see Nadine M. Weidman, *Constructing Scientific Psychology: Karl Lashley's Mind-Brain Debates* (Cambridge: Cambridge University Press, 1999), esp. chap. 3, and Frank A. Beach, "Karl Spencer Lashley, 1890–1958," *National Academy of Sciences Biographical Memoirs* (1961), http://www.nasonline.org/publications/biographical-memoirs/memoir-pdfs/lashley-karl.pdf (accessed June 1, 2019).
10. Franz and Lashley, "Retention of Habits," pp. 17 (first quotation), 10 (second quotation), 4 (third quotation), 17 (fourth and fifth quotations). See chapter 2, earlier, for a discussion of K. S. Lashley, "A Simple Maze: With Data on the Relation of the Distribution of Practice to the Rate of Learning," *Psychobiology* 1:5 (1918): 353–67.
11. Franz and Lashley, "Retention of Habits," pp. 8 (first quotation), 15 (second quotation), 17–18 (third quotation). The journal *Psychobiology* seems a perfect outlet for a surgical study, and it is intriguing that this coauthored essay was the lead article in the opening issue. The Johns Hopkins psychologist Knight Dunlap established and edited the periodical, which continued until 1920. Dunlap was the author of *An Outline of Psychobiology* (Baltimore: Johns Hopkins University Press, 1914).
12. Franz, "On the Functions of the Cerebrum," pp. 6 (first quotation), 13 (second quotation). In 1902 Franz noted his debt to C. Lloyd Morgan and, even more, Edward Thorndike for his associative theory of learning; see ibid, p. 6. The classic exposition was Edward L. Thorndike, "Animal Intelligence; an Experimental Study of the Associative Processes in Animals" (1898), in Thorndike, *Animal Intelligence: Experimental Studies* (New York: Macmillan, 1911), pp. 20–155.
13. The shift toward an organic, biologically grounded understanding of the mind is all the more significant because Lashley had recently collaborated with John B. Watson, when Watson still entertained the idea that instincts mix with learning to produce observed behavior.

Among their coauthored papers, see John B. Watson and K. S. Lashley, *Homing and Related Activities of Birds*, Papers from the Department of Marine Biology of the Carnegie Institution of Washington (Washington, D.C.: Carnegie Institution, 1915), 7: 5–84. For a later example of the direction of Lashley's thinking, see "The Thalamus and Emotion," *Psychological Review* 45:1 (1938): 42–61. Weidman makes an excellent case for these subtle transitions in theorizing—which call into question our retrospective impression that nature-nurture was a clear historical dichotomy—in *Constructing Scientific Psychology*, esp. chaps. 2–3.

14. For these professional connections, see Beach, "Lashley," pp. 176–78 (on Stone); anon., "Deans and Presidents," unpaginated typescript, biographical sketches of administrators, State University of New York Health Center, Syracuse, http://library.upstate.edu/collections/history/institution/ferguson/Deans%20and%20Presidents.pdf (accessed Dec. 6, 2013; not available online at this address June 1, 2019; print copy in the author's possession) (Jacobsen); Donald A. Dewsbury, "Frank Ambrose Beach, 1911–1988," *National Academy of Sciences Biographical Memoirs* (1998), pp. 5–6, http://www.nap.edu/html/biomems/fbeach.pdf (accessed June 1, 2019) (Beach); Peter M. Milner and Brenda Milner, "Donald Olding Hebb," *Biographical Memoirs of Fellows of the Royal Society* 42 (1996): 193–204 (Hebb); Lester R. Aronson and Ethel Tobach, "Preface," in T. C. Schneirla, *Selected Writings of T. C. Schneirla*, ed. Lester R. Aronson et al. (San Francisco: W. H. Freeman, 1972), p. ix (Maier and Schneirla); Miki Takasuna, "Six Researchers' Contributions to Comparative Psychology in Japan: 1900–1945," *Japanese Psychological Research* 47:2 (2005): 91 (Yoshioka); Stellar and Lindzey, "Morgan," pp. 343–45 (Morgan).
15. Stellar and Lindzey, "Morgan," p. 344 (quotation). For the chronology of Jacobsen and Carpenter in Yerkes's laboratory, see Jacobsen, Wolfe, and Jackson, "Experimental Analysis," p. 1, and Robert M. Yerkes, "Foreword," in C. R. Carpenter, "A Field Study of the Behavior and Social Relations of Howling Monkeys," *Comparative Psychology Monographs*, no. 48, 10:2 (Baltimore: Johns Hopkins University Press, 1934): 1–4. There was a kind of round robin migration through animal research sites, because Joseph Yoshioka worked first with Yerkes and then Lashley, so in reverse order than Jacobsen and Carpenter. See Joseph G. Yoshioka, "Learning versus Skill in Rats," *Pedagogical Seminary and Journal of Genetic Psychology* 41:2 (1932): 406, and Takasuna, "Comparative Psychology in Japan," p. 91. The group's ideas also gained visibility through a coauthored textbook, N. R. F. Maier and T. C. Schneirla, *Principles of Animal Psychology* (New York: McGraw-Hill, 1935). The authors saw the current scientific context as particularly attentive to the "treatment of psychological problems" (p. ix), and many of the experiments they chose for discussion involved disruptive surgeries, deprivation (of water, for example), or inflicted pain, e.g., pp. 271, 415.
16. Karl Lashley to Calvin Stone, Mar. 21, 1920, quoted in Beach, "Lashley," pp. 176 (first quotation), 177 (second and third quotations), 176 (fourth and fifth quotations), 177 (sixth, seventh, and eighth quotations); Franz and Lashley, "Retention of Habits," p. 3 (ninth quotation).
17. Walter B. Cannon to Knight Dunlap, Feb. 16, 1924, box 105, folder 1446, Walter B. Cannon Papers (H MS c40), Center for the History of Medicine, Francis A. Countway Library of Medicine, Harvard Medical School, Boston, Mass. Although Cannon did not identify Stone, he was named in Dunlap to Cannon, Feb. 14, 1924 (same folder). It is not clear which of Stone's papers eventually published in the *Journal of Comparative Psychology*, edited by Dunlap, provoked Cannon. See a bibliography of Stone's articles of the mid-1920s, in Ernest R. Hilgard, "Calvin Perry Stone, 1892–1954," *National Academy of Sciences Biographical Memoirs* (1994), pp. 411–12, http://www.nasonline.org/publications/biographical-memoirs/memoir-pdfs/stone-calvin-p.pdf (accessed June 1, 2019). Stone made no mention of palliative care, in any case, in connection with details of the surgery, in Calvin P. Stone, "The Retention of Copulatory Ability in Male Rats Following Castration," *Journal of Comparative Psychology* 7:5 (1927): 372.

18. Leonard Carmichael, "The Experimental Embryology of Mind," *Psychological Bulletin* 38:1 (1941): 8. The paper was Carmichael's presidential address to the American Psychological Association, delivered Sept. 6, 1940. An earlier series of Carmichael's papers focused on salamander and frog embryos, beginning with, "The Development of Behavior in Vertebrates Experimentally Removed from the Influence of External Stimulation," *Psychological Review* 33:1 (1926): 51–58.
19. Jacobsen, Wolfe, and Jackson, "Experimental Analysis," pp. 9 (first, second, and third quotations), 10 (fourth quotation). The authors recorded the amount of brain tissue removed from each chimpanzee on pp. 2–3. For Pavlov (1849–1936) on experimental neurosis, see "Experimental Neuroses" (1931) and "Concerning Human and Animal Neuroses" (1929), in Ivan Petrovitch Pavlov, *Lectures on Conditioned Reflexes: Conditioned Reflexes and Psychiatry* (vol. 2), ed. W. Horsley Gantt (1941; reprint ed., New York: International Publishers, 1963), pp. 73–74, 83–85. Pavlov seems to have discovered accidentally that training can cause abnormality in animals; he did not set out to produce dysfunction.
20. Calvin S. Hall, "The Inheritance of Emotionality," *Sigma Xi Quarterly* 26:1 (1938): 20 (first quotation), 17 (second quotation); Hall, "Emotional Behavior in the Rat, I," p. 390 (third quotation); Hall, "Emotional Behavior in the Rat, III," p. 351 (fourth quotation). Tolman's interest in instincts and drives makes it unsurprising that the idea of emotionality became prominent in his laboratory. See, e.g., Edward Chace Tolman, "Can Instincts Be Given Up in Psychology?" *Journal of Abnormal Psychology and Social Psychology* 17:2 (1922): 139–52; "The Nature of the Fundamental Drives," *Journal of Abnormal and Social Psychology* 20:4 (1926): 349–58, and his masterwork, *Purposive Behavior in Animals and Men* (1932; reprint ed., no place: Appleton, Century, Crofts, 1967). Before Hall, Yoshioka (1893–unknown), who completed a doctorate under Tolman in 1926, called attention to "emotional excitability" ("Learning versus Skill in Rats," p. 409). Still, Berkeley was not the sole site of explicit experimentation on the emotions. James R. Patrick, for example, who studied under J. F. Dashiell at the University of North Carolina in the late 1920s, oscillated between animal and human subjects. See, e.g., "The Effect of Emotional Stimuli on the Activity Level of the White Rat," *Journal of Comparative Psychology* 12:4 (1931): 357–63, and "Studies in Rational Behavior and Emotional Excitement. I. Rational Behavior in Human Subjects," *Journal of Comparative Psychology* 18:1 (1934): 1–22. In "Studies in Rational Behavior" (pp. 1–3), Patrick acknowledged his debt to G. V. Hamilton, Yerkes's one-time student who worked with both primates and humans. Because Yerkes once taught Tolman at Harvard as well, there was a logic to the converging intellectual lineages.
21. Yerkes and Yerkes, "Comparative Psychopathology," in Bentley and Cowdry, eds., *Problem of Mental Disorder*, pp. 327 (first quotation), title page (second quotation), 329 (third quotation).
22. Madison Bentley and E. V. Cowdry, "The Character of the Problem," in Bentley and Cowdry, eds., *Problem of Mental Disorder*, pp. 4 (first, second, third, and fourth quotations), 5 (fifth quotation); Yerkes and Yerkes, "Comparative Psychopathology," in ibid, p. 331 (sixth and seventh quotations).
23. Maier, *Abnormal Behavior in the Rat*, title page (first quotation), fig. 6 caption (second quotation), fig. 11 caption (third and fourth quotations), pp. 10 (fifth quotation), 53 (sixth quotation), 54 (seventh quotation). The account of the rat that collapsed appears on p. 44.
24. T. C. Schneirla, "Studies on Army Ants in Panama," *Journal of Comparative Psychology* 15:2 (1933): 267 (first quotation), 268 (second quotation), 276 (third quotation), 272 (fourth quotation), 274 (fifth, sixth, and seventh quotations). Italics in quotations appear in the texts unless otherwise noted. Schneirla's dissertation appeared as "Learning and Orientation in Ants," *Comparative Psychology Monographs* 6 (Baltimore: Johns Hopkins University Press, 1929): 1–143.

25. [C. R. Carpenter], "Report in Brief of a Survey Study of Orang Outans and Reservation Situation and Prospects; Atjeh, Sumatra," in journal "1-100 Field Notes of Sumatra" (hereafter Sumatra Journal), p. 74 (first quotation), box 5, folder "Research Observation Journals, Sumatra, July 20–Aug. 14, 1937?" [*sic*]; [C. R. Carpenter], "The Markle-Columbia Primate Expedition" (hereafter Markle-Columbia Primate Journal), pp. 6 (second quotation), 7 (third quotation), box 5, folder "Research Observation Journals, Markle-Columbia Primate Expedition, Background of Santiago Colony, 1938," group 19, Clarence Ray Carpenter Papers, Eberly Family Special Collections Library, Penn State University, University Park, Pa. All Carpenter papers are located in group 19. Mention of castration and lobotomy appears in Markle-Columbia Primate Journal, pp. 5, 7, respectively.
26. Carpenter published his dissertation as, "Psychobiological Studies of Social Behavior in Aves. I. The Effect of Complete and Incomplete Gonadectomy on the Primary Sexual Activity of the Male Pigeon," *Journal of Comparative Psychology* 16:1 (1933): 25–57, and "Psychobiological Studies of Social Behavior in Aves. II. The Effect of Complete and Incomplete Gonadectomy on Secondary Sexual Activity with Histological Studies," *Journal of Comparative Psychology* 16:1 (1933): 59–97.
27. [Carpenter], "Report in Brief," [Aug. 1937], Sumatra Journal, p. 80 (quotation), Carpenter Papers. Having occupied Manchuria since 1931, the Japanese military invaded and occupied Beijing beginning July 7, 1937. Possibly Carpenter had not heard this news, or perhaps these events seemed far from his work in Southeast Asia. On his way there, he had landed in Yokohama, Jan. 22, 1937; see Jan. 22, "Personal Diary: Dec. 28, 1936, California to Dec. 1937. Travel from California through Japan, China, Singapore, Thailand, etc." (hereafter East Asia Journal), unpaginated, box 5, folder "Research Observation Journals, Small Memo Books—1 on Gibbons, 3 on Japan, 1 Diary Dec. 1936-Dec. 1937, Siam 1937," Carpenter Papers.
28. "American Scientist Talks to Penang Rotarians about Origin of Man," *Straits Echo*, Sept. 1, 1938, clipping, in Markle-Columbia Primate Journal, p. 130 (quotations), Carpenter Papers. Note that the quotations are the words the newspaper chose to summarize Carpenter's speech.
29. "Indian Monkeys for United States. Cargo of 500. Experimental Colony to be Formed," *Bombay Times*, Oct. 3, 1938, clipping, and Wayfarer, "Stop Export of these Monkeys: In the Name of Humanity," no source, clipping, in Markle-Columbia Primate Journal, pp. 128 (first quotation), 132 (second quotation), Carpenter Papers. Some of the same words used by "Wayfarer" also appear in Bhagat Ram, secretary, Animals' Friends Society, Punjab, "Export of Monkeys," [*Sunday Standard*, Sept. 26, 1937], clipping, in Markle-Columbia Primate Journal, p. 129. For the American arrival, see "Monkeys Arrive Here for Breeding Purposes," *Boston Post*, Nov. 14 [1938], in Markle-Columbia Primate Journal, p. 133. Whatever the process by which the press picked up these scientific stories, such publicity was commonplace. See, e.g., "Scientists to Live with Jungle Apes: They Hope by Nine Months Study to Trace More of the 'Missing Links' to Man," clipping, *New York Times*, Dec. 16, 1936, in C. R. Carpenter, journal "Barro Colorado Island—Bard College (Howler) Expedition, 1935–1936; Santiago Island, Puerto Rico, June–July 1939" (hereafter BCI-SI Journal), p. 290, box 5, folder "Research Observation Journals, Barro Colorado Island—Bard College (Howler) Expedition, 1935–1936, Santiago Island, Puerto Rico, June–July 1939," Carpenter Papers. Note that despite Carpenter's title for the journal, the article's subject was fieldwork in Thailand.
30. Some of the most dedicated laboratory experimentalists seem least likely to have left travel records, either because they did not go abroad or did not maintain a systematic diary or correspondence while away. There is no record of international travel for Karl Lashley before World War II, for example. Reports of journeys produced by scholars who conducted fieldwork—including zoologist Harold Coolidge and ornithologists Frank Chapman, Ernst

Mayr, and Margaret Morse Nice—often do not have an explicit relationship to their laboratory studies, if they conducted them. From a professional viewpoint, fieldwork continued to have a questionable status in the specialty of animal behavior in the 1930s, as I explain in chapter 3. Travel accounts by researchers who also pursued controlled studies at home are available, however, for Walter Cannon, C. R. Carpenter, T. C. Schneirla, Calvin Perry Stone, and Robert Yerkes.

31. Margaret Morse Nice, "Birds of a Hungarian Lake," p. 6, typescript, box 12, folder 1, Margaret Morse Nice Papers (Collection 2993), Division of Rare and Manuscript Collections, Cornell University Library, Ithaca, N.Y. I have silently corrected small typographical errors (here "Angol" for "Anglo") in this quotation from Nice and other quotations of other writers.
32. [Calvin Perry Stone], "Letters to Family While on Sabbatical Leave in England and Europe, 1932" (hereafter Stone Sabbatical Letters), unpaginated typescripts, Dresden, June 28 (first quotation), Edinburgh, Apr. 26 (second and third quotations), Calvin Perry Stone Papers microfilm (SC0440), University Archives, Special Collections, Stanford University Library, Stanford, Calif. (hereafter Stone Papers microfilm).
33. Stone Sabbatical Letters, Oxford, Apr. 3 (first and second quotations), Berlin, June 6 (third quotation), Leipzig, June 26 (fourth quotation), [Leipzig], June 21 (fifth quotation), Glasgow, Apr. 2[8?] (sixth quotation), Stone Papers microfilm. See chapter 2, earlier, for his family background and wartime service.
34. Frank M. Chapman, *Life in an Air Castle: Nature Studies in the Tropics* (New York: D. Appleton Century, 1938), p. 3 (first quotation); [Carpenter], Jan. 19, East Asia Journal (second and third quotations), Carpenter Papers; Roberta Yerkes to Ada Yerkes and David Yerkes, n.d., series VI, box 161, folder 2524, Robert Mearns Yerkes Papers (MS 569), Manuscripts and Archives, Yale University Library, New Haven, Conn. (fourth and fifth quotations). On the trials of travel, see Chapman, *Life in an Air Castle*, p. 2 (tiny cabin); [Carpenter], Jan. 11–13, 21 (seasickness) and Feb. 3 (drunken passengers), East Asia Journal, Carpenter Papers; and Robert M. Yerkes to Ada Yerkes and David Yerkes, Aug. 17, 1927 (prospect of malaria), series VI, box 161, folder 2524, Yerkes Papers.
35. [Carpenter], Jan. 7 (first quotation), Jan. 23–24 (second quotation), Jan. 22 (third quotation), Jan. 30–31 (fourth and fifth quotations), Jan. 26 (sixth and seventh quotations), Jan. 31 (eighth, ninth, and tenth quotations), East Asia Journal, Carpenter Papers.
36. [Carpenter], Mar. 3, Feb. 17, journal "Research Observations Journals: Asiatic Primate Expedition, Field Observations in Siam, 1937" (hereafter Siam Journal), pp. 21 (first and second quotations), 1 (third and fourth quotations), box 5, folder "Research Observations Journals, Asiatic Primate Expedition, Field Observations in Siam, 1937," Carpenter Papers.
37. Harold J. Coolidge, Jr., and Theodore Roosevelt, Jr., *Three Kingdoms of Indo-China* (New York: Thomas Y. Crowell, 1933), p. 14 (first quotation); Chapman, *Life in an Air Castle*, p. 18 (second quotation). Coolidge and Roosevelt split up geographically in Indo-China, and, after coauthoring the introduction, also divided up the writing between them. All of my references and quotations appear either in coauthored or Coolidge-authored portions.
38. Cornelia James Cannon, "The G.P.U. in Action," p. 5, typescript, 1936, series II, subseries B, 12.10, Cannon Family Papers (MC 553), Arthur and Elizabeth Schlesinger Library, Harvard University, Cambridge, Mass., http://pds.lib.harvard.edu/pds/view/11822739 (accessed June 1, 2019). "The Editors," Harper and Brothers, to Mrs. Walter B. Cannon, Aug. 14, 1936, rejecting Cornelia's manuscript, is located in the same file.
39. [Cornelia James Cannon], "Letters Written and Journal Kept during Trip through Russia" (hereafter Cannon Russia Typescript), July 11, 1935, p. 6 (first quotation), July 12, p. 9 (second and third quotations), July 11, 1935, p. 6 (fourth quotation), July 12, p. 8 (fifth through eighth quotations), partly paginated typescript, 1936, series II, subseries B, 12.9, Cannon Family

Papers, http://pds.lib.harvard.edu/pds/view/11822745 (accessed June 1, 2019). Note that this typescript begins on p. 16 of the electronic file; the previous typescripts in the file cover the Cannons' experiences in Japan and China on their way to Russia. For accounts of their trip, see Elin L. Wolfe, A. Clifford Barger, and Saul Benison, *Walter B. Cannon: Science and Society* (Boston: Boston Medical Library in the Francis A. Countway Library of Medicine, 2000), chap. 17, and Marian Cannon Schlesinger, *Snatched from Oblivion: A Cambridge Memoir* (Boston: Little Brown, 1979), chap. 12. The personal reason that the Cannons circumnavigated the globe in order to arrive in Leningrad was to visit their daughter Wilma, newly married to the rising historian of China, John K. Fairbank, and living in Beijing for the sake of her husband's research. Examples of dangers encountered by scientists abroad that significantly exceeded inconvenience are as follows. One American member of the Kelley-Roosevelt Expedition of the Field Museum of Chicago died of malaria in Vientiane, French Indo-China, in 1929; see Coolidge and Roosevelt, *Three Kingdoms*, pp. 18, 171. In Dutch New Guinea to collect bird specimens in 1928, Ernst Mayr coped with headhunters, multiple diseases, poor weather, and afflicting insects; see Ernst Mayr, "II. My Dutch New Guinea Expedition, 1928," *Novitates Zoologicae* 36 (1930–1931): 20–26.

40. [Cornelia Cannon], Cannon Russia Typescript, July 11, 1935, p. 6, Cannon Family Papers.
41. Stone Sabbatical Letters, London, Apr. 15, 1932 (first, second, and third quotations), York, May 1 (fourth and fifth quotation), Leipzig, June 21 (sixth and seventh quotations), Heidelberg, July 22 (eighth quotation), Stone Papers microfilm. For Cannon's temperance pledge, along with the other officers in the Harvard-staffed Army medical unit, see Walter B. Cannon to Cornelia James Cannon, May 15, 1917, box 185, letter book 1, Cannon Papers. On the Cannons' beer drinking on the Trans-Siberian Railway, see Cornelia Cannon, Cannon Russia Typescript, July 10, 1935, p. 4, Cannon Family Papers.
42. [Carpenter], Feb. 3, East Asia Journal, Carpenter Papers (first and second quotations); Margaret Morse Nice, "Some Ornithological Experiences in Europe," p. 11 (third quotation), typescript, box 12, folder 1, Nice Papers. For Carpenter's other references to drinking on the same trip, see Jan. 20 (champagne), Jan. 24 and 26 (sake), East Asia Journal. For Nice's refusal of rum and acceptance of wine, see "Birds of a Hungarian Lake," pp. 4, 1.
43. Nice, "Birds of a Hungarian Lake," p. 6.
44. [Cornelia Cannon], Cannon Russia Transcript, July 12, 1935, p. 10 (first quotation), July 16, p. 16 (second, third, and fourth quotations), July 11, p. 6 (fifth, sixth, and seventh quotations), July 17, p. 18 (eighth and ninth quotations), Cannon Family Papers.
45. Cannon Schlesinger, *Snatched from Oblivion*, p. 160 (first quotation); Stone Sabbatical Letters, Leipzig, June 20 (second quotation), Leipzig, June 21 (third quotation), Stone Papers microfilm.
46. Walter B. Cannon, *Some Implications of the Evidence for Chemical Transmission of Nerve Impulses: Address at the Plenary Sitting of the Congress* (Moscow-Leningrad: State Publishing House for Biological and Medical Literature, 1935), p. 12 (first quotation); [Cornelia Cannon], Cannon Russia Typescript, Aug. 8, 1935, no page (second quotation), Aug. 10, no page (third quotation), Cannon Family Papers. The centerpiece of Pavlov's trip to the United States, a story often retold in scientific circles, was that Pavlov lost nearly all his money in a robbery in New York City. Along with the Rockefeller Foundation, Cannon orchestrated replacement of the stolen cash so that Pavlov could continue his trip. See Walter Bradford Cannon, *The Way of an Investigator: A Scientist's Experiences in Medical Research* (New York: Norton, 1945), pp. 185–86, and Cannon Schlesinger, *Snatched from Oblivion*, p. 219. Cannon Schlesinger identified "Prof. Pavlov" as the meaning of her mother's abbreviation, in ibid, p. 218.
47. Robert M. Yerkes [in Berlin] to [Ada, David, and Roberta Yerkes], June 16, 1929, series VI, box 163, folder 2566, Yerkes Papers (first and second quotations); Robert M. Yerkes, *Almost Human* (New York: Century, 1925), p. viii (third and fourth quotations); Stone

Sabbatical Letters, "Visit to Ernest Jones," Apr. 3, 1932, Stone Papers microfilm (fifth and sixth quotations); Nice, "Ornithological Experiences in Europe," p. 4 (seventh quotation). Stone's critique of McDougall was "William McDougall's Methodological Concept of Instinct," in Stuart A. Rice, ed., *Methods in Social Science: A Case Book* (Chicago: University of Chicago Press, 1931), pp. 186–94. McDougall also wrote a letter of introduction for Stone to the psychologist C. S. Myers in London; see Stone Sabbatical Letters, "Afternoon of April 15," Stone Papers microfilm. How Stone approached McDougall is an interesting question, because they never overlapped in an American university and so presumably were professional acquaintances rather than personal friends. The case casts light on the practice among American scientists of the old-fashioned scholarly custom of issuing letters of introduction. As I argue later, however, Stone's published criticism of McDougall was closer to veiled praise of the older theorist, who functioned as a stereotyped pariah for American empiricists. For his peers' criticism of McDougall, see David L. Krantz and David Allen, "The Rise and Fall of McDougall's Instinct Doctrine," *Journal of the History of the Behavioral Sciences* 3:4 (1967): 326–38, as well as my discussion of this scholarly interpretation in chapter 2.

48. Coolidge and Roosevelt, *Three Kingdoms*, pp. 18 (first quotation), 28 (second quotation), 96 (third quotation), 66 (fourth quotation), 67 (fifth quotation). Coolidge (1904–1985), who completed a Harvard BS degree in 1927, had done fieldwork in Africa (1926–1927) before this trip to French Indo-China (1928–1929). See his *Notes on the Gorilla*, offprint from *Report of the Harvard-African Expedition upon the African Republic of Liberia and the Belgian Congo* (Cambridge, Mass.: Harvard Museum of Comparative Zoology, 1930), pp. 623–35.
49. [Carpenter], July 25, Sumatra Journal, pp. 22 (first quotation), 20 (second quotation). Although there is some truth to the long-held historical argument that anthropological theories of culture displaced biological concepts of race between the world wars in Western Europe and the United States, many scientists continued to affirm the inheritance of physical, temperamental, and mental characteristics within human groups. This view was especially strong among researchers focused on the evolution of organisms, the intellectual circle analyzed in this book. For perspectives on scientific theories of race during the interwar period, see Daniel J. Kevles, *In the Name of Eugenics: Genetics and the Uses of Human Heredity* (New York: Knopf, 1985); Elazar Barkan, *The Retreat of Scientific Racism: Changing Concepts of Race in Britain and the United States between the World Wars* (Cambridge: Cambridge University Press, 1992); John W. Dower, *War without Mercy: Race and Power in the Pacific War* (New York: Pantheon, 1987); and Anne C. Rose, *Psychology and Selfhood in the Segregated South* (Chapel Hill: University of North Carolina Press, 2009).
50. [Cornelia Cannon], Cannon Russia Typescript, July 14, p. 13 (first five quotations), July 27, no page (sixth and seventh quotations), Cannon Family Papers. For Cornelia's eugenic ideas, see Maria I. Diedrich, *Cornelia James Cannon and the Future American Race* (Amherst: University of Massachusetts Press, 2010), esp. chap. 5, and my chapter 3, earlier.
51. Coolidge and Roosevelt, *Three Kingdoms*, p. 114.
52. Yerkes, *Almost Human*, p. xv (first quotation); [Robert M. Yerkes], journal "Daily Record, Anthropoid Work, Havana, Cuba, July–[Aug.], 1924. Robert M. Yerkes," July 18, 1924, p. 28 (second quotation), July 21, p. 38 (third quotation), July 23, p. 45 (fourth quotation), July 22, p. 42 (fifth quotation), series VI, box 172, folder 2666, Yerkes Papers.
53. Walter B. Cannon to Dr. O. H. Robertson, Sept. 20, 1923, box 40, folder 509 (first quotation); H. C. Chang to Walter B. Cannon, June 7, 1938, same folder (second quotation), Cannon Papers. One reason for Cannon's extended correspondence with faculty and students at Peking Union Medical College (established by missionaries in 1906 and taken over by the Rockefeller Foundation after 1915) was the institution's attempt to recruit him to the faculty immediately after World War I; see Wolfe, Barger, and Benison, *Cannon: Science and Society*, p. 64. For an excellent analysis of the college, see Mary Brown Bullock, *An*

American Transplant: The Rockefeller Foundation and Peking Union Medical College (Berkeley: University of California Press, 1980).

54. Frank M. Chapman, *My Tropical Air Castle: Nature Studies in Panama* (New York: D. Appleton, 1929), p. 15 (first quotation); Coolidge, *Three Kingdoms*, p. 118 (second quotation). For Carpenter's questionnaire, see "Questionnaire for Officers of Military Patrols of Atjey Relative to the Distribution and Habits of Orang Outans," in Sumatra Journal, pp. 83–84. In addition to Chapman, American researchers in my group who traveled to Barro Colorado Island for research before 1940 included Warder Clyde Allee, C. R. Carpenter, T. C. Schneirla, and William Morton Wheeler. The British entomologist and ornithologist W. H. Thorpe also conducted research on the island around 1932; see R. A. Hinde, "William Homan Thorpe. 1 April 1902–7 April 1986," *Biographical Memoirs of Fellows of the Royal Society* 33 (1987): 624.
55. [Cornelia Cannon], Cannon Russia Typescript, July 14, p. 14, Cannon Family Papers. For Cannon's praise of governments attentive to science, see *Chemical Transmission of Nerve Impulses*, p. 9. The fourth country he mentioned was Britain.
56. [Cornelia Cannon], Cannon Russia Typescript, July 30, no page (first and second quotations), July 25, p. 31 (third and fourth quotations), July 19, p. 20 (fifth quotation), July 25, p. 31 (sixth, and seventh quotations), Cannon Family Papers. My overall impression is that the Cannons went to Russia expecting to be pleased by Communist reforms, in part because of their loyalty to Pavlov and in part because they saw Soviet society as rationally planned. They reluctantly admitted the seriousness of Stalinist problems. Just days before these expressions of anxiety, for example, Cornelia wrote: "Father flourishes and is as enthusiastic and surprised over everything as I am" (Cannon Russia Typescript, July 19, p. 25). Their favorable view of the Soviet system was not unique among American scientists. Richard S. Lyman, a psychiatrist trained at Johns Hopkins by Adolf Meyer, wrote to Meyer, Oct. 22, 1933, that "I am frankly Russophile and almost enthusiastic about the Soviet experimentation," having visited Pavlov's laboratory in Leningrad in 1930 (folder I/2450/5, Adolf Meyer Papers, Alan Mason Chesney Medical Archives, Johns Hopkins University Medical Institutions, Baltimore, Md.). To set Lyman's praise in biographical context, see Anne C. Rose, "Racial Experiments in Psychiatry's Provinces: Richard S. Lyman and His Colleagues in China and the American South, 1932–51," *History of Psychiatry* 23:4 (2012): 419–36.
57. Nice, "Ornithological Experiences in Europe," p. 7 (first and second quotations); Stone Sabbatical Letters, Stone to Ruth Stone, June 26, 1932 (third quotation), Stone [in Dresden] to "Dear Family," June 28 (fourth and fifth quotations), "Visiting the Laboratories, Institute of Psychology, Berlin," June 6 (sixth quotation), Stone Papers microfilm. At the Institute of Psychology, Stone did meet Lewin's gentile colleague, Wolfgang Köhler, who less than a year later published an isolated protest to Nazi policies that restricted Jews in the academy; see Köhler, "Conversations in Germany" (issued Apr. 28, 1933), in Klaus Hentschel, ed., *Physics and National Socialism: An Anthology of Primary Sources, Science Networks-Historical Studies*, vol. 18 (Basel: Birkhaüser, 1996), pp. 36–40. Köhler emigrated to the United States two years later. Although Stone commented in his letter home of June 28 on the international politics surrounding the war debt, he apparently did not see that public events were raising life-transforming personal issues for his German peers.
58. Margaret Morse Nice to Warder Clyde Allee, May 14, 1949, box 20, folder 12, Warder Clyde Allee Papers, Special Collections Research Center, University of Chicago Library, Chicago, Ill. (quotation). Allee replied to Nice that although he did not want to judge scientists by their politics, "I do want to be less politically naïve than Dr. Lorenz is reported to have been" (May 24, 1949, same folder). The paper coauthored by Nice and Lorenz was "The Maturation of Some Activities in Young Redstarts and Serins," typescript, box 12, folder 2, Nice Papers. The same folder contains their extended correspondence about this joint study.

59. Margaret Morse Nice and Joseph J. Hickey, "Christmas 1946," typescript, pp. 1 (first quotation), 2 (second and third quotations), box 5, folder 11, Allee Papers.
60. On Cannon's activities to assist scientists abroad, see Cannon, *Way of an Investigator*, pp. 164 (Soviet Union), 163 (China), 161–62 (Spain), and Anne C. Rose, "The Translator's Preface: Language Boundaries of English-Speaking Life Scientists, 1914–1945," typescript in the author's possession.
61. Carmichael, "Experimental Embryology of Mind," p. 17.
62. Leonard Carmichael, "Heredity and Environment: Are They Antithetical?" *Journal of Abnormal and Social Psychology* 20:3 (1925): 245–60; Calvin P. Stone, "Wildness and Savageness in Rats of Different Strains," in Karl S. Lashley, ed., *Studies in the Dynamics of Behavior* (Chicago: University of Chicago Press, 1932), pp. 3–55; Hall, "Inheritance of Emotionality" (1938). Yerkes's earlier experiment on temperament, now repeated by Stone in an effort to achieve more precision of measurement, was "The Heredity of Savageness and Wildness in Rats," *Journal of Animal Behavior* 3:4 (1913): 286–96.
63. Tolman, *Purposive Behavior in Man and Animals*, pp. 4 (first quotation), 3 (second quotation); Calvin P. Stone, "Maturation and 'Instinctive' Functions," in Fred A. Moss, ed., *Comparative Psychology* (1934), rev. ed. (New York: Prentice-Hall, 1942), p. 42 (third quotation).
64. Calvin P. Stone, "Multiply, Vary, Let the Strongest Live and the Weakest Die—Charles Darwin," *Psychological Bulletin* 40:1 (1943): 1–24; Hall, "Inheritance of Emotionality," p. 20. For Hall's series of papers dating back to 1934, see n. 8, earlier. For Cannon's experiments that disrupted animal emotions, see *Bodily Changes in Pain, Hunger Fear and Rage: An Account of Recent Researches into the Function of Emotional Excitement* (1915; reprint ed., New York: D. Appleton, 1920).
65. Stone, "McDougall's Concept of Instinct," pp. 187 (first quotation), 188 (second quotation), 189 (third quotation).
66. Leonard Carmichael to Wallace Craig, Feb. 3, 1936 (first and fourth quotations); Wallace Craig to Leonard Carmichael, Feb. 15, 1936 (second and third quotations), folder "Wallace Craig, 1936"; Leonard Carmichael to Robert M. Yerkes, June 23, 1936 (fifth quotation), folder "Robert M. Yerkes 7," Leonard Carmichael Papers (Mss.B.C212), American Philosophical Society, Philadelphia, Pa. As Carmichael rose in the academy, he tried quite valiantly to funnel financial support from various agencies to Craig, before Craig's death in 1954; see folder "Wallace Craig," no date, Carmichael Papers. Craig occasionally issued short essays demonstrating his enduring curiosity and wit, e.g., "The Twilight Song of the Wood Peewee: A Preliminary Statement," *Auk* 43:2 (1926): 150–52, and "The Music of the Wood Pewee's Song and One of Its Laws," *Auk* 50:2 (1933): 174–78. But he produced no major studies, experimental or theoretical, after World War I.
67. Robert M. Yerkes, "Primate Coöperation and Intelligence," *American Journal of Psychology* 50:1/4 (1937): 265.
68. [Warder Clyde Allee], "Group Action among Animals and Its Social Implications," undated typescript, p. 1, box 1, folder 1 (first and second quotations), Allee Papers; W. C. Allee, "Where Angels Fear to Tread: A Contribution from General Sociology to Human Ethics," *Science*, new series 97:2528 (1943): 520 (third and fourth quotations); [Warder Clyde Allee], "Mass Action among Animals and Its Social Implications," undated typescript, p. 5, box 1, folder 1, Allee Papers (fifth quotation). It is possible that these two typescripts ("Group Action" and "Mass Action") were different versions of the same paper.
69. Yerkes, "Primate Coöperation and Intelligence," pp. 255 (first quotation), 263 (second quotation), 262 (third quotation), 260 (fourth quotation).
70. Tolman, *Purposive Behavior in Man and Animals*, pp. 425 (first quotation), 424 (second and third quotations); T. C. Schneirla, "A Consideration of Some Conceptual Trends in

Comparative Psychology" (1952), in *Selected Writings of T. C. Schneirla*, ed. Aronson et al., p. 912 (fourth quotation). On American controversy of the era about what science measures, see Christopher D. Green, "Of Immortal Mythological Beasts: Operationism in Psychology," *Theory and Psychology* 2:3 (1992): 291–320. An external reason for these reexaminations of the authority of science was the development of quantum mechanics and the uncertainty principle by Werner Heisenberg (1901–1976) in the second half of the 1920s, for which he received the Nobel Prize for Physics in 1932. John Dewey appropriated Heisenberg's ideas for his own pragmatist view of the instrumentality of knowledge. Part of Dewey's argument was that science, understood as an evolving method, should formulate social values that change as circumstances require. See John Dewey, *The Quest for Certainty: A Study of the Relation of Knowledge and Action* (New York: Minton Balch, 1929). On the intellectual struggles over scientific certainty in the United States between the wars, see Mark C. Smith, *Social Science in the Crucible: The American Debate over Objectivity and Purpose* (Durham, N.C.: Duke University Press, 1994).

71. Margaret Morse Nice, "Some Ornithological Experiences in Europe," *Bird-Banding* 4:3 (1933): 147 (first, second, and third quotations), 154 (fourth and fifth quotations). This article is the published version of the manuscript cited in n. 42, earlier. Nice frequently admitted to feeling disoriented in a new physical setting, with unfamiliar birds. See also her admission of "inexperience," in "Birds of a Hungarian Lake," p. 2.
72. Walter Cannon, *Chemical Transmission of Nerve Impulses*, pp. 6 (first quotation), 4 (second and third quotations).
73. Robert M. Yerkes, "Primate Coöperation and Intelligence," pp. 269 (first quotation), 270 (second and third quotations), 254 (fourth quotation), 270 (fifth, sixth, seventh, eighth, and ninth quotations).
74. Cannon, "Body Physiologic and Body Politic," p. 1 (first quotation); Walter B. Cannon, "'Voodoo' Death," *American Anthropologist*, new series, 44:2 (1942): 176 (second quotation). For his correspondence with the journal, see Walter B. Cannon to Ralph Linton, Oct. 14, 1941, and Linton to Cannon, Oct. 16, 1941, box 67, folder 904, Cannon Papers.
75. Cannon, *Bodily Changes in Pain, Hunger Fear and Rage*, title page (first quotation); Cannon, "'Voodoo' Death," p. 175 (second quotation); Ralph S. Banay, MD, to Walter Cannon, July 15, 1941, box 67, folder 903, Cannon Papers (third quotation); Cannon, "'Voodoo' Death," pp. 176 (fourth, fifth, and sixth quotations), 175 (seventh, eighth, and ninth quotations). Cannon collected information about cases of physiologically unexplained mortality from an amazing variety of individuals around the world, entirely filling at least five folders (898–901, 903) in box 67, Cannon Papers. As the scholar who popularized the notion of homeostasis (see, e.g., *The Wisdom of the Body* [New York: Norton, 1932]), Cannon now doggedly pursued anecdotes about why the body fatally loses its equilibrium without discernable cause.
76. Lauriston Sharp to Walter B. Cannon, Aug. 10, 1940, box 67, folder 903, Cannon Papers.
77. Ibid.
78. Cannon Schlesinger, *Snatched from Oblivion*, p. 63 (first quotation); Cannon, *Chemical Transmission of Nerve Impulses*, p. 8 (second and third quotations); Cannon, *Bodily Changes in Pain, Hunger Fear and Rage*, title page (fourth quotation).

Chapter 5

1. Theodosius Dobzhansky, *Genetics and the Origin of Species* (New York: Columbia University Press, 1937), p. 149 (first quotation); T. [*sic*] Dobzhansky to "Dear Friends," Nov. 20, 1948,

folder "Dobzhansky-Mayr Correspondence, Aug. 1944–Nov. 1949," Theodosius Dobzhansky Papers (Mss.B.D65), American Philosophical Society, Philadelphia, Pa. (second and third quotations).

2. Dobzhansky, *Genetics and Origin of Species*, preface, n.p. (first quotation), p. 13 (second quotation). On Dobzhansky's fieldwork, see his lively global narratives, 1925–1960, in *The Roving Naturalist: Travel Letters of Theodosius Dobzhansky*, ed. Bentley Glass (Philadelphia: American Philosophical Society, 1980), as well as Francisco J. Ayala, "Theodosius Dobzhansky, 1900–1975," *National Academy of Sciences Biographical Memoirs* (1985), esp. p. 178, http://www.nasonline.org/publications/biographical-memoirs/memoir-pdfs/dobzhansky-theodosius.pdf (accessed June 1, 2019). Darwin's book is *The Expression of the Emotions in Man and Animals* (1872; reprint ed., Chicago: University of Chicago Press, 1965).
3. Daniel S. Lehrman, "A Critique of Lorenz's Theory of Instinctive Behavior," *Quarterly Review of Biology* 28:4 (1953): 345 (first quotation); Charlotte Sleigh, *Six Legs Better: A Cultural History of Myrmecology* (Baltimore: Johns Hopkins University Press, 2007), p. 175 (second quotation); Edward O. Wilson, *The Insect Societies* (Cambridge: Harvard University Press, 1971), p. 460 (third and fourth quotations). The massive amount of writing devoted to the so-called modern synthesis or evolutionary synthesis, a mid-twentieth-century intellectual movement oriented toward population genetics, has obscured the continuing strength of holistic studies of single organisms. Ernst Mayr, a chief architect of the evolutionary synthesis, offers a retrospective view in "Prologue: Some Thoughts on the History of the Evolutionary Synthesis," in Ernst Mayr and William B. Provine, eds., *The Evolutionary Synthesis: Perspectives on the Unification of Biology* (Cambridge: Harvard University Press, 1980), pp. 1–48. Sleigh analyzes postwar disciplinary conflicts at Harvard between practitioners of microscopic and macroscopic approaches, in *Six Legs Better*, pp. 211–18. Lily E. Kay explains how the availability of Rockefeller Foundation support for genetics at Caltech energized molecular biology beginning around 1930, in *The Molecular Vision of Life: Caltech, The Rockefeller Foundation, and the Rise of the New Biology* (New York: Oxford University Press, 1993). The high profile of human psychotherapy after the war, in contrast, helped validate homeostatic assessment of single organisms of all kinds. See, e.g., Marga Vicedo, *The Nature and Nurture of Love: From Imprinting to Attachment in Cold War America* (Chicago: University of Chicago Press, 2013). The near parity at this time of examinations of populations and individuals is visible in back-to-back conferences on these respective subjects hosted by Princeton University in 1947: Conference on Genetics, Paleontology, and Evolution (Jan. 2–4) and Symposium on Heredity and Environment (Apr. 10). Despite their similar titles, the first consisted of evolutionary biologists (including Mayr, J. B. S. Haldane, and Sewall Wright) and the second of psychologists (including Leonard Carmichael, Karl Lashley, and Calvin Stone). For the respective proceedings, see Glenn L. Jepsen, Ernst Mayr, and George Gaylord Simpson, eds., *Genetics, Paleontology, and Evolution* (Princeton: Princeton University Press, 1949), and the entire November 1947 number of the *Psychological Review* 54:6 (1947): 297–352.
4. Widely varying estimates of total military and civilian war fatalities are the result of the immeasurable nature of many variables: deliberate bombing of civilian targets; ethnic cleansing; and famine, disease, and scarce medical care due to war-related civil disruption. What a scholar considers to be a "war-related" outcome also affects total numbers. One historian estimates the war dead as thirty-seven million, and another as fifty-five million; Wikipedia lists a figure as high as eighty-five million. See, respectively, James L. Stokesbury, *A Short History of World War II* (New York: William Morrow, 1980), p. 378; John W. Dower, *War without Mercy: Race and Power in the Pacific War* (New York: Pantheon, 1986), p. 295; and "World War II Casualties," http://en.wikipedia.org/wiki/World_War_II_casualties (accessed June 1, 2019).

5. Technological innovation was a hallmark of World War II. On land, World War I resembled an old-fashioned struggle organized around masses of contending infantry, with modest tank and air support. During World War II, physical scientists and engineers continuously upgraded equipment ranging from an individual's rifle to the atomic bomb. The National Roster of Scientific and Specialized Personnel—in effect a military placement service for experts—had recruited nearly 40,000 chemists, 10,000 civil engineers (the highest number among multiple engineering specialties), and 6,000 physicists by September 1941 (Leonard Carmichael, "The National Roster of Scientific and Specialized Personnel: Third Progress Report," *Science*, new series 95:2456 [Jan. 23, 1942]: 88). The same report noted the recruitment of 3,000 psychologists. Historians often identify World War II as an influential turning point for human psychology and psychiatry, both intellectually and institutionally; see, e.g., Ellen Herman, *The Romance of American Psychology: Political Culture and the Age of Experts* (Berkeley: University of California Press, 1995), and Frank Biess and Daniel M. Gross, eds., *Science and Emotions after 1945: A Transatlantic Perspective* (Chicago: University of Chicago Press, 2014). The evidence presented in this chapter suggests instead steady academic development comparatively detached from war and politics among animal behavior experts. One Cold War conference—titled "Group Processes," sponsored by the Josiah Macy, Jr., Foundation, and held at Cornell University, Ithaca, New York, Sept. 26–30, 1954—gathered politically diverse international figures otherwise connected by their scholarship. The Austrian Konrad Lorenz and Dutch-born Niko Tinbergen, the first a former Nazi Party member and second a victim of Nazi occupation, met along with American psychologists Frank Beach and Daniel Lehrman. Others included émigré and American-born biologists, psychoanalysts, and anthropologists: Ernst Mayr, Erik Erikson, Frieda Fromm-Reichmann, and Margaret Mead. For a transcript, see Bertram Schaffner, ed., *Group Processes: Transactions of the First Conference* (New York: Josiah Macy, Jr., Foundation, 1955).
6. Tania Munz, *The Dancing Bees: Karl von Frisch and the Discovery of the Honeybee Language* (Chicago: University of Chicago Press, 2016), p. 122 (first quotation); Ute Deichmann, *Biologists under Hitler*, trans. Thomas Dunlap (Cambridge: Harvard University Press, 1996), p. 179 (second quotation). In his 1957 memoir, von Frisch dated a crucial insight into bee communication to experiments performed in spring 1945, precisely the moment when Allied troops were advancing across Germany-occupied Europe. He recorded his reaction to the appearance of American troops in his retrospective account, but highlighted scientific discovery, with little mention of the pervasive physical devastation and human dislocation of the war's end. I take this in part to be a Cold War literary strategy to put his professional accomplishment in the foreground, but also a revealing sign of his absorption in research. See Karl von Frisch, *A Biologist Remembers*, trans. Lisbeth Gombrich (Oxford: Pergamon Press, 1967), esp. pp. 142–43. On the violence and suffering in the immediate postwar period, see Keith Lowe, *Savage Continent: Europe in the Aftermath of World War II* (New York: Picador, 2013). Debate about the sincerity and depth of Lorenz's Nazism continues, but the evidence of his willful collaboration is undeniable; see Deichmann, *Biologists under Hitler*, pp. 179–205.
7. "The Nobel Prize in Physiology or Medicine, 1933," http://www.nobelprize.org/nobel_prizes/medicine/laureates/1933/ (accessed June 1, 2019) (first quotation); Thomas H. Morgan, "The Relation of Genetics to Physiology and Medicine," Nobel Lecture, June 4, 1934, pp. 315 (second and third quotations), 313 (fourth quotation), http://www.nobelprize.org/nobel_prizes/medicine/laureates/1933/morgan-lecture.html (accessed June 1, 2019). On Morgan's successive study of various organisms, see A. H. Sturtevant, "Thomas Hunt Morgan, 1866–1945," *National Academy of Sciences Biographical Memoirs* (1959), esp. pp. 286–93, http://www.nasonline.org/publications/biographical-memoirs/memoir-pdfs/morgan-thomas-hunt.pdf (accessed June 1, 2019).

8. Ernst Mayr to Theodosius Dobzhansky, Feb. 28, 1939, box D65.m, folder "Mayr, Ernst, 1937–1947," Dobzhansky Papers. Mayr's first major book, adapted from his Jesup lectures at Columbia University in 1941, aimed to integrate genetics, systematics, and behavioral analysis; see *Systematics and the Origin of Species: From the Viewpoint of a Zoologist* (New York: Columbia University Press, 1942). See also Jürgen Haffer's excellent study, *Ornithology, Evolution, and Philosophy: The Life and Science of Ernst Mayr, 1904–2005* (Berlin: Springer, 2008), esp. pt. 2. On changes in the research focus of American experimental geneticists, see Theophilus Painter, "John Thomas Patterson, 1878–1960," *National Academy of Sciences Biographical Memoirs* (1965), pp. 230–31, http://www.nasonline.org/publications/biographical-memoirs/memoir-pdfs/patterson-john-t.pdf; Theodosius Dobzhansky, "Leslie Clarence Dunn, 1893–1974," *National Academy of Sciences Biographical Memoirs* (1974), pp. 81–83, http://www.nasonline.org/publications/biographical-memoirs/memoir-pdfs/dunn-leslie.pdf; and "Background Note," Dobzhansky Papers, http://www.amphilsoc.org/mole/view?docId=ead/Mss.B.D65-ead.xml (all accessed June 1, 2019). Perhaps because these scientists experienced significant scholarly transformation during their careers, they all retained catholic interests. For example, Patterson analyzed Indian tools collected near his home in Austin, Texas; see J. T. Patterson, "The Corner-Tang Flint Artifacts of Texas," *University of Texas Bulletin*, no. 3618, May 8, 1936 (offprint). As co-investigator with his son, a sociologist, Dunn examined the interaction of social and genetic factors in the history of Roman Jewry; see Leslie C. Dunn and Stephen P. Dunn, "The Jewish Community of Rome," *Scientific American* 196:3 (1957): 118–28. For the Rockefeller commitment to genetics in the 1930s, see Kay, *Molecular Vision of Life*, chaps. 3–4.
9. W. S. Hunter, "Summary Comments on the Heredity-Environment Symposium," *Psychological Review* 54:6 (1947): 348 (first quotation), 349 (second quotation); Calvin P. Stone, "Methodological Resources for the Experimental Study of Innate Behavior as Related to Environmental Factors, ibid, p. 345 (third and fourth quotations). The papers by Carmichael and Lashley, respectively, were "The Growth of the Sensory Control of Behavior before Birth," pp. 316–24, and "Structural Variation in the Nervous System in Relation to Behavior," pp. 325–34. The remaining papers were Frank A. Beach, "Evolutionary Changes in the Physiological Control of Mating Behavior in Mammals," pp. 297–315, and Clifford T. Morgan, "The Hoarding Instinct," pp. 335–41. Note that all were published in the same journal issue. On Lashley's influence as a teacher, see chapter 4, earlier.
10. The structure of the Group Processes conference was so dialogical that the identification of specific papers seemed almost misleading. For example, the topic of Frank Beach was "Ontogeny and Living Systems," in Schaffner, ed., *Group Processes*, pp. 9–74; by the second page (p. 10) of the transcribed session, however, Beach was so often interrupted by lengthy comments (all recorded verbatim as if part of his paper) that he could barely present his prepared remarks. With this caveat about format in mind, the other major papers were by Lorenz, "Morphology and Behavior Patterns in Closely Allied Species," pp. 168–220, and Tinbergen, "Psychology and Ethology as Supplementary Parts of a Science of Behavior," pp. 75–167. Although Lehrman was one of six keynote speakers, his paper, "The Perception of Animal Behavior," pp. 259–67, was little more than a note, perhaps a reflection of his junior status as a recent PhD who had completed his degree just months before. No matter whether they were Americans who emphasized environmental variables or Europeans who stressed set behavior patterns, they alluded to genetics. In his talk on individual development, Beach explained that "the first system I would like to discuss is the gene" (p. 11). Apologetically, Lorenz opted to use the wording " 'inherited coordinations,' " despite "the geneticists' objection to this term" (p. 169).
11. Frank A. Beach, "The Snark Was a Boojum" (1950), in Donald A. Dewsbury, ed., *Foundations of Comparative Psychology* (New York: Van Nostrand Reinhold, 1984), p. 67 (first quotation);

Morgan, "Hoarding Instinct," p. 339 (second quotation); Lashley, "Structural Variation in the Nervous System," p. 330 (third quotation).

12. Abstract of Charles H. Sawyer, "Reflex Induction of Ovulation in the Estrogen-Treated Rabbit by Artificial Vaginal Stimulation," in "Abstracts: American Association of Anatomists, Sixty-Second Annual Session . . . 1949," *Anatomical Record* 103:3 (1949): 502 (first quotation); Frank A. Beach, "'Psychosomatic' Phenomena in Animals," *Psychosomatic Medicine* 14:4 (1952): 263 (second and third quotations); Austin H. Riesen, "The Development of Visual Perception in Man and Chimpanzee," *Science*, new series 106:2744 (Aug. 1, 1947): 108 (fourth quotation); Norman R. F. Maier, *Frustration: The Study of Behavior without a Goal* (1949; reprint ed., Westport, Conn.: Greenwood Press, 1982), 28 (fifth quotation); Harry F. Harlow, "Mice, Monkeys, Men, and Motives," *Psychological Review* 60:1 (1953): 27 (sixth and seventh quotations). If more examples of induced pain are needed, see also Beach, "'Psychosomatic' Phenomena in Animals," esp. p. 266, and Maier, *Frustration*, esp. p. 27. Harlow's judgment was hardly an offhand comment, appearing as it did in his presidential address (Sept. 3, 1951) to the Division of Experimental Psychologists of the American Psychological Association.
13. Beach, "'Psychosomatic' Phenomena in Animals," pp. 273 (first, second, and third quotations), 267 (fourth quotation), 261 (fifth quotation), 268 (sixth, seventh, and eighth quotations). On Beach, see Donald A. Dewsbury, "Frank Ambrose Beach, 1911–1988," *National Academy of Sciences Biographical Memoirs* (1998), http://www.nasonline.org/publications/biographical-memoirs/memoir-pdfs/beach-frank-a.pdf (accessed June 1, 2019). For a clear statement of William McDougall's view of emotion relatively late in his career, see his *Outline of Psychology* (New York: Scribners, 1923), p. 326; note that his ideas evolved, but did not fundamentally change, beginning with his signature book, *An Introduction to Social Psychology* (1908; reprint ed., London: Methuen, 1960). See Anne C. Rose, "William McDougall, American Psychologist: A Reconsideration of Nature-Nurture Debates in the Interwar United States," *Journal of the History of the Behavioral Sciences* 52:4 (2016): 325–48.
14. Howard T. Odum, "The Bird Navigation Controversy," *Auk* 65:4 (1948): 590 (first quotation); J. B. Rhine, "The Present Outlook on the Question of PSI in Animals," *Journal of Parapsychology* 15:4 (1951): 244 (second and third quotations); J. G. Pratt, "The Homing Problem in Pigeons," *Journal of Parapsychology* 17:1 (1953): 54 (fourth quotation). Italics in quotations appear in the original texts unless otherwise noted. Odum's references to "geophysical ornithology" (p. 584), physicists working in the field (p. 590), the applicability of relativity theory (p. 586), and US Army sponsorship (p. 585) help to explain why debate about homing reappeared at this time. Simultaneously, parapsychology took a distinct turn toward possible occult capacities in disparate live subjects, another indication of the vigor of holistic biology. See, e.g., Pierre Duval and Evelyn Montredon, "ESP Experiments in Mice" (1968), Robert Brier, "PK Effect on a Plant-Polygraph System" (1969), and Jean Barry, "Retarding Fungus Growth by PK" (1968), in J. B. Rhine, ed., *Progress in Parapsychology* (Durham, N.C.: Parapsychology Press, 1971), pp. 17–27, 102–17, and 118–21.
15. Frank A. Beach and Julian Jaynes, "Effects of Early Experience upon the Behavior of Animals," *Psychological Bulletin* 51:3 (1954): 243. Leonard Carmichael's initial work appeared as "The Development of Behavior in Vertebrates Experimentally Removed from the Influence of External Stimulation," *Psychological Review* 33:1 (1926): 51–58. Apprehensive around 1960 that social deprivation experiments were under attack, Konrad Lorenz defended them as reliable tests for distinguishing innate and learned behavior and, in the process, implied that the technique had been his discovery in the first place. See *Evolution and Modification of Behavior* (Chicago: University of Chicago Press, 1965), chap. 7, the English translation of a

1961 German-language essay examined in context by Richard W. Burkhardt, Jr., *Patterns of Behavior: Konrad Lorenz, Niko Tinbergen, and the Founding of Ethology* (Chicago: University of Chicago Press, 2005), pp. 428–29. Lorenz's pioneering scientific papers were far more observational than experimental, however, making it questionable whether he deserved credit for the deprivation method. For a sample of his early work, see "Contributions to the Study of the Ethology of Social Corvidae" (1931), and "Companions as Factors in the Bird's Environment: The Conspecific as the Eliciting Factor for Social Behaviour Patterns" (1935), in *Studies in Animal and Human Behaviour*, trans. Robert Martin (Cambridge: Harvard University Press, 1970), 1: 1–56, 101–258. Ironically, *Evolution and Modification of Behavior* favorably cited American practitioners of deprivation such as Beach and Harry Harlow (though not Carmichael), without acknowledging New World scholars as having a hand its genesis (pp. 111–12).

16. Frank A. Beach, "Comparison of Copulatory Behavior of Male Rats Raised in Isolation, Cohabitation, and Segregation," *Journal of Genetic Psychology* 60:1 (1942): 132. Notes on the affective bond of animal mothers and offspring appeared in John B. Watson, *Animal Education: An Experimental Study of the Psychological Development of the White Rat, Correlated with the Growth of Its Nervous System* (Chicago: University of Chicago Press, 1903), pp. 63–64, and Robert M. Yerkes, "Maternal Instinct in a Monkey," *Journal of Animal Behavior* 5:5 (1915): 403–05.
17. Zing Yang Kuo, "Studies on the Basic Factors in Animal Fighting: VII. Inter-Species Coexistence in Mammals," *Journal of Genetic Psychology* 97 (1960): 211 (quotations). The accompanying articles were "Studies on the Basic Factors in Animal Fighting: V. Inter-Species Coexistence in Fish," *Journal of Genetic Psychology* 97 (1960): 181–94, and "Studies on the Basic Factors in Animal Fighting: VI. Inter-Species Coexistence in Birds," *Journal of Genetic Psychology* 97 (1960): 195–209. Note that volume 97 of the journal contained only one issue (hence unnumbered). Correspondence among Kuo and Carmichael, Carmichael and the journal editor (Carl Murchison), and readers requesting offprints from Carmichael—as the papers instructed interested readers to do—are found in folders "Zing Yang Kuo—Articles on Animal Aggression," 1–3, Leonard Carmichael Papers (Mss.B.C212), American Philosophical Society. See also Geoffrey H. Blowers, "'To Be a Big Shot or To Be Shot': Zing-Yang Kuo's Other Career," *History of Psychology* 4:4 (2001): 367–87.
18. Harry F. Harlow and Robert R. Zimmermann, "The Development of Affectional Responses in Infant Monkeys," *Proceedings of the American Philosophical Society* 102:5 (1958): 503 (first quotation); Joseph B. Sidowski and Donald B. Lindsley, "Harry Frederick Harlow, 1905–1981," *National Academy of Sciences Biographical Memoirs* (1989), p. 226 (second quotation), http://www.nasonline.org/publications/biographical-memoirs/memoir-pdfs/harlow-harry-f.pdf (accessed June 1, 2019); Harry F. Harlow, "Mice, Monkeys, Men, and Motives," pp. 24 (third quotation) and 29 (fourth quotation). Using the bibliography of Harlow's authored and coauthored publications, 1932–1978, in the NAS memoir by Sidowski and Lindsley (pp. 236–57), I count 294 total publications and thirty-three involving surgical ablations, 1939–1973. I did not include studies that used drugs or radiation to immobilize brain areas. Only one of the thirty-three surgical studies, issued in 1939, was sole-authored by Harlow; I infer that he viewed such protocols as a reliable experimental base for doctoral dissertations and hence encouraged his students to perform them. For a sample of photos of solitary infants, see Harlow and Zimmermann, "Affectional Responses in Infant Monkeys," passim. The hydraulic ("tissue-tension") model that Harlow and other American psychologists deemed inadequate was typically traced by them to Curt P. Richter, "Animal Behavior and Internal Drives," *Quarterly Review of Biology* 2:3 (1927): 307–43. Despite the slipperiness of Harlow's concept of love in the 1950s and 1960s, he was unusual among American psychologists in giving

at least sporadic attention to the emotions over many years. See, e.g., Harry F. Harlow and Ross Stagner, "Psychology of Feelings and Emotions. II. Theory of Emotions," *Psychological Review* 40:2 (1933): 184–95, and Harry F. Harlow, Margaret Kuenne Harlow, and Donald R. Meyer, "Learning Motivated by a Manipulation Drive," *Journal of Experimental Psychology* 40:2 (1950): 228–34, as well as Deborah Blum, *Love at Goon Park: Harry Harlow and the Science of Affection* (Cambridge, Mass.: Perseus, 2002).

19. Dobzhansky, *Genetics and Origin of Species*, p. 303.
20. Harlow and Zimmermann, "Affectional Responses in Infant Monkeys," p. 509.
21. T. C. Schneirla, "'Cruel' Ants and Occam's Razor," *Journal of Comparative Psychology* 34:1 (1942): 80 (first and second quotations); T. C. Schneirla, "Bees," *Ecology* 32:3 (1951): 564 (third quotation); Konrad Lorenz, *The Natural Science of the Human Species: An Introduction to Comparative Behavioral Research, The "Russian Manuscript" (1944–1948)*, ed. Agnes von Cranach, trans. Robert D. Martin (Cambridge: MIT Press, 1996), p. 94 (fourth quotation). Schneirla's "Bees" essay was a review of *Bees: Their Vision, Chemical Senses, and Language* (1950) by the 1973 Nobel laureate Karl von Frisch. Even if an emphasis on individual behavior had the capacity to obscure evolutionary relationships, investigators of nonhuman animals were more likely than human-focused psychologists to think about descent. Receiving the Nobel Prize in Physiology or Medicine, Lorenz delivered his Nobel lecture (Dec. 12, 1973) on the distinction between traits that are merely analogous (structurally similar) and homologous (linked by common ancestry). See "Analogy as a Source of Knowledge," http://www.nobelprize.org/nobel_prizes/medicine/laureates/1973/lorenz-lecture.pdf (accessed June 1, 2019). Similarly, in a volume of original essays memorializing Schneirla, his one-time student Daniel S. Lehrman addressed the relationship between phylogeny and ontogeny—the development of species and individuals. See "Semantic and Conceptual Issues in the Nature-Nurture Problem," in Lester R. Aronson et al., eds., *Development and Evolution of Behavior: Essays in Memory of T. C. Schneirla* (San Francisco: W. H. Freeman, 1970), pp. 17–52.
22. Lorenz, *Natural Science of Human Species*, p. 89; Calvin S. Hall and Paul H. Whiteman, "The Effects of Infantile Stimulation upon Later Emotional Stability in the Mouse," *Journal of Comparative and Physiological Psychology* 44:1 (1951): 61.
23. Ernst Mayr, "Bird Conservation Problems in the Southwest Pacific," *Audubon Magazine* 47 (1945): 282.
24. Clarence R. Carpenter [*sic*] to B. V. Moore, June 1, 1945, box 39, folder "Assignment Orders, 1943–46," group 19, Clarence Ray Carpenter Papers, Eberly Family Special Collections Library, Penn State University, University Park, Pa. (first quotation); Stokesbury, *Short History of World War II*, p. 179 (second quotation). All further Carpenter citations in this chapter are also found in group 19, box 39.
25. Robert M. Yerkes, "Psychology and Defense," *Proceedings of the American Philosophical Society* 84:4 (1941): 527 (first quotation), 531 (second quotation), 533 (third quotation), 540 (fourth quotation), 539 (fifth quotation), 540 (sixth quotation).
26. "Address by President Leonard Carmichael at Convocation, October 2, 1940, 11:45 a.m. Neo-Patriotism," typescript, pp. 6 (first quotation), 8 (second quotation), 16 (third quotation), 18 (fifth quotation), folder "Speech: Neo-Patriotism Convocation, 2 Oct. 1940," Carmichael Papers; Leonard Carmichael, "Man and Society in War and Peace," *Christian Leader* 125 (Oct. 16, 1943): 616 (fourth quotation), 615 (sixth and seventh quotations), 616 (eighth quotation). The latter article was initially a talk given at the Congress Square Church, Portland, Maine, Sept. 21, 1943. In the earlier Tufts address, Carmichael cited a long list of modern cultural ills: modernist literature (including revived interest in Herman Melville and Walt Whitman, along with the popularity among literary insiders of James Joyce, whose writings Carmichael

called "somewhat pornographic"), p. 5; Freudian psychoanalysis, pp. 6–7; atheism, p. 7; and theories of relativity and quantum mechanics, p. 8. Carmichael's National Roster reports are listed in the full bibliography of his writings, in Carl Pfaffmann, "Leonard Carmichael, 1898–1973," *National Academy of Sciences Biographical Memoirs* (1980), pp. 41–43, http://www.nasonline.org/publications/biographical-memoirs/memoir-pdfs/carmichael-leonard.pdf (accessed June 1, 2019).

27. Walter B. Cannon, "The Body Physiologic and the Body Politic," *Science*, new series, 93:2401 (1941): 7 (first quotation), 1 (second, third, and fourth quotations), 10 (fifth quotation), 6 (sixth quotation), 6 (seventh quotation); Edward Chace Tolman, *Drives toward War* (New York: Appleton-Century, 1942), p. 95 (eighth and ninth quotations); Edward Chace Tolman, "Motivation, Learning, and Adjustment," *Proceedings of the American Philosophical Society* 84:4 (1941): 554 (tenth quotation).
28. Lucretius Smith, "The Genetics Controversy," *Soviet Russia Today* 8 (1940): 10 (first and second quotations), 13 (third and fourth quotations); Calvin Perry Stone to Herbert Hoover, Oct. 13, 1941, Calvin Perry Stone Papers microfilm (SC0440), University Archives, Special Collections, Stanford University Library, Stanford, Calif. (fifth through eighth quotations). Morgan is identified as "Lucretius Smith" in Sturtevant, "Morgan," p. 324. Despite Morgan's caution during wartime about openly defending Soviet science, his respect for the Soviet Union as a rationally planned society offering scientists prestige and funding was commonplace among Anglo-American peers during the 1930s. Consider Walter B. Cannon's praise of Soviet state-supported research in his address at the International Physiological Congress, Leningrad, 1935, *Some Implications of the Evidence for Chemical Transmission of Nerve Impulses: Address at the Plenary Sitting of the Congress* (Moscow-Leningrad: State Publishing House for Biological and Medical Literature, 1935), esp. p. 9. Similarly, Julian Huxley made multiple trips to Russia at Soviet invitation between 1931 and 1945, described in Ronald W. Clark, *Sir Julian Huxley, F. R. S.* (London: Phoenix House, 1960), pp. 43–46, 90, and chronicled firsthand in Huxley, *A Scientist among the Soviets* (London: Chatto and Windus, 1932). For an excellent general study, see Nikolai Krementsov, *International Science between the World Wars: The Case of Genetics* (London: Routledge, 2005).
29. W. C. Allee, "Where Angels Fear to Tread: A Contribution from General Sociology to Human Ethics," *Science*, new series 97:2528 (1943): 525 (first quotation); Warder Clyde Allee to Margaret Morse Nice, May 24, 1949 (second quotation), box 20, folder 12, Warder Clyde Allee Papers, Special Collections Research Center, University of Chicago Library, Chicago, Ill. Nice first broached the subject of Rockefeller support for Lorenz in a letter to Allee, Apr. 27, 1949. It was not until Aug. 3 that Allee agreed to write to Alan Gregg there; on Sept. 1 Allee informed Nice that "the Rockefeller people have not even acknowledged the receipt of my letter," and "they usually do acknowledge my rare communications immediately" (same folder). Given Rockefeller's substantial earlier efforts to rescue international scientists exposed to political threats, the visible reluctance all around to sponsor Lorenz was the result, presumably, of Lorenz's one-time Nazi Party membership, as well as renewed American isolationism during the Cold War and especially fear of scrutiny by the House Un-American Activities Committee. For a summary of Rockefeller rescue of social scientists, see Darwin H. Stapleton, "Intellectuals Flee from Fascism: Rockefeller Support of Social Scientists, 1933–1945" (2005), pp. 5–10, http://www.rockarch.org/publications/resrep/stapleton1.pdf (accessed June 1, 2019). Because Rockefeller funding took many forms, however, much international rescue occurred via programs not specifically designed to provide safety to individuals. Consider scholars from around the world who attended the Rockefeller-funded year-long seminar at Yale, "The Impact of Culture on Personality," 1932–1933, and subsequently declined to return to their home countries; see Anne C. Rose, *Psychology and*

Selfhood in the Segregated South (Chapel Hill: University of North Carolina Press, 2009), pp. 97–98, 120.

30. "Motion Picture Unit Shoots Jungle Training Film Here," *The Runway* 1:10 (Dec. 31, 1943): front page (quotations), folder "'Land and Live in the Jungle' Training Film Production, 1943–44, Carpenter Papers. For later projects, see Clarence R. Carpenter and George A. Levi, "Preliminary Project Proposal: Night Vision of Flying Personnel," typescript, June 8, 1944; [C. R. Carpenter], "Report on Vision Problems Peculiar to Artic, Desert, and Tropic Regions," typescript, June 14, 1944, folder "'Land and Live'"; and [C. R. Carpenter?], *Handling and Release of Pigeons from Aircraft, AAF School of Applied Tactics, Orlando, Florida*, folder "Handling and Release of Pigeons from Aircraft," Carpenter Papers.
31. Clarence R. Carpenter to Lt. Col. H. O. Russell, Feb. 2, 1944, folder "Assignment Orders, 1943–46," Carpenter Papers. For a case of an assignment that took far longer than planned due to problems, see Special Order, no. 295, Nov. 24, 1943, and Special Order, no. 20, Jan. 20, 1944, which, respectively, extended the period devoted to the jungle survival film from 60 to 125 days and then from 125 to 155 days; folder "Assignment Orders," Carpenter Papers. Military orders requiring travel stateside among Orlando, New York, New Haven, and Culver City, Calif., with occasional stops at home in State College, Pa., are found in the same folder.
32. Special Order, no. 12, June 12, [1945], folder "Assignment Orders," Carpenter Papers (quotation). An order marked 300.4, July 26, 1946, identified the upcoming date of discharge as Sept. 15, 1946 (same folder). B. V. Moore and M. R. Trabue, respectively Carpenter's department head and dean, together wrote unsuccessfully to the Commanding General, Army Air Force, Orlando, June 6, 1945, to secure Carpenter's discharge before European deployment, folder "Wartime Correspondence with Penn State Personnel, 1943–1945," Carpenter Papers. Although Carpenter listed six authored and coauthored Air Force manuals on his academic vitae, his publications otherwise skipped from "Societies of Monkeys and Apes," *Biological Symposia* 8 (1942): 177–204, to Carpenter and Clarence E. Glick, "What 2,775 Soldier Students Thought of Biarritz," *Educational Record* (Oct. 1946): 477–92; see Clarence Ray Carpenter Vitae, typescript, p. 2, Eberly Special Collections Library. On the interface of colonialism and science in the World War II era, see esp. Frederick Cooper, "Development, Modernization, and the Social Sciences in the Era of Decolonization: The Examples of British and French Africa," *Revue d'Histoire des Sciences Humaines* 1:10 (2004): 9–38, and Helen Tilley, *Africa as a Living Laboratory: Empire, Development, and the Problem of Scientific Knowledge, 1870–1950* (Chicago: University of Chicago Press, 2011).
33. Carpenter was born on Nov. 28, 1905, and Harlow on Oct. 31, 1905. Other animal behavior researchers about the same age whose careers continued unabated include (in order of birth date) Norman R. F. Maier (1900–1977), T. C. Schneirla (1902–1968), Frank A. Beach (1911–1988), and Clifford T. Morgan (1915–1976). For biographical information on figures not otherwise discussed here, see Eliot Stellar and Gardner Lindzey, "Clifford T. Morgan, 1915–1976," *American Journal of Psychology* 91:2 (1978): 343–48, and Donald A. Dewsbury, "On Publishing Controversy: Norman R. F. Maier and the Genesis of Seizures," *American Psychologist* 48:8 (1993): 869–77.
34. Biographical information on Beach and Schneirla may be found in Dewsbury, "Beach"; Ethel Tobach and Lester A. Aronson, "T. C. Schneirla: A Biographical Note," in Aronson et al., eds., *Development and Evolution of Behavior*, pp. xi–xviii; and Sleigh, *Six Legs Better*, pt. 3 passim. For a bibliography of Schneirla's works, see Lester R. Aronson et al., eds., *Selected Writings of T. C. Schneirla* (San Francisco: W. H. Freeman, 1972), pp. 1017–24. As a recent PhD (Harvard, 1940), Beach succeeded the amphibian specialist Gladwyn Kingsley Noble (1894–1940) as chair of what Noble had founded in 1928 as the Department of Experimental Biology (see esp. Dewsbury, "Beach," pp. 7–8). Noble—and even earlier, the ornithologist Frank

M. Chapman—had made the American Museum of Natural History a serious biological research site. See Walter L. Necker, "Gladwyn Kingsley Noble, 1894–1940: A Herpetological Bibliography," *Herpetologica* 2:2 (1940): 47–51, 53–55, and Robert Cushman Murphy, "Frank Michler Chapman, 1864–1945," *Auk* 67 (1950): 307–15. Schneirla spent the remainder of his career at the museum and was fully committed to the dual goals of scholarship and public education; see T. C. Schneirla, "The Study of Animal Behavior: Its History and Relation to the Museum" (issued originally in two parts, 1958 and 1959), in Aronson et al., eds., *Selected Writings of Schneirla*, pp. 90–127. On Schneirla's funded research in Mexico during the war, see Tobach and Aronson, "Schneirla," in Aronson et al., eds., *Development and Evolution of Behavior*, pp. xiv–xv. On the inclusion of his name on the House Un-American Activities Committee list of suspected communists (1947, 1951, 1953, and 1957), see Sleigh, *Six Legs Better*, pp. 171–72.

35. For biographical information, see Jay S. Rosenblatt, "Daniel Sanford Lehrman, June 1, 1919–August 27, 1972," *National Academy of Sciences Biographical Memoirs* (1995), http://www.nasonline.org/publications/biographical-memoirs/memoir-pdfs/lehrman-daniel-s.pdf (accessed June 1, 2019), and Rae Silver and Jay S. Rosenblatt, "The Development of a Developmentalist, Daniel S. Lehrman," *Developmental Psychobiology* 20:6 (1987): 563–70. The 1995 publication date of the NAS memoir does not appear on the website due to a posting error; it was provided in an email to me from Tony Ziselberger, Jan. 21, 2014. Lehrman's first research project appeared as G. K. Noble and D. S. Lehrman, "Egg Recognition by the Laughing Gull," *Auk* 57:1 (1940): 22–43. On Jewish scientists and other professionals, see Steve Nadis and Shing-Tung Yau, *A History in Sum: 150 Years of Mathematics at Harvard, 1825–1975* (Cambridge: Harvard University Press, 2013), pp. 80–85 (on anti-Semitic hiring between the wars), as well as Dan A. Oren, *Joining the Club: A History of Jews and Yale*, 2d ed. (New Haven: Yale University Press, 2001); Susanne Klingenstein, *Jews in the American Academy, 1900–1940: The Dynamics of Intellectual Assimilation* (New Haven: Yale University Press, 1991); and Lila Corwin Berman, *Speaking of Jews: Rabbis, Intellectuals, and the Creation of an American Public Identity* (Berkeley: University of California Press, 2009). Dewsbury notes Beach's difficulty with foreign languages, in "Beach," p.7.
36. Ernst Mayr, *Birds of the Southwest Pacific: A Field Guide to the Birds of the Area between Samoa, New Caledonia, and Micronesia* (New York: Macmillan, 1945), title page (first quotation), Robert Cushman Murphy [Chair, Department of Birds, American Museum of Natural History], "Preface," in ibid, p. ix (second and third quotations); Mayr, "Bird Conservation Problems," p. 282 (fourth, fifth, and sixth quotations). For details of Mayr's citizenship struggle, see Haffer, *Ornithology, Evolution, and Philosophy*, chap. 6 and esp. pp. 250–52. Haffer explains that although Mayr worked in the United States beginning in 1931, he did not apply for citizenship until the war broke out, in part—as Mayr later recalled—because he felt personally loyal to Germany. As late as 1940, moreover, his one-time doctoral mentor in Berlin, Erwin Stresemann, urged Mayr to return to Germany. One reason Mayr and his wife Gretel avoided deportation was that their daughters, Christa and Susanne, were born in the United States in 1936 and 1937, respectively, and hence were citizens (Haffer, *Ornithology, Evolution, and Philosophy*, pp. 104, 242). A chronological list of Mayr's publications may be found in Haffer, *Ornithology, Evolution, and Philosophy*, pp. 406–53. For his routine and continuous work in systematics, see, e.g., John T. Zimmer and Ernst Mayr, "New Species of Birds Described from 1938 to 1941," *Auk* 60:2 (1943): 249–62.
37. Ernst Mayr to Theodosius Dobzhansky, May 14, 1947, box D65.m, folder "Mayr, Ernst, 1937–1947," Dobzhansky Papers.
38. Theodosius Dobzhansky to John Thomas Patterson, May 6, 1936, folder "John Thomas Patterson," Dobzhansky Papers (first and second quotations); 1962 interview quoted in Kay,

Molecular Vision of Life, p. 84 (third through sixth quotations). Further details of the negotiation between Dobzhansky and the University of Texas, where it is clear that university administrators were willing to bend state law in order to hire an eminent scientist, may be found in Patterson to Dobzhansky, Apr. 22 and May 15, 1936, as well as Dobhansky's withdrawal of his application, Dobzhansky to Patterson, May 28, 1936, all in folder "Patterson," Dobzhansky Papers.

39. Theodosius Dobzhansky to Ernst Mayr, Aug. 15, 1945, box D65.m, folder "Dobzhansky-Mayr Correspondence, Aug. 1944-Nov. 1949," Dobzhansky Papers (first through fourth quotations); untitled typescript marked (in hand) "Please Return," including quotations from Niko Tinbergen to Margaret Morse Nice, June 3, 1945, box 5, folder 11, Margaret Morse Nice Papers (Collection 2993), Division of Rare and Manuscript Collections, Cornell University Library, Ithaca, N.Y. (fifth through seventh quotations). Nice coauthored a circular letter to American ornithologists with Joseph J. Hickey, Nov. 14, 1946, titled "Christmas 1946," outlining the needs of European colleagues and requesting donations, box 5, folder 11, Nice Papers. On Yoshioka and Kuo, respectively, see Miki Takasuna, "Six Researchers' Contributions to Comparative Psychology in Japan: 1900–1945," *Japanese Psychological Research* 47:2 (2005): 91 (Yoshioka); Blowers, "To Be a Big Shot," pp. 374–87 (Kuo). The last article by Yoshioka that I could locate was Joseph G. Yoshioka and Harold E. Jones, "An Analysis of Children's Maze Learning, in Terms of Stages of Learning," *Journal of Genetic Psychology* 67:2 (1945): 203–14. The affiliation of the authors was the Institute of Child Welfare, University of California, the institution where Yoshioka earned his doctorate in psychology in the late 1920s.
40. Dobzhansky to Ernst Mayr, Aug. 15, 1945 (quotation). Dobzhansky and Mayr's coauthored papers were "Th. [*sic*] Dobzhansky and Ernst Mayr, "Experiments on Sexual Isolation in *Drosophila*: I. Geographic Strains of *Drosophila willistoni*," *Proceedings of the National Academy of Sciences* 30:9 (1944): 238–44, and Ernst Mayr and Th. [*sic*] Dobzhansky, "Experiments on Sexual Isolation in *Drosophila*: IV. Modifications of the Degree of Isolation between *Drosophila pseudoobscura* and *Drosophila persimilis* and of Sexual Preferences in *Drosophila prosaltans*," *Proceedings of the National Academy of Sciences* 31:2 (1945): 75–82.
41. Dewsbury, "On Publishing Controversy," p. 871.
42. On the deployment of physical scientists and engineers, see n. 5, earlier. For casualty figures, consider comparatively reliable numbers provided by James L. Stokesbury. Stokesbury estimates twenty-two million military casualties (those who died or were wounded, imprisoned, or missing) during World War I. Counting deaths alone during World War II, he cites at least seventeen million military fatalities and twenty million civilian fatalities (*A Short History of World War I* [New York: William Morrow, 1981], p. 310, and *Short History of World War* II, p. 378). Stokesbury apparently did not see a need to estimate civilian casualties for World War I, a notoriously stationary war fought most memorably on the French-Belgian border and to a lesser extent in eastern Europe, the Mideast, and the Pacific. Wikipedia proposes ten million military and seven million civilian deaths alone during World War I, numbers which include deaths from influenza both among military personnel and civilians ("World War I Casualties," http://en.wikipedia.org/wiki/World_War_I_casualties [accessed June 1, 2019). In all these figures, what one considers war-related harm greatly affects one's calculations.
43. Hunter, "Summary Comments," p. 348.
44. Konrad Z. Lorenz, *King Solomon's Ring: New Light on Animal Ways*, trans. Marjorie Kerr Wilson (New York: Thomas Y. Crowell, 1952), p. 77. A German-language edition had appeared in 1949. On Lorenz's experience as a POW, see Agnes von Cranach, "Editor's Foreword," in Lorenz, *Natural Science of the Human Species*, pp. ix-xv. Lorenz wrote to

Margaret Nice, May 10, 1949, that he had a "beautill [*sic*] nervous breakdown" after his return home (box 1, folder 17, Nice Papers). British theorizing that flagged the intellectual utility of evolution includes R. A. Fisher, *The Genetical Theory of Natural Selection: A Complete Variorum Edition* (1930; reprint ed., Oxford: Oxford University Press, 1999); J. B. S. Haldane, *The Causes of Evolution* (1932; reprint ed., Ithaca: Cornell University Press, 1966); and especially Julian Huxley, *Evolution: The Modern Synthesis* (1942; reprint ed., New York: Harper and Brothers, 1943).

45. L. C. Dunn and Theodosius Dobzhansky, *Heredity, Race, and Society*, rev. ed. (New York: New American Library, 1952), pp. 83 (first quotation), 56 (second and third quotations), 124 (fourth quotation), 83 (fifth quotation), 99 (sixth quotation), 92 (seventh quotation), 101 (eighth quotation), 63 (ninth quotation).
46. Frank A. Beach, "Locks and Beagles," *American Psychologist* 24:11 (1969): 972.
47. Frank A. Beach, "Current Concepts of Play in Animals," *American Naturalist* 79:785 (1945): 540 (first quotation), 531 (second quotation); Frank A. Beach and Julian Jaynes, "Effects of Early Experience," pp. 239 (third quotation), 240 (fourth quotation). Beach's 1952 paper was " 'Psychosomatic' Phenomena in Animals."
48. Beach, "Locks and Beagles," p. 988.
49. Frank A. Beach, "Of Course, Animals Can Think," *Natural History* 56:3 (1947): 144 (first quotation), 116 (second quotation).
50. Beach, "Snark Was a Boojum," in Dewsbury, *Foundations of Comparative Psychology*, p. 65 (first quotation); Frank A. Beach, "The Descent of Instinct," *Psychological Review* 62:6 (1955): 403 (second and third quotations), 404 (fourth and fifth quotations), 407 (sixth quotation).
51. On these travels, see Burkhardt, *Patterns of Behavior*, pp. 291–92 (Tinbergen), 398–403 (Lorenz), 390–98 (Lehrman and Schneirla in Paris), and 403–07 (Beach and Lehrman in Groningen), as well as resulting publications that began as public lectures, including Karl von Frisch, *Bees*, esp. pp. xi–xii; N. [*sic*] Tinbergen, *The Study of Instinct* (1951; reprint ed., Oxford: Oxford University Press, 1969), esp. pp. xiii–xiv; and Schaffner, ed., *Group Processes*.
52. Lehrman, "Semantic and Conceptual Issues," in Aronson et al., eds., *Development and Evolution of Behavior*, pp. 18 (first quotation), 22 (second and third quotations). For the full citation of Lehrman's "Critique of Lorenz's Theory," see n. 3, earlier.
53. Tinbergen, *Study of Instinct*, pp. xiii (first quotation), 6 (second and third quotations); Lehrman, "Critique of Lorenz's Theory," p. 337 (fourth quotation); Lehrman quoted in Burkhardt, *Patterns of Behavior*, p. 399 (fifth quotation).
54. Lehrman, "Critique of Lorenz's Theory," pp. 344 (first quotation), 344 (second quotation), 352 (third quotation), 356 (fourth quotation); T. C. Schneirla, "A Consideration of Some Conceptual Trends in Comparative Psychology," in Aronson et al., eds., *Selecting Writings of Schneirla*, pp. 896 (fifth quotation), 897 (sixth quotation); T. C. Schneirla, "Instinct and Aggression," *Natural History* 75 (1966), pp. 18 (seventh and eighth quotations), 16 (ninth quotation). Schneirla's essay anthologized by Aronson began as a paper presented at the 1951 meeting of the American Psychological Association, subsequently published in 1952 in the *Psychological Bulletin*.
55. Schneirla, "Study of Animal Behavior," in Aronson et al., *Selecting Writings of Schneirla*, p. 116 (first quotation); Konrad Lorenz, "Morphology and Behavior Patterns in Closely Allied Species," in Schaffner, ed., *Group Processes*, p. 169 (second quotation); Beach, "Ontology and Living Systems, in ibid, p. 10 (third quotation); Schneirla, "Consideration of Some Conceptual Trends," in Aronson et al., *Selecting Writings of Schneirla*, pp. 897 (fourth quotation), 912 (fifth quotations); T. C. Schneirla, "The Relationship between Observation and Experimentation in the Field Study of Behavior," in ibid, p. 9 (sixth quotation). The

title by Schneirla first appeared in 1950 in the *Annals of the New York Academy of Sciences*. Prominent figures at the 1947 heredity-environment symposium at Princeton (Beach, Carmichael, Lashley, C. T. Morgan, and Stone) also stressed that the nature-nurture debate was over; see W. S. Hunter, "Summary Comments," pp. 348–52.

56. T. C. Schneirla, "A Comparison of Species and Genera in the Ant Subfamily *Dorylinae* with Respect to Functional Patterns (1) (2)," *Insectes Sociaux* 4:3 (1957): 263 (first quotation), 287 (second quotation), 263 (third and fourth quotations), 265 (fifth, sixth, and seventh quotations), 281 (eighth quotation), 287 (ninth quotation), 283 (tenth quotation), 293 (eleventh quotation). Schneirla's dissertation, completed for a doctorate awarded by the University of Michigan in 1928, was based on experimentation rather than fieldwork; it appeared as "Learning and Orientation in Ants," 6 *Comparative Psychology Monographs* (Baltimore: Johns Hopkins University Press, 1929): 1–143. His first field study was, "Studies on Army Ants in Panama," *Journal of Comparative Psychology* 15:2 (1933): 267–99.
57. For laboratory components, see "A Comparison in the Ant Subfamily *Dorylinae*," pp. 265, 275 (histology), 270 (light). Among twenty-eight references listed in the bibliography (pp. 297–98), four were written in French, three in German, and one in Spanish (the last on Argentinean ants). It is my impression that as American-born animal scientists increasingly received all their training in the United States, their command of languages other than English diminished. Assuming Schneirla had some facility in three foreign languages, he seems exceptional.
58. Beach, "Locks and Beagles," pp. 982 (first quotation), 985 (second quotation).
59. Beach, "Locks and Beagles," pp. 989 (first and second quotations), 981 (third quotation).
60. Press Release, Nobel Prize in Physiology or Medicine, 1973, http://www.nobelprize.org/nobel_prizes/medicine/laureates/1973/press.html (accessed June 1, 2019).
61. Lorenz, *King Solomon's Ring*, pp. 73 (first quotation), 57 (fifth quotation); Konrad Lorenz, *On Aggression*, trans. Marjorie Kerr Wilson (New York: Harcourt, Brace, and World, 1966), pp. 61 (second quotation), 206 (third and fourth quotations). For Lorenz's emphasis on human aggression, see *On Aggression*, chap. 14. Although neither Schneirla nor Lehrman was alive when Lorenz won the Nobel Prize, American discomfort with Lorenz's so-said pessimism and determinism continued; see, e.g., Samuel S. Kim, "The Lorenzian Theory of Aggression and Peace Research: A Critique," *Journal of Peace Research* 13:4 (1976): 253–76. Both German and British experts on animal behavior, in contrast, were committed intellectually to inherited ritual patterns and produced an expansive body of writing at odds with the American preference for open-ended experimentation. See, e.g., Konrad Lorenz, *Evolution and Modification of Behavior*; K. Z. Lorenz [*sic*], "Evolution of Ritualization in the Biological and Cultural Spheres," *Philosophical Transactions of the Royal Society of London*, series B (Biological Sciences) 251:772 (1966): 273–84; and Julian Huxley, "Introduction" to "A Discussion of Ritualization of Behaviour in Animals and Man," special issue, ibid, pp. 249–71.
62. Daniel S. Lehrman, "The Physiological Basis of Parental Feeding Behavior in the Ring Dove (*Streptopelia risoria*)," *Behaviour* 7:4 (1955): 263.

Chapter 6

1. Rae Silver and Jay Rosenblatt, "The Development of a Developmentalist: Daniel S. Lehrman," *Developmental Psychobiology* 20:6 (1987): 563 (first and second quotations); "Introduction," Ethel Tobach, John Gianutsos, Howard R. Topoff, and Charles G. Gross, eds., *The Four Horsemen: Racism, Sexism, Militarism, and Social Darwinism* (New York: Behavioral Publications, 1974), p. 18 (third quotation). For the details of Lehrman's death, see Jay S.

Rosenblatt, "Daniel Sanford Lehrman, June 1, 1919–August 27, 1972," *National Academy of Sciences Biographical Memoirs* (1995), http://www.nasonline.org/publications/biographical-memoirs/memoir-pdfs/lehrman-daniel-s.pdf (accessed June 1, 2019). The 1995 publication date of the NAS memoir does not appear on the website due to a posting error; it was provided in an email to me from Tony Ziselberger, Jan. 21, 2014. Lehrman's famous paper was "A Critique of Konrad Lorenz's Theory of Instinctive Behavior," *Quarterly Review of Biology* 28:4 (1953): 337–63.

2. Donald R. Griffin, *The Question of Animal Awareness: Evolutionary Continuity of Mental Experience* (New York: Rockefeller University, 1976), pp. 25 (first and second quotations), title page (third quotation), 78 (fourth quotation).
3. Frank A. Beach, "Historical Origins of Modern Research on Hormones and Behavior," *Hormones and Behavior* 15:4 (1981): 363.
4. George B. Schaller, *The Mountain Gorilla: Ecology and Behavior* (Chicago: University of Chicago Press, 1963), p. 3.
5. Griffin, *Animal Awareness*, pp. 5 (first and second quotations); Edward O. Wilson, *Sociobiology: The New Synthesis* (Cambridge: Harvard University Press, 1975), p. 23 (third and four quotations); Beach, "Historical Origins of Modern Research on Hormones and Behavior," pp. 328–33. Griffin's capstone early paper on bat navigation, coauthored by a fellow Harvard doctoral student, was Donald R. Griffin and Robert Galambos, "The Sensory Basis of Obstacle Avoidance by Flying Bats," *Journal of Experimental Zoology* 86:3 [1941]: 481–506. Griffin's undergraduate and graduate studies at Harvard were highly interdisciplinary—involving biology, physics, physiology, and psychology—but he was not permitted to use the bat navigation research for his doctorate in biology. To satisfy his department, Griffin completed a paper on the more conventional subject of bird migration. In the complex world of Harvard science, Karl Lashley directed Griffin's dissertation, even though Lashley himself had not researched bird orientation since his coauthored papers with John B. Watson in the 1910s and, more recently, mainly trained students to measure sensations and cognition using deprivation techniques. Lashley's joint appointment in the departments of psychology and biology made his supervision of Griffin possible. The same bat studies became Galambos's dissertation. Notably, Griffin earned a Junior Fellowship, which exempted him from many departmental requirements. For these institutional details, see Charles G. Gross, "Donald R. Griffin, 1915–2003," *National Academy of Sciences Biographical Memoirs* (2005), pp. 6–10, http://www.nasonline.org/publications/biographical-memoirs/memoir-pdfs/griffin-donald.pdf (accessed June 1, 2019), and James A. Simmons, "Donald R. Griffin: 1915–2003," *Bat Research News* 45:1 (2004), http://www.batresearchnews.org/Miller/Griffin.html (accessed June 1, 2019). Griffin's selective memory of Lashley in print is confirmed by his private correspondence with Lashley's would-be biographer, Darryl Bruce. Contacted by Bruce, Griffin replied, "I knew Lashley only as a graduate student at Harvard from 1938–1942 and occasional visits with him later" (Griffin to Bruce, Dec. 21, 1983, box 7, folder 67, Donald R. Griffin Papers, RG 450G875, FA 164, Rockefeller Archive Center, Sleepy Hollow, N.Y. [hereafter RAC]). In fact, they corresponded often until Griffin received a final letter (May 22, 1958) just prior to Lashley's death. Griffin contacted Lashley whenever he needed a reference from "the man under whom one did graduate work" (Griffin to Lashley, June 29, 1945). Lashley showed continuing interest in navigation and homing, warmly praising Griffin's developing work (e.g., Lashley to Griffin, Jan. 14, 1948, May 22, 1958). I infer that Griffin's enthusiasm for ethology in the 1970s obscured his memory of Lashley's influence on him. All correspondence with and about Lashley is located in box 7, folder 67, Griffin Papers.
6. Michael W. Fox, *Inhumane Society: The American Way of Exploiting Animals* (New York: St. Martin's, 1990), p. 13 (first quotation); Marc Bekoff, *The Emotional Lives of Animals: A Leading Scientist Explores Animal Joy, Sorrow, and Empathy—and Why They Matter* (Novato, Calif.:

New World Library, 2007), p. 1 (second quotation). Although I have argued that animal research in the United States never focused on a few species, there is much scholarship on a few historical preferences. See, e.g., Bonnie Tocher Clause, "The Wistar Rat as a Right Choice: Establishing Mammalian Standards and the Ideal of a Standardized Mammal," *Journal of the History of Biology* 26:2 (1993): 329–49, and Robert E. Kohler, *Lords of the Fly: Drosophila Genetics and the Experimental Life* (Chicago: University of Chicago Press, 1994).

7. Julian Huxley, *Evolution: The Modern Synthesis* (1942; reprint ed., New York: Harper and Brothers, 1943), title page (first quotation), John B. Calhoun, "Population Density and Social Pathology," *Scientific American* 306:2 (1962): 141 (second and third quotations), 145 (fourth quotation), 144 (fifth quotation); Donald A. Dewsbury, "Patterns of Copulatory Behavior in Male Mammals," *Quarterly Review of Biology* 47:1 (1972): 23 (sixth quotation), 24 (seventh quotation). The paragraph on human male copulation appears in ibid, p. 19. Vassiliki Betty Smocovitis sets Huxley's manifesto in historical perspective, in *Unifying Biology: The Evolutionary Synthesis and Evolutionary Biology* (Princeton: Princeton University Press, 1996), pt. 2. Calhoun had been designing increasingly elaborate living structures over many years; see, e.g., John B. Calhoun, "A Method for Self-Control of Population Growth among Mammals Living in the Wild," *Science*, new series, 109:2831 (1949): 333–35. His ambitious experiments influenced urban sociology and popular culture; see Edmund Ramsden and Jon Adams, "Escaping the Laboratory: The Rodent Experiments of John B. Calhoun and Their Cultural Influence," *Journal of Social History* 42:3 (2009): 761–92. Arriving at Johns Hopkins in 1946 as a researcher in the School of Hygiene and Public Health, Calhoun became part of an ongoing project begun during World War II to control Baltimore's escalating rat population, the apparent result of war production causing urban growth. See Ramsden and Adams, "Escaping the Laboratory," p. 764, and Christine Keiner, "Wartime Rat Control, Rodent Ecology, and the Rise and Fall of Chemical Rodenticides," *Endeavor* 29:3 (2005): 119–25.

8. Donald R. Griffin, *Listening in the Dark: The Acoustic Orientation of Bats and Men* (New Haven: Yale University Press, 1958), pp. 77 (first quotation), 386 (second, third, fourth, and fifth quotations); Gross, "Griffin," p. 14 (sixth, seventh, eighth, and ninth quotations). It seems significant that Griffin held back from publicizing the concept "echolocation" in *Listening in the Dark* (1958), based on his Trumbull Lectures at Yale in 1955. Writing prior to a time when announcing a big theory was commonplace in the United States, Griffin did not introduce the cross-species phenomenon he named "echolocation" until page seventy-seven of a four-hundred-page book, choosing instead the more descriptive phrase "acoustic orientation" for his subtitle. He had devoted the first quarter of his book mainly to facts of bat anatomy and physiology. A decade before the Yale lectures, Griffin had devised the term "echolocation" to designate the navigation process: Donald R. Griffin, "Echolocation by Blind Men, Bats and Radar," *Science*, new series, 100:2609 (1944): 589–90. His caution in announcing a theory, however, makes sense in light of later criticism of his claim for animal awareness.

9. Griffin, *Listening in the Dark*, p. 379 (first, second, and third quotations); Peter Singer, *Animal Liberation: A New Ethics for Our Treatment of Animals* (New York: New York Review of Books/Random House, 1975), p. ix (fourth and fifth quotations). Those arguing the cause of animals in captivity include Singer, *Animal Liberation*, esp. chaps. 2–3; Marian Stamp Dawkins, *Animal Suffering: The Science of Animal Welfare* (London: Chapman and Hall, 1980), esp. chaps. 3–4; Fox, *Inhumane Society*; and Marian Stamp Dawkins, "The Science of Animal Suffering," *Ethology* 114:10 (2008): esp. 941–43. For experimental evidence on one food-producing animal, see Victoria Braithwaite, *Do Fish Feel Pain?* (Oxford: Oxford University Press, 2010).

10. R. P. Hanson and Lars Karstad, "Feral Swine in the Southeastern United States," *Journal of Wildlife Management* 23:1 (1959): 71 (first quotation), 68 (second, third, and fourth quotations).
11. George B. Schaller, *The Year of the Gorilla* (Chicago: University of Chicago Press, 1964), p. 1. Schaller's explanation of Coolidge's invitation also appears on p. 1. Coolidge (1904–1985) did fieldwork in Africa in 1926–1927, just before completing a BS degree at Harvard. He published the results as *Notes on the Gorilla*, offprint from *Report of the Harvard-African Expedition upon the African Republic of Liberia and the Belgian Congo* (Cambridge, Mass.: Harvard Museum of Comparative Zoology, 1930), pp. 623–35. *The Mountain Gorilla* (1963) was Schaller's scholarly presentation of the gorilla research. Earlier, he had coauthored articles with Emlen (cited in *Mountain Gorilla*, p. 411, and including J. T. Emlen, Jr., and G. Schaller, "In the Home of the Mountain Gorilla," *Animal Kingdom* 63:3 [1960]: 98–108). On Emlen, see Wesley E. Lanyon, Stephen T. Emlen, and Gordon H. Orians, "In Memoriam: John Thompson Emlen, Jr., 1908–1997," *Auk* 117:1 (2000): 223. Griffin noted the hospitality of the Creole Petroleum Company in *Listening in the Dark*, p. 287; an explanation of the fieldwork is the subject of chap. 11. Standard Oil of New Jersey owned Creole Petroleum, and at the time of Griffin's visit, the subsidiary was one of the largest corporate oil-producers in the world.
12. Konrad Z. Lorenz, *King Solomon's Ring: New Light on Animal Ways*, trans. Marjorie Kerr Wilson (New York: Thomas Y. Crowell, 1952), p. viii.
13. In the post–World War II period, émigré scholars, besides Schaller, who are considered in this chapter include (in alphabetical order): Michael W. Fox (veterinary science, birthplace Britain), Thomas Nagel (philosopher, birthplace Belgrade, then-Yugoslavia), and Peter Singer (philosopher, birthplace Australia, and an occasional American resident). Although they ended up in the United States for various reasons—most often war-related displacement—and at various ages, one attraction seems the reputation of the American academy as a well-supported enterprise. In the Netherlands in 1947, Niko Tinbergen was discouraged, as paraphrased here by his correspondent Ernst Mayr, by "the general lethargy of his country and his inability to get anybody out of his rut" (Mayr to Frank Beach, Sept. 9, 1947, HUGFP 14.7, box 5, folder 202, Ernest Mayr Papers, Harvard University Archives, Cambridge, Mass.). Tinbergen pursued an appointment in the United States for several years (see Tinbergen to Mayr, HUGFP 14.7, Sept. 5, 1945, box 3, folder 142, Mayr Papers). Mayr, Beach, W. C. Allee, Margaret Morse Nice, and perhaps others worked behind the scenes to secure him an American position, until he settled at Oxford in 1949.
14. Jane van Lawick-Goodall, *In the Shadow of Man* (Boston: Houghton Mifflin, 1971), pp. 138 (first, second, and third quotations), 139 (fourth and fifth quotations), 268 (sixth quotation), 78 (seventh quotation). European scientists certainly argued abstractly, too. But they seemed less inhibited about interspersing affecting narratives with strident theorizing. Consider, for example, the juxtaposition of Konrad Lorenz's popular books—*King Solomon's Ring* (1952), *Man Meets Dog*, trans. Marjorie Kerr Wilson (1953; reprint ed., New York: Penguin, 1964), and *On Aggression*, trans. Marjorie Kerr Wilson (New York: Harcourt, Brace, and World, 1966)—with such difficult polemics such as *Evolution and Modification of Behavior* (Chicago: University of Chicago Press, 1965). Eventually, Americans—albeit often outside the academy—began composing first-person reports of their lives with animals. A steady stream of such books has followed Dian Fossey, *Gorillas in the Mist* (Boston: Houghton Mifflin, 1983).
15. George Gaylord Simpson, "The Crisis in Biology," *American Scholar* 36:3 (1967): 370 (first quotation); 371 (second quotation); Richard C. Lewontin, "Is Nature Probable or Capricious?" *BioScience* 16:1 (1966): 25 (third quotation); Theodosius Dobzhansky, "Are Naturalists Old-Fashioned?" *American Naturalist* 100:915 (1966): 550 (fourth quotation),

544 (fifth quotation). Dobzhansky noted (p. 544) that he borrowed the term "compositionist" from George Gaylord Simpson, *This View of Life: The World of an Evolutionist* (New York: Harcourt, Brace and World, 1964). In this 1967 paper, Simpson echoed his own earlier argument for evolutionary uniqueness, initially offered as a critique of the then-popular expectation that alien life forms might be found in space, "The Nonprevalence of Humanoids," *Science*, new series, 143:3608 (1964): 769–75. In both cases, Simpson argued that identical biochemistry would likely not produce equivalent life, because the conditions of evolutionary descent would not be replicated.

16. Peter Tompkins and Christopher Bird, *The Secret Life of Plants* (New York: Harper and Row, 1973), pp. xiii (first quotation), xi (second quotation), xiv (third and fourth quotations). Italics in quotations appear in the original texts unless otherwise noted.
17. Griffin, *Animal Awareness*, p. 73 (first quotation); Singer, *Animal Liberation*, p. 263 (second quotation). Marian Dawkins was not as dismissive of plants as others who were concerned with animal sentience, but she was just as definite: "Emotion seems to be a property of a highly developed nervous system. No plant has a structure which has anything like the complexity of an insect or even a jellyfish. As well as this, plants show no behavioural evidence of suffering or trying to escape. A plant does not uproot itself and run away or try to shake off someone who is cutting off one of its branches" (*Animal Suffering*, p. 119).
18. Karl von Frisch, *Bees: Their Vision, Chemical Senses, and Language* (Ithaca: Cornell University Press, 1950), p. 85 (first quotation); Karl von Frisch, "Decoding the Language of the Bee," Nobel Lecture in Physiology and Medicine, Dec. 12, 1973, https://www.nobelprize.org/uploads/2018/06/frisch-lecture.pdf (accessed Sept. 24, 2019), p. 84 (second quotation); Richard Dawkins, *The Selfish Gene* (New York: Oxford University Press, 1976), pp. 49 (third and fourth quotations), 2 (fifth quotation), ix (remaining quotations).
19. Wilson, *Sociobiology*, pp. 13 (first quotation), 14 (second quotation), 10 (third and fourth quotations), 4 (fifth quotation).
20. Edward O. Wilson, *Naturalist* (Washington, D.C.: Island Press, 1994), pp. 330 (first quotation), 308 (second quotation), 349 (third quotation). Wilson devotes chap. 17 to an account of the controversy.
21. Marshall Sahlins, *The Use and Abuse of Biology: An Anthropological Critique of Sociobiology* (Ann Arbor: University of Michigan Press, 1976), p. 11 (first quotation); Ernst Mayr to Michael Ruse, HUGFP 74.7, Oct. 24, 1977, box 25, folder 1277, Mayr Papers (second quotation). It is difficult to tell whether Wilson and Sahlins referred so little to work in each other's fields because scholarship in both disciplines had become too voluminous for nonspecialists to keep up, because the academy placed less and less value on being a generalist, because of an absence of theoretical consensus, or because the occasion was public debate. Sahlins's book was an extended essay of 120 pages, perhaps making it understandable that the only sociobiologists he cited in a short bibliography (pp. 115–20) in addition to Wilson were two theorists of altruism, W. D. Hamilton and Robert Trivers. The bulk of his volume was an exposition of kinship systems (pp. 17–67), a subject he knew well. In contrast, Wilson compiled a sixty-four-page double-column bibliography (pp. 599–663), the vast majority of citations pertaining to animals other than humans. It read as an impressive record of biology's accomplishments. Among well-known anthropologists in Sahlins's bibliography—Franz Boas, Mary Douglas, E. E. Evans-Pritchard, Claude Lévi-Strauss, Bronislaw Malinowski, and H. R. Radcliffe-Brown—only Lévi-Strauss appeared on Wilson's list, along with Ashley Montagu, not mentioned by Sahlins. Wilson's omissions were noteworthy, because he argued that humans are animals and that biology should legitimately supersede the social sciences in studying them. In the absence of critical examination of the deficiency of cultural anthropology and related social sciences, Wilson asked the reader to trust his judgment. With

similar lopsidedness, however, Sahlins's high wall between animal and human societies suggested his disregard of impressive evidence of communication and cooperation within animal species, rightly highlighted by Wilson.

22. Margaret Floy Washburn, *The Animal Mind: A Text-Book of Comparative Psychology*, 2d ed. (New York: Macmillan, 1917), p. 24. Among later scholars of animal consciousness, M. Dawkins had the strongest historical memory of forebears with whom she did not wholly agree, including G. J. Romanes, John B. Watson, and Robert Yerkes (*Animal Suffering*, pp. 11, 14, 15). She did not mention Washburn.
23. Thomas Nagel, "What Is It Like to Be a Bat?," *Philosophical Review* 83:4 (1974): 445 (first quotation), 438 (second quotation), 439 (third quotation). Ever attentive to evolutionary theory, though not persuaded by its materialist premises, Nagel has recently expanded his early argument for consciousness into a closely reasoned proposal for idealism; see Thomas Nagel, *Mind and Cosmos: Why the Materialist Neo-Darwinian Conception of Nature Is Almost Certainly False* (Oxford: Oxford University Press, 2012).
24. Dobzhansky, "Are Naturalists Old-Fashioned?," p. 542 (first and second quotations); Wilson, *Naturalist* (third quotation), title page.
25. Griffin, *Animal Awareness*, p. vii.
26. Marc Bekoff, "The Development of Social Interaction, Play, and Metacommunication in Mammals: An Ethological Perspective," *Quarterly Review of Biology* 47:4 (1972): 427 (first and second quotations), 421 (third and fourth quotations), 413 (fifth and sixth quotations). Discussion of the comparative psychology approach of Lehrman and Schneirla is found on p. 413. Beach appears as the lead authority on play, p. 416.
27. Bekoff, "Development of Social Interaction," pp. 412 (first and second quotations), 418 (photograph), 421 (photograph); Lorenz, *Evolution and Modification of Behavior*, p. 42 (third quotation); Lorenz, *On Aggression*, p. 61 (fourth quotation). Ritualization was a popular concept and, in a prominent example, the subject of a London conference in 1966; see, e.g., Julian Huxley, "A Discussion on Ritualization of Behaviour in Animals and Man," *Philosophical Transactions of the Royal Society of London*, ser. B (Biological Sciences) 251:772 (1966): 247–526. Participants included not only ethologists such as Lorenz, Tinbergen, and Robert Hinde, but also existential psychiatrists Erik Erikson and R. D. Laing, anthropologist Victor W. Turner, sociologist Edward Shils, and many others.
28. Marc Bekoff, "Social Communication among Canids: Evidence for the Evolution of a Stereotyped Mammalian Display," *Science*, new series, 197:4308 (1977): 1097 (first, second, and third quotations), 1099 (fourth quotation). Endnotes appear on p. 1099. In relatively brief notes for a short *Science* article, Bekoff cited neither Lorenz nor Tinbergen, but, in addition to the Dawkinses, Robert Hinde and Julian Huxley, among ethologists, as well as names often mentioned by Lorenz as his predecessors: Lorenz's teacher Oskar Heinroth and the American biologist C. O. Whitman. Farther afield, but still more theoretically minded than American psychologists, E. O. Wilson and semiotician Thomas A. Sebeok also appeared as relevant authors. It is hard to know whether deliberate efforts to bridge the gap between developmental psychology and ethology facilitated persuasion of the Americans. Robert A. Hinde, for example, aimed for reconciliation, in *Animal Behaviour: A Synthesis of Ethology and Comparative Psychology* (New York: McGraw-Hill, 1966). The book fairly surveyed the subject of individual motivation, although its prominent evolutionary framework of analysis and his methodological endorsement of modeling attested to Hinde's ethological roots.
29. Donald R. Griffin, "Foreword," von Frisch, *Bees*, p. v (quotations). Von Frisch explained Griffin's assistance with the American tour, pp. xi–xii.
30. Donald R. Griffin to Karl Lashley, July 22, 1949 (first quotation), Griffin to Lashley, Aug. 18, 1949 (second quotation), box 7, folder 67; Griffin to John Tyler Bonner, Apr. 30, 1957 (third,

fourth, and fifth quotations), box 1, folder 4; Griffin to Karl von Frisch, Oct. 16, 1965 (sixth quotation), box 13, folder 133; Robert M. Yerkes to Griffin, Jan. 27, 1949 (seventh quotation), box 14, folder 138, Griffin Papers. The contact at Yale Press that Yerkes mentioned was his daughter, Roberta, an assistant editor. It may have been Wallace Craig, now long inactive as a researcher, who introduced Griffin to von Frisch's writing. The single letter by Craig in the Griffin Papers, dated July 15, 1948, indicates that the Craigs and Griffins had lived at the same address in Cambridge (10 Bowdoin St.), presumably an apartment house, before Griffin left Harvard for Cornell in 1946. Craig's 1948 letter congratulates the Griffins on the birth of their twins and also reveals that Craig kept up with Griffin's work on bird behavior. Craig reported that he himself was writing a "long paper" (box 2, folder 12). Over the years, Griffin's correspondence with ethologists was most extensive with von Frisch (box 13, folder 133). Letters exchanged with Lorenz were warm but less frequent (box 7, folder 66). Correspondence with Tinbergen was limited to the late 1940s (box 13, folder 126).

31. Schaller, *Mountain Gorilla*, pp. 233 (first and second quotations); Lanyon, Emlen, and Gordon, "John Emlen," p. 225 (third quotation). Schaller's close analysis of chest-beating appears in *Mountain Gorilla*, pp. 221–35. Although he seemed genuinely impressed by Yerkes's work, Schaller commented that gorilla studies are "a highly inbred literature giving the appearance of truth through mere repetition" (*Mountain Gorilla*, p. 19). Yerkes's gorilla material composed a large section of the overview he coauthored with his wife, Robert W. Yerkes and Ada W. Yerkes, *The Great Apes: A Study of Anthropoid Life* (New Haven: Yale University Press, 1929), pt. V. On his own, Yerkes previously published his experiments involving a captive gorilla, "The Mind of a Gorilla," *Genetic Psychology Monographs* 2:1 and 2 (1927): 1–193. Yerkes did not systematically analyze gorillas in the wild, making one brief trip to Africa in 1929 to arrange for captive subjects (see chapter 3, earlier). On the eclectic biology of Emlen, he met the Oxford evolutionary biologist David Lack, when Lack visited the University of California at Davis in 1939, and unnamed ethologists during an African sabbatical in 1953–1954 (Lanyon et al., "John Emlen," p. 223). He was Lack's junior coauthor of "Observations on Breeding Behavior in Tricolored Red-Wings," *Condor* 41:6 (1939): 225–30. His solo publications did not routinely reference European ethologists until the mid-1950s, however. Contrast, e.g., John T. Emlen, Jr., "Flocking Behavior in Birds," *Auk* 69:2 (1952): esp. 169–70, with extended consideration of Gustav Kramer, Lack, Lorenz, and Tinbergen, in John T. Emlen, Jr., "The Study of Behavior in Birds," in *Recent Studies in Avian Biology*, ed. Albert Wolfson (Urbana: University of Illinois Press. 1955), esp. 143–53. On this transitional period, see also Sumner W. Matteson, "John T. Emlen Jr.: A Naturalist for all Seasons, Part 2: Of Adventure, Innovation, and Conscience, 1934–1959," *Passenger Pigeon* 60:3 (1998): 229–39, and Stephen T. Emlen, "Taking My Cues from Nature's Clues," in Lee C. Drickamer and Donald A. Dewsbury, eds., *Leaders in Animal Behavior: The Second Generation* (Cambridge: Cambridge University Press, 2010), esp. 245–46. Emlen was respected by both animal psychologists and evolutionary biologists. When Frank Beach hoped to hire him at the American Museum of Natural History in 1943, and Beach asked his colleague, Ernst Mayr, for Mayr's opinion. Mayr wrote back positively, although the recruitment did not come to fruition; see Mayr to Beach, HUGFP 14.7, May 25, 1943, box 2, folder 72, Mayr Papers. On Richter, Harry Harlow attacked Richter's "idealized theoretical paper" in Harlow's 1951 presidential address to the division of experimentalists of the American Psychological Association, "Mice, Monkeys, Men, and Motives," *Psychological Review* 60:1 (1953): 23 (quotation). Richter's foundational article was Curt P. Richter, "Animal Behavior and Internal Drives," *Quarterly Review of Biology* 2:3 (1927): 307–43.

32. For Bekoff's acknowledgment of Fox, see Bekoff, "Development of Social Interaction," p. 427; for reference to his doctoral thesis, see Bekoff, "Social Communication in Canids,"

p. 1097, n. 14. For evidence that Bekoff's work was done in the Department of Psychology, where Fox had an appointment, see Michael W. Fox, "Behavior Genetics of F1 and F2 Coyote-Dog Hybrids," *Applied Animal Ethology* [after 1984 titled *Applied Animal Behaviour Science*] 1:2 (1975): 185 (Fox's academic affiliation), 194 (full citation for Bekoff's dissertation). Bekoff listed Fox's most important early publications, in Bekoff, "Development of Social Interaction," p. 429. Fox joined the Humane Society of the United States in 1976; see https://www.activistfacts.com/person/3365-michael-w-fox/ (accessed June 1, 2019). Robert Trivers's correspondence with Ernst Mayr, 1966–1987, may be found in HUGFP 14.7 (various folders), Mayr Papers. Mayr's letter requesting help on behalf of Lorenz was Mayr to Frank Beach, Aug. 3, 1965, HUGFP 74.7, with a reply from Beach to Mayr, Aug. 10, 1965, box 11, folder 862, Mayr Papers.

33. Robert L. Trivers, "Foreword," R. Dawkins, *Selfish Gene*, p. vi (first and second quotations); Ernst Mayr to Robert Trivers, Nov. 17, 1978 (third and fourth quotations), HUGFP 74.7, box 26, folder 1300, Mayr Papers. Trivers's reply to Mayr, Nov. 28, 1978, is in the same box.
34. George Gaylord Simpson, *Tempo and Mode in Evolution* (New York: Columbia University Press, 1944), esp. chap. 7; see notes 15–17, earlier, for 1960s citations. Note that the concepts of physics that Simpson applied to evolution in the 1940s were highly innovative and accommodated nature's unpredictability. In effect, he envisioned quantum leaps in adaptation if an organism had already accumulated enough small genetic mutations to be ready to exploit a fortuitous change in the environment: "quantum evolution occurs only if there is preadaptation for an effectively possible new mode of life" (*Tempo and Mode*, p. 212). In contrast, he identified speciation and phyletic evolution as more ordinary and ongoing kinds of evolutionary transformation, perhaps analogous to the constancy of Newtonian physics.
35. Ethel Tobach, "Social Darwinism Rides Again," in Tobach et al., eds., *Four Horsemen*, pp. 103 (first and second quotations), 110 (third quotation), 107 (fourth quotation).
36. Frank A. Beach, "Behavioral Endocrinology: An Emerging Discipline," *American Scientist* 63:2 (1975): 178 (first quotation), 179 (second quotation), 181 (third quotation); Beach, "Historical Origins of Modern Research on Hormones and Behavior," pp. 325 (fourth quotation), 351 (fifth quotation); Frank Beach to Ernst Mayr, Sept. 8, 1976, HUGFP 74.7, box 24, folder 1246, Mayr Papers (sixth quotation). Beach's benchmark work on hormones in 1948 took more explicit account of evolution than his later writing. *Hormones and Behavior: A Survey of Interrelationships between Endocrine Secretions and Patterns of Overt Response* (New York: Paul B. Hoeber, 1948) was organized around species comparisons. See also chapter 5, earlier, on his early comparative subjects such as animal play.
37. John Emlen, "Behavior in Birds," in *Avian Biology*, ed. Wolfson, p. 123 (quotations). For Stephen T. Emlen's account of his scientific development, see "Taking My Cues from Nature's Clues," in Drickamer and Dewsbury, eds., *Leaders in Animal Behavior*, esp. pp. 258, 264, 274. For samples of the work of Emlen's teachers, see Nelson G. Hairston, Frederick E. Smith, and Lawrence B. Slobodkin, "Community Structure, Population Control, and Competition," *American Naturalist* 94:879 (1960): 421–25, and L. B. Slobodkin, F. E. Smith, and N. G. Hairston, "Regulation in Terrestrial Ecosystems, and the Implied Balance of Nature," *American Naturalist* 101:918 (1967): 109–124.
38. M. Dawkins, *Animal Suffering*, title page (first quotation); Singer, *Animal Liberation*, pp. 65–66 (second quotation); Fox, *Inhumane Society*, pp. xvii (third and fourth quotations), 9 (fifth quotation). The phrase "holocaust of the animal kingdom" also appears on p. 9. Analogies to Nazi genocide are common in this literature; see also Singer, *Animal Liberation*, pp. 81 (Nazi medical experiments) and 270 (Hitler). David Fraser emphasizes the activism of the 1970s as a social catalyst for animal welfare science, "Animal Behavior, Animal Welfare and the Scientific Study of Affect," *Applied Animal Behaviour Science*, 118:2–3 (2009): 108–117.

I thank Marian Dawkins for pointing out to me the importance of William Homan Thorpe (1902–1986), the pioneering and eclectic British ethologist who contributed "The Assessment of Pain and Distress in Animals," Appendix III, to the F. W. Rogers Brambell, chair, *Report of the Technical Committee to Enquire into the Welfare of Animals Kept under Intensive Livestock Husbandry Systems . . . Presented to Parliament by the Secretary of State for Scotland and Minister of Agriculture, Fisheries and Food by Command of Her Majesty* (1965), pp. 71–79. See also W. H. Thorpe, "The Concepts of Learning and Their Relation to Those of Instinct," *Physiological Mechanisms in Animal Behaviour*, Symposia of the Society for Experimental Biology, no. IV (Cambridge: Cambridge University Press, 1950), pp. 387–408, and R. A. Hinde, "William Homan Thorpe. 1 April 1902–7 April 1986," *Biographical Memoirs of Fellows of the Royal Society* 33 (1987): 620–39.

39. Rupshi Mitra and Robert M. Sapolsky, "Effects of Enrichment Predominate over Those of Chronic Stress on Fear-Related Behavior in Male Rats," *Stress* 12:4 (2009): 305 (first quotation); Robert M. Sapolsky, "Genetic Hyping," *The Sciences* 40:2 (2000): 15 (second and third quotations). An example of Sapolsky's early work is Robert M. Sapolsky and Glen E. Mott, "Social Subordinance in Wild Baboons Is Associated with Suppressed High Density Lipoprotein-Cholesterol Concentrations: The Possible Role of Chronic Social Stress," *Endocrinology* 121:5 (1987): 1605–10. His seminal work for a wide audience is *A Primate's Memoir* (New York: Scribner, 2001). His synthetic mindset also echoed the combined nature-nurture orientation of the Hungarian émigré to Canada, Hans Selye (1907–1982), who Sapolsky admired. See Selye's account of the unconventional and controversial route that led him to a theory of acclimation to stress, *The Story of the Adaptation Syndrome* (Montreal: Acta, 1952).
40. Ernst Mayr and William B. Provine, eds., *The Evolutionary Synthesis: Perspectives on the Unification of Biology* (Cambridge: Harvard University Press, 1980); Beach, "Historical Origins of Modern Research on Hormones and Behavior." Carl N. Degler summarizes rising esteem of Darwinian evolution, broadly understood, in the later twentieth century, *In Search of Human Nature: The Decline and Revival of Darwinism in American Social Thought* (New York: Oxford University Press, 1991), pt. 3. There has been analysis on the American psychological tradition that Lashley so much shaped, but this legacy seems dwarfed by the evolutionary revival. On Lashley, see the nuanced study by Nadine M. Weidman, *Constructing Scientific Psychology: Karl Lashley's Mind-Brain Debates* (Cambridge: Cambridge University Press, 1999).
41. Wilson, *Sociobiology*, pp. 7 (Wheeler), 8 (Allee), 19 (Simon), 22 (Lorenz); Griffin, *Animal Awareness*, pp. 13 (Tolman), 32 (Chomsky), 34 (Lorenz); Simpson, "Crisis in Biology," p. 365 (Wöhler). Rediscovery of Edward Tolman (1886–1959) accompanied the return of the idea of consciousness, now understood biologically. Although Tolman worked hard between the wars to accommodate the empiricism of his peers in psychology, his work continued to bear marks of his training under more philosophically (and sometimes idealistically) minded mentors at Harvard, including Hugo Münsterberg, Herbert Langfeld, Ralph Barton Perry, and Edwin Holt (see chapter 2, earlier). Most often cited in the 1970s was Tolman's summary paper, "Cognitive Maps in Rats and Men," *Psychological Review* 55:4 (1948): 189–208. Marian Dawkins specifically named Tolman as not being among the conventional "Behaviourists" in her book, *Animal Suffering* (p. 15), but then did not cite his work in a long bibliography of more than two hundred items. Perhaps she did not see him as deeply influential.
42. Calhoun, "Population Density and Social Pathology," pp. 143 (first and third quotations), 146 (second quotation).
43. Tinbergen, *Study of Instinct*, p. 1 (first quotation), 5 (second quotation); Lorenz, *On Aggression*, pp. 85 (third quotation), 89 (fourth and fifth quotations), 150 (sixth quotation).

44. Robert L. Trivers, "The Evolution of Reciprocal Altruism," *Quarterly Review of Biology* 46:1 (1971): 48 (first quotation), 49 (second quotation), 49 (third quotation). Trivers's reliance on animal examples of self-sacrificing behavior in fish (pp. 39–43) and birds (pp. 43–45) makes it easy to miss that he shifts focus to humans when discussing the catalyzing function of private emotions (see the section titled "The Psychological System Underlying *Human* Reciprocal Altruism," pp. 47–54, emphasis added). The linear structure of the essay reflects the basic sociobiological assumption that animal behavior is the evolutionary underpinning of human activities; but Trivers's choice of wording stops short of an unequivocal endorsement that animals, too, have emotions.
45. Trivers, "Reciprocal Altruism," p. 48. For McDougall's idea of emotional signals, see William McDougall, *Outline of Psychology* (New York: Scribners, 1923), esp. p. 326. The emphasis on emotions as communicators was more explicit in 1923 than in McDougall's foundational work, though not substantially different; see William McDougall, *An Introduction to Social Psychology* (1908; reprint ed., London: Methuen, 1960).
46. Wilson, *Sociobiology*, p. 4.
47. Griffin, *Animal Awareness*, title page (quotation). Another philosopher of the period much concerned with emotions in the evolution of mind, though less focused on animals, is Susanne K. Langer, *Mind: An Essay on Human Feeling*, vols. 1–3 (Baltimore: Johns Hopkins University Press, 1967–1982).
48. Griffin, *Animal Awareness*, p. 5 (first, second, third, and fourth quotations); William James, *The Principles of Psychology*, vol. 1 (1890), in *The Works of William James*, electronic edition, ed. Frederick H. Burkhardt, Fredson Bowers, and Ignas K. Skrupskelis (Cambridge: Harvard University Press, 1975–1988), pp. 219 (fifth quotation), 380 (sixth quotation), 429 (seventh quotation).
49. Arturo Rosenblueth, Norbert Wiener, and Julian Bigelow, "Behavior, Purpose and Teleology," *Philosophy of Science* 10:1 (1943): 18 (first quotation), 23 (second quotation); Tolman, "Cognitive Map," p. 192 (third, fourth, and fifth quotations). See n. 41, earlier, for the revival of interest in Tolman's idea of active cognition. Although strict behaviorists and traditional evolutionists alike rejected consideration of purposiveness in relation to changes in life forms, other twentieth-century intellectuals have struggled with the significance of intentionality in the minds of organisms and the presence of ends (most important, survival) in the evolutionary process. Whereas George Gaylord Simpson dismissed religious explanations of evolutionary change, he took seriously apparent goal-seeking in evolution, especially in connection with the influence of motivation on behavior. See Simpson, *Tempo and Mode in Evolution*, chap. 5, and esp. pp. 152, 167. In constructing a naturalistic idealism, Thomas Nagel recently made the existence of purposiveness in individuals and seeming orderliness of unfolding natural processes, sometimes named "emergent evolution," a central argument in *Mind and Cosmos*. These recent discussions give the cautious restoration of individual goal-seeking in post–World War I American psychology (see chapter 2) much broader historical and metaphysical applications. At that time, some Anglo-American scientists advocated emergent evolution, including C. Lloyd Morgan, *Emergent Evolution: The Gifford Lectures* (New York: Henry Holt, 1923); William Morton Wheeler, "Emergent Evolution and the Social," *Science*, new series, 64:1662 (1926): 433–40; and William Morton Wheeler, *Emergent Evolution and the Development of Societies* (New York: Norton, 1928).
50. M. Dawkins, *Animal Suffering*, pp. 15 (first and second quotations), 23 (third quotation), 119 (fourth quotation), 117 (fifth quotation). She discussed Cannon on p. 56.
51. M. Dawkins, "Science of Animal Suffering," pp. 940 (first and second quotations), 941 (third quotation), 942 (fourth quotation). The experiment on elephant preferences appears on

p. 941. She expanded her argument in *Why Animals Matter: Animal Consciousness, Animal Welfare, and Human Well-Being* (Oxford: Oxford University Press, 2012).

52. M. Dawkins, *Animal Suffering*, p. 24 (first quotation); Singer, *Animal Liberation*, pp. 235 (second quotation), x (third quotation); Wilson, *Sociobiology*, p. 3 (fourth quotation). An example of analysis of proto-morality is Frans de Waal, *Primates and Philosophers: How Morality Evolved*, ed. Stephen Macedo and Josiah Ober (Princeton: Princeton University Press, 2006).
53. Sapolsky, *Primate's Memoir*, p. 303.
54. Fox, *Inhumane Society*, p. 92.
55. Wilson, *Naturalist*, p. 6.

Bibliographic Notes

I have organized my comments by the kind of source: manuscript collections, biographies and memoirs, journals, primary-source monographs, and secondary sources. Under each heading I emphasize materials on American history, but also mention international studies for the sake of comparison. Full bibliographic information for all titles may be found in the notes.

Archives

Unpublished materials reveal the private evolution of scientific ideas and the network of personal communication among my principal figures. Although there are no major manuscript collections for some leading American scientists who studied animals (notably Karl Lashley, Edward Tolman, and Frank Beach), existing archives fill many gaps.

The most useful collections for the early twentieth century are the Robert M. Yerkes Papers (Manuscripts and Archives, Yale University Library) and the Walter B. Cannon Papers (Center for the History of Medicine, Countway Library, Harvard Medical School). Both contain extensive professional correspondence. Separate collections of family materials support these archives: Robert and Ada Yerkes Papers (Whitney Library, New Haven Museum) and Cannon Family Papers (Schlesinger Library, Harvard University). For the years during and after World War II, the most helpful collections are the Ernst Mayr Papers (Harvard University Archives) and the Donald R. Griffin Papers (Rockefeller Archive Center). Along with the professional correspondence found in both, the highlights of the Griffin Papers are research notes and photographs.

Other collections of similar breadth were secondary to my investigation because the figures' research subjects or career choices made them less central to American animal behavior science. Travel diaries and field notes, devoted to ant and primate investigations, respectively, are exciting components of the William Morton Wheeler Papers (Harvard University Archives) and Clarence Ray Carpenter Papers (Eberly Family Special Collections Library, Penn State University). Collections that intriguingly document the integration of personal beliefs, professional activities, and scientific thinking include the Margaret Morse Nice Papers (Division of Rare and Manuscript Collections, Cornell University Library), Theodosius Dobzhansky Papers (American Philosophical Society), and Warder Clyde Allee Papers (Special Collections Research Center, University of Chicago Library). The William McDougall Papers (Rubenstein Rare Book and Manuscript Library, Duke University), mainly chronicling McDougall's years at Duke (1927–1938), also contain notes on his children's development in pre–World War I Britain, as well as research and family photographs from the same period. The Leonard B. Carmichael Papers (American Philosophical Society) show the path taken by one psychologist who became a university and public administrator.

Additional collections provide indirect insights into individuals. These include the papers of the physicist Richard Chace Tolman, brother of the psychologist Edward C. Tolman (Archives, California Institute of Technology), clippings in the Margaret Floy Washburn Folders (Archives and Special Collections Library, Vassar College), and letters by and reminiscences of Calvin Perry Stone (microfilm, Special Collections and University Archives, Stanford University).

Biographies and Memoirs

The ongoing series of National Academy of Sciences Biographical Memoirs, begun in 1877, publishes memorial essays by colleagues. The collection includes substantial pieces on virtually all of the Americans I cover. The papers are noteworthy for their factual detail, personal insights, and bibliographic lists. The intimacy of writers and subjects encourages admiration but not bias, perhaps due to an ethos of scientific objectivity. Taken together, the memorials attest to the intricate intellectual and personal networks connecting American scholars of animal behavior.

Full-length scholarly biographies of these Americans are less common, but there are some excellent sources: Nadine Weidman, *Constructing Scientific Psychology: Karl Lashley's Mind-Brain Debates*; Marilyn Bailey Ogilvie, *For the Birds: American Ornithologist Margaret Morse Nice*; and the two-volume biography by Saul Benison et al., *Walter B. Cannon: The Life and Times of a Young Scientist*, and Elin L. Wolfe et al., *Walter B. Cannon: Science and Society*. The Cannon family also provides a window on the wider Harvard community (including Robert Yerkes, William Morton Wheeler, and Ralph Barton Perry) through the biography of Cannon's wife by Maria I. Diedrich, *Cornelia James Cannon and the Future American Race*, and the memoir by his daughter, Marian Cannon Schlesinger, *Snatched from Oblivion: A Cambridge Memoir*. Margaret Morse Nice remembered her life as a female ornithologist, most active professionally 1930–1950, in *Research Is a Passion with Me*. Two exceptional memoirs give insight into American animal behavior research in the second half of the twentieth century: E. O. Wilson, *Naturalist*, and Robert Sapolsky, *A Primate's Memoir*.

For international perspectives, there are outstanding biographies of European animal behavior scientists, although, in my view, these exaggerate European leadership and repeat truisms about American contemporaries. Most noteworthy are Richard W. Burkhardt, Jr., *Patterns of Behavior: Konrad Lorenz, Niko Tinbergen, and the Founding of Ethology*, and Tania Munz, *The Dancing Bees: Karl von Frisch and the Discovery of the Honeybee Language*. An excellent lens on European-American interactions is the memoir by Karl von Frisch, *A Biologist Remembers*. Pioneering work on Asian contributors to American animal behavior science invites further investigation: Geoffrey H. Blowers, "'To Be a Big Shot or To Be Shot': Zing-Yang Kuo's Other Career" (*History of Psychology*) and Miki Takasuna, on Joseph G. Yoshioka, "Six Researchers' Contributions to Comparative Psychology in Japan, 1900–1945" (*Japanese Psychological Research*).

Journals and Key Articles

Historically, only a handful of American journals specialized in the science of animal minds, and scholars in the field consequently published widely in other periodicals associated with the disciplines of psychology and biology. Robert Yerkes was the prime mover behind successive animal behavior journals: *Journal of Comparative Neurology and Psychology* (1904–1910), *Journal of Animal Behavior* (1911–1921), and *Journal of Comparative Psychology* (1921–1946, 1983–present). Research reports too long for the latter two outlets appeared in collateral book series issued contemporaneously with the journals: Behavior Monographs and Comparative Psychology Monographs.

Psychology journals favored by animal behavior scholars include the *American Journal of Psychology* (begun 1887), *Psychological Review* (1894), *Psychological Bulletin* (1904), *Journal of Abnormal and Social Psychology* (1906, issued with variable titles), *Journal of Experimental Psychology* (1916), *Pedagogical Seminary and Journal of Genetic Psychology* (1924), and *American Psychologist* (1946). Here, their articles competed for space with papers on human psychology, a situation that spurred the launch of specialty journals focused on animals. American biology journals that circulated work by animal behavior scientists include the *American Naturalist* (1867), *Biological Bulletin* (1897), *Journal of Experimental Zoology* (1904), and *Quarterly Review of Biology* (1926). Essays specifically about birds often appeared in *The Auk* (1884), the journal of the American Ornithological Society, and less frequently in *Audubon Magazine* (1887), issued by the National Audubon Society. Especially when scholars aimed at broad audiences, they published in *Proceedings of the American Philosophical Society* (1838) and *Science* (1880).

A short list of journal articles that encapsulated evolving scientific concerns stand out as especially noteworthy: Karl Lashley, "A Simple Maze" (*Psychobiology*, 1918); Edward Tolman, "Can Instincts Be Given Up in Psychology?" (*Journal of Abnormal and Social Psychology*, 1922); Leonard Carmichael, "Heredity and Environment" (*Journal of Abnormal and Social Psychology*, 1925); Leonard Carmichael, "The Experimental Embryology of Mind" (*Psychological Bulletin*, 1941); Daniel Lehrman, "A Critique of Lorenz's Theory of Instinctive Behavior" (*Quarterly Review of Biology*, 1953); Frank Beach and Julian Jaynes, "Effects of Early Experience upon the Behavior of Animals" (*Psychological Bulletin*, 1954); Frank Beach, "The Descent of Instinct" (*Psychological Review*, 1955); Robert Trivers, "The Evolution of Reciprocal Altruism" (*Quarterly Review of Biology*, 1971); and Marc Bekoff, "The Development of Social Interaction, Play, and Metacommunication in Mammals" (*Quarterly Review of Biology*, 1972). These crisp reflections on debated issues were not necessarily the most influential publications. The steady stream of empirical studies by Robert Yerkes hit few high notes, for example, but as a body of work profoundly shaped the study of animal psychology.

Theories

The same American endorsement of empiricism that affected the frequency and impact of polemical or summary articles made the production of book-length theories secondary to reports of practical experiments. Scholars nonetheless periodically integrated their findings into general propositions. As time went on, the researchers seemed increasingly inclined to do so. Outstanding examples include Margaret Washburn, *The Animal Mind* (1908, 1917, 1926, 1936); Edward Thorndike, *Animal Intelligence* (1911); Walter Cannon, *Bodily Changes in Pain, Hunger, Fear and Rage* (1915); Robert Yerkes, *Almost Human* (1925); Warder Clyde Allee, *Animal Aggregations* (1931); Walter Cannon, *The Wisdom of the Body* (1932); Edward Tolman, *Purposive Behavior in Animals and Men* (1932); Norman R. F. Maier and T. C. Schneirla, *Principles of Animal Psychology* (1935); Theodosius Dobzhansky, *Genetics and the Origin of Species* (1937); Margaret Morse Nice, *Studies in the Life History of the Song Sparrow* (1937, 1943); Ernst Mayr, *Systematics and the Origin of Species* (1942); Robert Yerkes, *Chimpanzees* (1943); Frank Beach, *Hormones and Behavior* (1948); Donald Griffin, *Listening in the Dark* (1958); George Schaller, *The Mountain Gorilla* (1963); Edward O. Wilson, *Sociobiology* (1975); and Donald Griffin, *The Question of Animal Awareness* (1976). Two British books that figured prominently in American discussions are William McDougall, *Introduction to Social Psychology* (1908) and Marian Stamp Dawkins, *Animal Suffering* (1980).

Secondary Works on Animals and Emotions

I have relied on excellent bodies of scholarship that constitute fairly separate subfields in the history of science and culture.

Histories of Anglo-American attitudes toward animals include Keith Thomas, *Man and the Natural World: A History of Modern Sensibility*; James Turner, *Reckoning with the Beast: Animals, Pain, and Humanity in the Victorian Mind*; and Janet M. Davis, *The Gospel of Kindness: Animal Welfare and the Making of Modern America*. Critical studies of animal-human relations that are less historically focused include Eileen Crist, *Images of Animals: Anthropomorphism and the Animal Mind*; Marian Stamp Dawkins: *Why Animals Matter: Animal Consciousness, Animal Welfare, and Human Well-Being*; and John Marzluff, *Welcome to Subirdia: Sharing Our Neighborhoods with Wrens, Robins, Woodpeckers, and Other Wildlife*.

Evolutionary thinking, in the form of the "modern synthesis" from World War II onward, was a vehicle for the "nature" side of nature-nurture debate and drew attention to the emotions as a component of biology. Although recent analyses of human feelings highlight the agency of culture in forming emotions, the case for animals rode this historic wave of resurgent evolutionary theory. Especially useful are Vassiliki Betty Smocovitis, *Unifying Biology: The Evolutionary Synthesis and Evolutionary Biology*, and Hamilton Cravens: *The Triumph of Evolution: American Scientists and the Heredity-Environment Controversy, 1900–1941*. Because the Americans I study were unconventional evolutionists from 1900 on—and not the simple behaviorists often described—I have been aided by parallel work on international debates about evolution: for Germany, Anne Harrington, *Reenchanted Science: Holism in German Culture from Wilhelm II to Hitler*, and Daniel Schacter, *Stranger behind the Engram: Theories of Memory and the Psychology of Science*; for the Soviet Union, Nikolai Krementsov, *International Science between the World Wars: The Case of Genetics*, and Loren Graham, *Lysenko's Ghost: Epigenetics and Russia*; and for transatlantic connections, Charlotte Sleigh, *Six Legs Better: A Cultural History of Myrmecology*.

The emotions have commanded the attention of recent scholars ranging from philosophers to neuroscientists. Historians have not been left out. In between perspectives on the distant past (Thomas Dixon, *From Passions to Emotions: The Creation of a Secular Psychological Category*) and on recent decades (Ruth Leys, *The Ascent of Affect: Genealogy and Critique*), early-to-mid-twentieth-century America has been neglected. This book aims to fill the gap.

Index

For the benefit of digital users, indexed terms that span two pages (e.g., 52–53) may, on occasion, appear on only one of those pages.